International Nuclear Science, Technology and Engineering Conference (iNuSTEC2018)

Sustainable Nuclear Economy for the Nation Progress

IOP Conference Series: Materials Science and Engineering
Volume 555

Skudai, Malaysia
23 - 25 November 2018

Editors:

Khaidir Hamzah
Nahrul Khair Alang Md Rashid
Faridah Mohamad Idris
Abdul Aziz Mohamed
Mohd Syukri Yahya
Faizal KP Kunchi Mohamed

ISBN: 978-1-5108-9386-3
ISSN: 1757-8981

Printed from e-media with permission by:

Curran Associates, Inc.
57 Morehouse Lane
Red Hook, NY 12571

Some format issues inherent in the e-media version may also appear in this print version.

Printed by Curran Associates, Inc. (2019)

For permission requests, please contact the Institute of Physics at the address below.

Institute of Physics
Dirac House, Temple Back
Bristol BS1 6BE UK

Phone: 44 1 17 929 7481
Fax: 44 1 17 920 0979

techtracking@iop.org

Additional copies of this publication are available from:

Curran Associates, Inc.
57 Morehouse Lane
Red Hook, NY 12571 USA
Phone: 845-758-0400
Fax: 845-758-2633
Email: curran@proceedings.com
Web: www.proceedings.com

Table of contents

Volume 555

International Nuclear Science, Technology and Engineering Conference 2018

23–25 November 2018, Universiti Teknologi Malaysia, Skudai, Johor, Malaysia

Accepted papers received: 22 March 2019
Published online: 19 July 2019

Preface

Papers

IOP Conf. Series: Materials Science and Engineering **555** (2019) 011001 doi:10.1088/1757-899X/555/1/011001

IOP Conference Series: Materials Science and Engineering

International Nuclear Science, Technology and Engineering Conference 2018 (iNuSTEC2018)

"Sustainable Nuclear Economy For The Nation Progress"

Universiti Teknologi Malaysia
Skudai, Johor, Malaysia
23-25 November 2018

Editors

Khaidir Hamzah
Universiti Teknologi Malaysia (UTM)
Nahrul Khair Alang Md Rashid
Universiti Teknologi Malaysia (UTM)
Faridah Mohamad Idris
Malaysian Nuclear Agency (Nuclear Malaysia)
Abdul Aziz Mohamed
Universiti Tenaga Nasional Malaysia (UNITEN)
Mohd Syukri Yahya
Universiti Tenaga Nasional Malaysia (UNITEN)
Faizal KP Kunchi Mohamed
Universiti Kebangsaan Malaysia (UKM)

All papers have been peer-reviewed

Supporting/Sponsoring Organisations

Ministry of Education Malaysia; Ministry of Science Technology and Innovation Malaysia; Malaysian Nuclear Agency; Hitachi-GE Nuclear Energy, Ltd; ASME Malaysia Section; Malaysian Welding and Joining Society; Malaysia Nuclear Power Corporation; Reactor Interest Group, Asia-Oceania Neutron Scattering Association, Japan Atomic Energy Society.

IOP Conf. Series: Materials Science and Engineering **555** (2019) 011001 doi:10.1088/1757-899X/555/1/011001

PREFACE

All praises to God the Al-Mighty for giving us the strengths and the His blessing in making Nuclear Science, Technology and Engineering Conference 2018 (iNuSTEC2018) a success.

The history in organizing Nuclear Science, Technology and Engineering Conference started back in 2010. During that year Then, it was called NuSTEC2010 and was organized only at national and regional levels. From that year onwards, we succeeded in organizing our second (iNuSTEC2011), third (iNuSTEC2012), fourth (international iNUSTEC2013), fifth (NuSTEC2014), sixth (international iNuSTEC2015), seventh (international iNuSTEC2016) and eighth (international iNuSTEC2017) conferences. Based on these achievements, we have decided to reconvene the conference sustainability by organizing regional/international iNuSTEC on annual basis.

The iNuSTEC2018 was held from 23- 25 November 2018 at Universiti Teknologi Malaysia (UTM) in Skudai, Johor, Malaysia. The conference continued the tradition of continuous effort by Malaysian Nuclear society to bring together the nuclear communities in an annual event to discuss and to discourse on progress in nuclear researches and related fields. The conference particularly encouraged the interaction of research students, research institutions and academics with the more established community. iNuSTEC2018 also played an essential role in introducing and fostering new research talents and in advancing the future profile of nuclear science, technology and engineering regionally and globally.

iNuSTEC2018 comprised of the conference parallel sessions and Nuclear Youth competition UTM had also organized a post-conference workshop on Introduction to MNCP on the 26 to 27th of November 2018. At the national forefront of science and energy development, we were honoured to be able to highlight the usage of science and engineering in the nuclear fields and to promote the use of nuclear science and technology for peaceful use through these conference, workshop and side-meetings.

We would like to acknowledge the Ministry of Energy, Science, Technology, Environment and Climate Change Malaysia (MESTECC), Ministry of Education Malaysia (MOE), Malaysian Nuclear Agency (Nuklear Malaysia), International Atomic Energy Agency (IAEA), Hitachi-GE Nuclear Energy Ltd., Japan Atomic Energy Agency (JAEA), Japan Atomic Energy Society (JAES), Tokyo Institute Technology, China National Nuclear Corporation (CNNC), Nusantara Technologies Sdn Bhd, Asia-Oceania Neutron Scattering Association (AONSA), ASME Section Malaysia, Malaysian Welding and Joining society, Institute of Nuclear Engineers (INE MNS), Nuclear Engineering Student Society (NESS), NGOs, universities and companies for their moral and financial support. Special thanks to iNuSTEC2018 organizing committees for their effort and commitment during this conference. We look forward to meeting all of you in 2019 – iNuSTEC2019 @ UKM.

Assoc Prof Dr Khaidzir Hamzah (UTM)
Dr Faridah Mohamad Idris (MNS)
Chairs of iNuSTEC2018
Skudai, Johor
Malaysia

IOP Conf. Series: Materials Science and Engineering **555** (2019) 011001 doi:10.1088/1757-899X/555/1/011001

MESSAGE FROM: PRESIDENT, MALAYSIAN NUCLEAR SOCIETY (MNS)

On behalf of the Malaysian Nuclear Society (MNS), I would like to express our thanks to all participants of the International Nuclear Science, Technology and Engineering Conference 2018 (iNuSTEC2018). MNS is honoured to receive supports from distinguished international and national experts in the field of nuclear (neutron) science and engineering as International Advisory and National Advisory Panels for iNuSTEC2018. We are indeed pleased to have Organizing Committees of iNuSTEC2018 and the Secretariats working with along to ensure the success of iNuSTEC2018. At the forefront of nuclear science and engineering MNS collaborates with universities and related NGOs to bring the success of iNuSTEC2018. MNS members shall strive to make it better and stronger in future.

In conjunction with this conference, MNS is also honoured to organise 3rd Nuclear Youth Competition 2018 (NYC2018) to help bring the young nuclear generation into the main stream of nuclear development and issues. A post conference workshop on introduction to MCNP simulation had been organised to give essential knowledge for Malaysian nuclear researchers in using computers as a tool of a simulation method.

This conference brings together experienced and young researchers to discuss and debate latest development in all areas of nuclear science, engineering and technology applications for friendly purposes at international level. This would spur and catalyse further research and development leading to the sustainable and improvement of the quality of life without compromising the quality of the environment in Malaysia and in the world.

We would like to express our gratitude to UTM for hosting iNuSTEC2018, NYC2018 and the post conference workshop and, to all the personnel involved for their unwavering efforts to make this event a reality. I wish to also record my appreciations and thanks to all sponsors for their generous support. May all these efforts bring benefits to all of us and I congratulate everyone involved in making this conference successful event.

"Sustainable Nuclear Science and Technology for Well-Being of Mankind"

Thank you.

Assoc Prof Dr Abdul Aziz Mohamed
President, MNS

IOP Conf. Series: Materials Science and Engineering **555** (2019) 011001 doi:10.1088/1757-899X/555/1/011001

CONTENTS

IOP Conf. Series: Materials Science and Engineering **555** (2019) 011001 doi:10.1088/1757-899X/555/1/011001

PROCEEDING EDITORIAL COMMITTEE

Abdul Aziz Mohamed (MNS/ UNITEN)
Faridah Mohamad Idris (MNS/Nuklear Malaysia)
Khaidzir Hamzah (UTM)
Nahrul Khair Alang Md Rashid (UTM)
Faizal KP Kunchi Mohamed (MNS/UKM)
Julie Andrianny Murshidi (MNS/Nuklear Malaysia)
Zulhairun Abdul Karim, (UTM)
Abu Hassan Husin (UiTM)
Mohd Syukri Yahya (MNS/ UNITEN)
Julia Abdul Karim (MNS/Nuklear Malaysia) Safwan Shalbi, (UTM/INE MNS)
Muhammad Aliff Ashraff Bin Rosdi, (UTM/INE MNS)
Asyraf Arif Bin Abu Bakar, (UNITEN/ INE MNS)
Abdul Hannan Damahuri, (UNITEN/INE MNS)

IOP Conf. Series: Materials Science and Engineering **555** (2019) 011001 doi:10.1088/1757-899X/555/1/011001

iNuSTEC2018 ORGANISING COMMITTEE

Patrons:
Prof Datuk Ir Dr Wahid Omar
Vice Chancellor, Universiti Teknologi Malaysia (UTM)

Advisors:
Professor Dato' Ir Dr Mohamed Rafiq bin Dato' Abdul Kadir
Dean, College of Engineering, Universiti Teknologi Malaysia
Prof Dr Mohd Ghazali bin Mohd Nawawi
Chairman, School of Chemical and Energy Engineering, Universiti Teknologi Malaysia
Prof Dr Nahrul Khair Alang Md Rashid (Universiti Teknologi Malaysia)
Advisor, Malaysian Nuclear Society

Chairman:
Assoc Prof Dr Khaidzir Hamzah
Nuclear Engineering Undergraduate, UTM

Co-chairman:
Assoc Prof Abdul Aziz Mohamed
President, Malaysian Nuclear Society/ UNITEN

Deputy Chairman:
Dr Faridah Mohamad Idris, MNS/ Nuklear Malaysia

Co-deputy Chairman:
Dr Jasman Zainal, UTM

Secretary:
Dr Zulhairun Abdul Karim, UTM

Treasurer:
Dr Julie Andrianny Murshidi, MNS Treasurer

Committee Members:
Dr Goh Pei Sean, UTM
Dr Wan Norharyati Wan Salleh, UTM
Assoc Prof Muneer Aziz Mohamed Saleh, UTM
Assoc Prof Dr Nasri A. Hamid, MNS/UNITEN
Assoc Prof Dr Irman Abdul Rahman, MNS/UKM
Assoc Prof Dr Suhairul Hashim, UTM
Dr Julia Abdul Karim, MNS/ Nuklear Malaysia
Dr Mohd Yuhyi Mohd Tadza, MNS/UMP
Dr Muhammad Rawi Mohamed Zin, MNS/ Nuklear Malaysia

Dr Hassan Mohamed, MNS/UNITEN
Ir Dr Abu Bakar Hasan, MNS/USIM
Prof Dr Hasan Abu Kasim, MNS/UM
Assoc Prof Dr Norhasliza Yusof, MNS/UM

IOP Conf. Series: Materials Science and Engineering **555** (2019) 011001 doi:10.1088/1757-899X/555/1/011001

MALAYSIA NUCLEAR SOCIETY

Azraf Azman, MNS/ Nuklear Malaysia
Shalina Sheikh Muhammad, MNS/ Nuklear Malaysia
Dr Hafizal Yazid, MNS/Nuklear Malaysia
Rafhayudi Jamro, MNS/ Nuklear Malaysia
Mohd Rizal Mamat @ Ibrahim, MNS/Nuklear Malaysia
Anwar Abdul Rahman, MNS/Nuklear Malaysia
Dr Zalina Laili, MNS/Nuklear Malaysia
Nazaratul Ashifa Bt. Abdullah Salim, MNS/Nuklear Malaysia
Anas Muhamad Pauzi, MNS/UNITEN
Dr Mazleha Maskin, MNS/Nuklear Malaysia

INTERNATIONAL ADVISORY COMMITTEE

Prof Emeritus Noramly Muslim, MNS Honorary Fellow
Prof Deok Jung Lee, Ulsan National Institute of Science and Technology, Rep of Korea
Prof Masaki Saito, Tokyo Institute of Technology, Japan
Professeur Émérite Philippe Quentin, Université Bordeaux, France
Dato' Raja Aziz Raja Adnan, IAEA, Vienna, Austria
Prof Si Hwan Kim, Hanyang University, Rep of Korea
Prof Sung-Min Choi, Korea Advanced Institute of Science and Technology, Rep of Korea
Assoc Prof Dr Zhe Gao, Tsinghua University, China

NATIONAL ADVISORY COMMITTEE

Prof Dr Mohd Zamri Yusoff, Universiti Tenaga Nasional
Prof Dr Hj Khudzir Hj Ismail, Universiti Teknologi MARA
Prof Emeritus Dato' Wira Ir Dr Md Noor Salleh, International Islamic University Malaysia
Prof Dr Mohamad Kassim, Universiti Kebangsaan Malaysia
Prof Dr Shahidan Radiman, Universiti Kebangsaan Malaysia
Dr Mohd Abd Wahab Bin Yusof, Malaysian Nuclear Agency
Prof Dr Amran Abdul Majid, Malaysian Radiation Protection Association
Assoc Prof Dr Abu Hassan bin Husin, Universiti Teknologi MARA
Prof Dato' Dr Mohd Sofian Azirun, Universiti Malaya
Prof Ir Dr Abu Bakar Hasan, Universiti Sains Islam Malaysia
Assoc Prof Dr Khaidzir Hamzah, Universiti Teknologi Malaysia
Prof Ir Dr Nahrul Khair bin Alang Md Rashid, Universiti Teknologi Malaysia
Prof Dr Wan Ahmad Tajuddin Wan Abdullah, Universiti Malaya
Dato' Dr Ghazali bin Dato' Mohd Yusoff, Nusantara Technologies Sdn Bhd
Prof Ir Megat Johari Bin Megat Mohd Noor, BEM Board Member/UTM

IOP Conf. Series: Materials Science and Engineering **555** (2019) 011001 doi:10.1088/1757-899X/555/1/011001

SCIENTIFIC AND PUBLICATION

Prof Ir Dr Nahrul Khair Alang Md Rashid, UTM/ MNS (Chairman)
Assoc Prof Dr Khaidzir Hamzah, UTM
Assoc Prof Dr Abdul Aziz Mohamed MNS/UNITEN
Assoc Prof Dr Faizal K.P. Kunchi Mohamed, MNS/UKM
Dr Jasman Zainal, UTM
Dr Faridah Mohamad Idris, MNS/Nuklear Malaysia
Dr Julie Andrianny Murshidi, MNS/Nuklear Malaysia
Dr Mohd Syukri Yahya, MNS/UNITEN
Dr Hassan Mohamed, MNS/UNITEN
Dr Julia Abdul Karim MNS/Nuklear Malaysia
Dr Hafizal Yazid MNS/Nuklear Malaysia
Assoc Prof Dr Ahmad Taufek bin Abdul Rahman UiTM)
Assoc Prof Dr Irman Abdul Rahman, MNS/ UKM
Dr Norsyahidah Mohd Hidzir, UKM
Dr Syazwani Mohd Fadil, UKM
Safwan Shalbi, INE MNS
Muhammad Aliff Ashraff Bin Rosdi, INE MNS
Asyraf Arif Bin Abu Bakar, INE MNS
Abdul Hannan Damahuri, INE MNS
Muhammad Syahir Sarkawi, INE MNS
Nursyazwani Mohd Ali, INE MNS
Muhammad Arif Sazali, INE MNS
Nurlaila Syamsul Bahri, INE MNS

NUCLEAR ENGINEERING STUDENT SOCIETY (NESS) SECRETARIAT

Oon Jun Teck
Mohd Umar Sarhan Bin Khairul Anuar
Khor Wei Gene
Khor Joo Wei
Nur Fatihah Bt Mazlan
Rafieza Bt Rosli
Yap Wee Kin
Ainur Najwa Bt Mohamad Pouzi
Nur Aisyah Binti Mohd Kamal
Munira Wahida Bt Alias
Nur Alia Bt Md Yunus
Jekan A/L Asokumaran
Bryan Yong Wei Hoong
Abu Huzaifah Bin Bukhari
Amier Gaddafi Bin Che Mohamad Zaki
Muhammad Azri Bin Mohd Ali
Muhammad Izzudin Bin Zainal
Devintharaaj A/L Kumar
Siti Naqibah Hasan Azhari
Fatin Izyan Bazilah Rosli
Muhammad Irfan Mustaqim Bin Azzian
Ammar Bin Ismail
Ahmad Azizan Qusyairi Bin Mat Sariba
Hanis Syahirah Bt Arifin
Nurizzati Binti Mohd Nordin
Nurhasyemi Binti Suhaimi
Asyraf Bin Abd Razak
Muhammad Amir Bin Tukiran
Naqeeb Bin Hakam Ad-Lib
Muhammad Nasiruddin Bin Zulkurnain
Dzul Qarnain Bin Zulkifli
Kishoor A/L Ravindran
Mohammad Zaim Zalhazmi Bin Zahar
Syahirah Nadiah Binti Abd Rahman
Foo Wei Yee

ACKNOWLEDGEMENT

iNuSTEC2018 was organized by:

Universiti Teknologi Malaysia (UTM)

and co-organized by:

Institute of Nuclear Engineers (INE-MNS)
Nuclear Engineering Student Society (NESS

In Collaboration with

Malaysian Nuclear Agency (Nuklear Malaysia)
Universiti Teknologi MARA (UiTM)
Universiti Kebangsaan Malaysia (UKM)
Universiti Tenaga Nasional (UNITEN)
Universiti Malaya (UM)
Universiti Sains Islam Malaysia (USIM)
Universiti Tun Hussein Onn Malaysia (UTHM)
Universiti Islam Antarabangsa Malaysia (IIUM)

Supported by:

ASME Malaysia Section
Malaysian Welding and Joining Society

IOP Conf. Series: Materials Science and Engineering **555** (2019) 011001 doi:10.1088/1757-899X/555/1/011001

IOP Conf. Series: Materials Science and Engineering **555** (2019) 011002 doi:10.1088/1757-899X/555/1/011002

IOP Conf. Series: Materials Science and Engineering **555** (2019) 011003 doi:10.1088/1757-899X/555/1/011003

Peer review statement

All papers published in this volume of *IOP Conference Series: Materials Science and Engineering* have been peer reviewed through processes administered by the proceedings Editors. Reviews were conducted by expert referees to the professional and scientific standards expected of a proceedings journal published by IOP Publishing.

IOP Conf. Series: Materials Science and Engineering **555** (2019) 012001 doi:10.1088/1757-899X/555/1/012001

Criticality safety analysis of spent fuel pool for TRIGA Puspati Reactor using MCNP5

Nurlaila Syamsul Bahri[1,a)], Jasman Zainal[1], Khaidzir Hamzah[1], Muneer Aziz Mohammad Saleh[1], Mohamad Hairie Rabir[2]

[1]Department of Nuclear Engineering, School of Chemical and Energy Engineering, Faculty of Engineering, Universiti Teknologi Malaysia, 81310, Johor, Malaysia.
[2]Nuclear & Reactor Physics Section, Reactor Technology Centre, Technical Support Division, Malaysian Nuclear Agency, Bangi, 43000 Kajang, Selangor, Malaysia

[a]jasman@utm.my

Abstract. Spent fuel pool uses water to temporarily cool spent nuclear fuels before final disposal. The Malaysian Nuclear Agency is recently working on its spent fuel pool facility to store irradiated TRIGA fuel rods. These irradiated fuels still contain significant amount of fissile material which are capable to induce criticality in the fuel storage facility. This study is therefore performed to investigate the safety criticality analysis for the proposed fuel storage rack design of 5x5 arrays. MCNP5 code was used to model the 5x5 array of fuel storage rack in order to determine the optimal fuel pitch distance. The analysis was based on the standard UZrH1.6 TRIGA fuel cladded with stainless steel used in PUSPATI TRIGA Reactor, which is 19.9% enriched 235U and 20% weight of Uranium. By varying the pitch distance between 4 cm to 14 cm with no neutron absorber included, the optimum fuel rack design was determined where sub-criticality condition can be maintained.

1. Introduction

PUSPATI TRIGA Reactor (RTP) is a 1MWth research reactor installed in Bangi, Malaysia, which achieved its first criticality in 1982. After many maintenance and upgrade along its safe operation years, RTP is still in service to produce radioisotopes, neutrons for research, education and training activities [1]. Soon enough, the fuel will reach its maximum burnup limit which will be categorized as spent, and thus, the fuel should be removed out from reactor operation.

Spent nuclear fuel is a high-level waste that requires a special designated interim storage facility before being sent to disposal facility of reprocessing facility. Even though it is considered as spent, the spent nuclear fuels are still generate heat from the decay process of actinides and fission products while being stored in interim storage facility. The interim storage facility must be to ensure the facility will always be in subcritical state at any conditions. The goal of this study is to perform the safety criticality analysis of RTP TRIGA fuel rod and determine the optimum 5x5 fuel array rack design, based on the standard and requirements provided by IAEA.

2. Storage rack modelling using Monte Carlo N-Particle (MCNP) code

In this study, MCNP code was used in modelling several designs of the 5x5 array storage rack with varied fuel pitch distances. The fuel data was based on the standard $UZrH_{1.6}$ TRIGA fuel with 19.9% enriched 235U and composed of 20% weight of Uranium and cladded with stainless steel 304 (SS304).

The dimensions of the fuel are as described in Figure 1. Results of interest are infinite (k_{inf}) and effective (k_{eff}) multiplication factors as a function of fuel pitch distances.

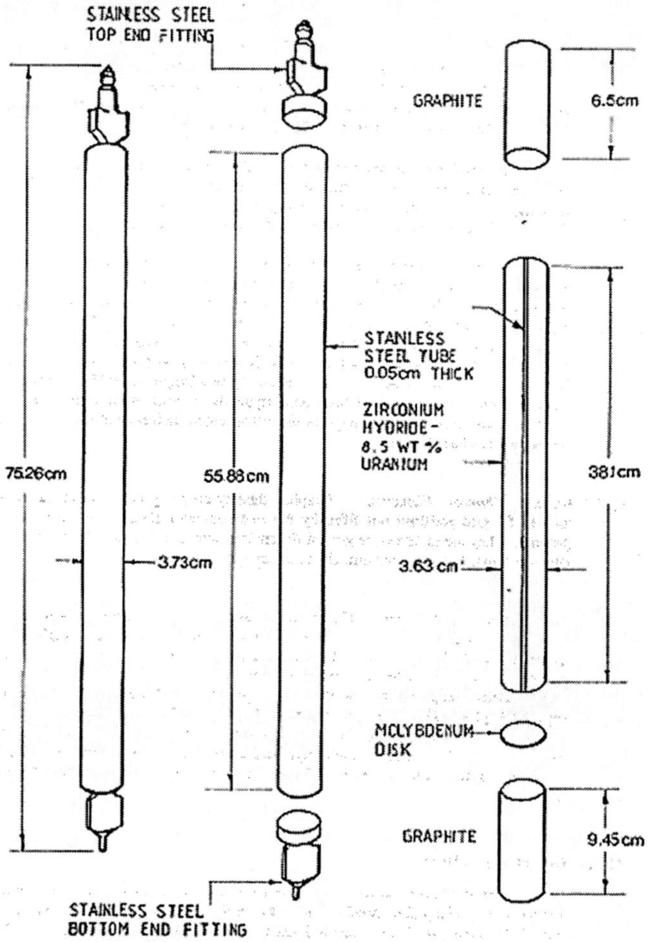

Figure 1. TRIGA fuel dimension [2].

The design of fuel racks with 5x5 array proposed in this study is based on the ability of the coolant to circulate around the fuels in the rack. The configuration of fuel array design should be taken seriously as it could be lead to insufficient coolant flow which results in cladding failures and subsequently overheating. These may also lead to critical or supercritical conditions in the fuel storage pool [3].

3. Results and discussions

Figure 3 shows cross section of the MCNP model of the RTP fuel rod, while Figure 4 shows top and side view of 5x5 array of the storage rack. The criticality safety calculation was performed for fuel pitch distances ranging from 4 cm to 14 cm. Conservative and best estimate analysis approach using infinite and finite lattice model respectfully were considered. In infinite lattice model, the rack side boundaries were assumed to be reflective.

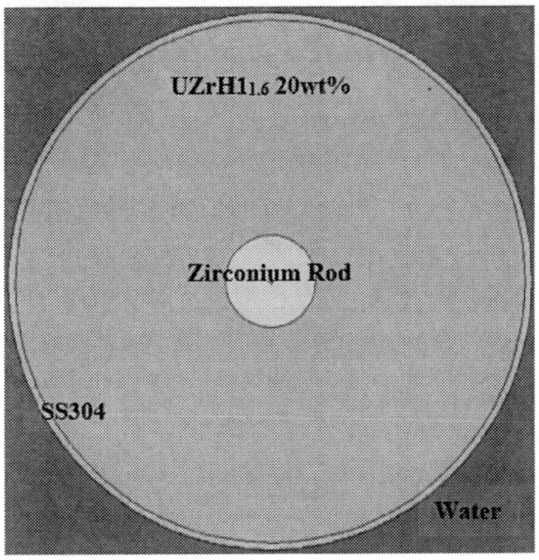

Figure 2. Cross section of RTP fuel rod.

IAEA requires that multiplication factors of the research reactor fuel storage to be lower than 0.80 under normal situations and lower than 0.95 under accidental conditions [4]. These values must be calculated by assuming the worst-case scenarios. Thus it was assumed that; a) the fuel element was all fresh (it has its original enrichment), b) the uranium weight of 20% type is used instead of 12% and 8.5% fuel type in RTP, and c) the geometric cross-sectional configuration is the same. The following is the results of the multiplication factor as a function of fuel pitch distances as calculated using MCNP5 code.

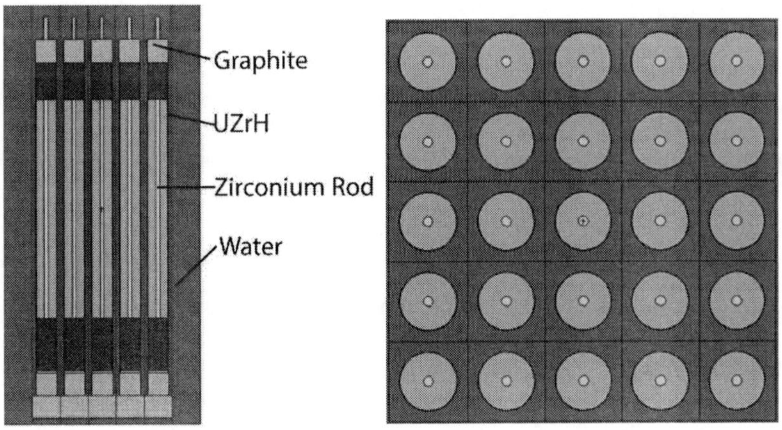

Figure 3. Top and side view of 5x5 array.

IOP Conf. Series: Materials Science and Engineering **555** (2019) 012001 doi:10.1088/1757-899X/555/1/012001

Table 1. Results of multiplication factor of 20%wt Uranium content RTP fresh fuel with different assembly pitch.

Assembly Pitch (cm)	K_{inf}	K_{eff}
4	1.235348 ± 0.004797	0.66209 ± 0.008235
6	1.192878 ± 0.004112	0.70123 ± 0.005841
8	1.096726 ± 0.004287	0.701812 ± 0.005292
10	0.97445 ± 0.004485	0.67054 ± 0.004567
12	0.848506 ± 0.004722	0.620182 ± 0.004293
14	0.731256 ± 0.00448	0.562762 ±0.004108

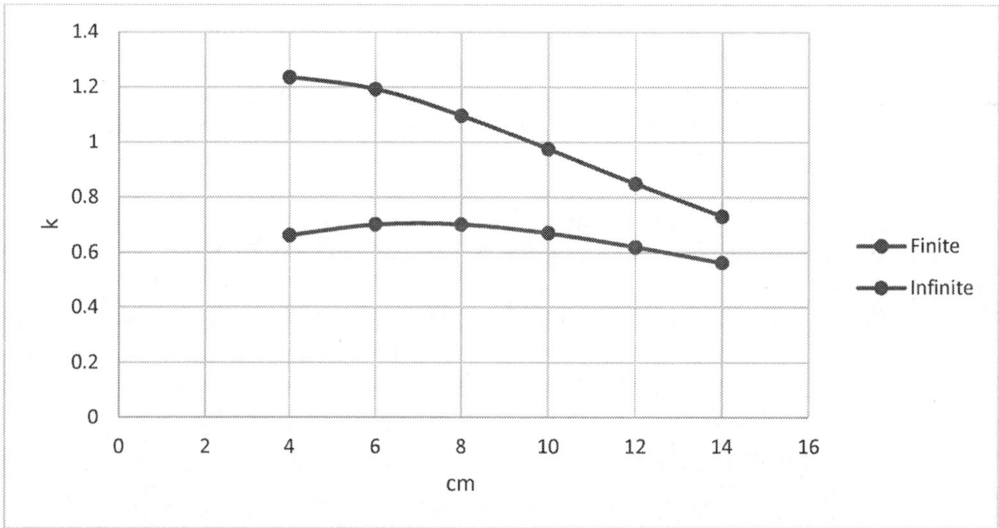

Figure 4. Multiplication factor of 20%wt Uranium content RTP fresh fuel with different assembly pitch.

Figure 4 shows the k_{inf} decreases as the assembly pitch increases. The k_{inf} value which indicates the maximum criticality can be achieved at a fuel pitch distance of 4 cm which is the closest distance the fuel rods can get. If the fuel pitch is less than 10 cm, the 5x5 fuel assembly will be supercritical.

The criticality calculation for the finite lattice (for best estimation approach), shows a slightly different trend of k_{eff} as a function of fuel pitch distance. All range of pitch distances for the 5x5 fuel assembly give k_{eff} values below 0.8. This result may be due to the modelling method in which there is empty space outside the 5x5 fuel assembly that leads to significant neutron leakage.

4. Conclusion

The subcritical state with k_{inf} value below 1.0 for the 5x5 array of fuel rods can be achieved at fuel pitch distance of 10 cm and bigger while k_{inf} 0.80 can be achieved at pitch distance of at minimum 13 cm in an infinite lattice model. The k_{eff} however was found to be within the safety margin, where at all pitch distance, the obtained k_{eff} are below 0.8. The safest rack design could thus be based on the conservative analysis.

5. References

[1] Malaysian Nuclear Agency, TRIGA PUSPATI Reactor, Available: http://www.nuclearmalaysia.gov.my/RnD/energy/reactor/reactorTech
[2] International Atomic Energy Agency, Nuclear Fuel Cycle. Available from: https://www.iaea.org/OurWork/ST/NE/NEFW
[3] Experience with Spent Fuel Storage at Research and Test Reactors *Proc. of an Advisory Group meeting* (Vienna Austria) p 41.
[4] Mahmoud R F, Shaat M K, Nagy M E, Agamy S.A, Abdelrahman A. A 2014 Burn-Up Credit in Criticality Safety of PWR Spent Fuel *Nuclear Eng. and Design,* 280, 628-633. Elsevier Ltd

Acknowledgement

This research is fully supported by a research grant provided by Universiti Teknologi Malaysia (UTM) through project vote number Q.J130000.2646.15J71. The authors gratefully acknowledge all helps and comments during the process of this research as it is going.

IOP Conf. Series: Materials Science and Engineering **555** (2019) 012002 doi:10.1088/1757-899X/555/1/012002

Estimation of fuel temperature reactivity coefficient of a research reactor using ANFIS

Nur Syazwani Mohd Ali[1], Khaidzir Hamzah[1], Faridah Mohamad Idris[2], Nahrul Khair Alang Md Rashid[1], Mohamad Sabri Minhat[2], Mohamad Hairie Rabir[2]

[1]Department of Energy Engineering, Faculty of Chemical and Energy Engineering, Universiti Teknologi Malaysia,81310 Johor Bahru, Johor, Malaysia.
[2]Malaysian Nuclear Agency, Bangi, 43000 Kajang, Selangor, Malaysia

nsyazwani68@liveutm.onmicrosoft.com

Abstract. Fuel Temperature reactivity coefficient is one of the important operational and safety parameters in nuclear reactors to monitor for normal operation and transient safety analysis in research as well as in power reactors. The aim of this work is to estimate the temperature reactivity coefficient by using adaptive neuro-fuzzy inference system (ANFIS) as a preliminary assessment for developing an online monitoring system in research reactors such as the Reactor TRIGA PUSPATI (RTP). The input for the learning mechanism is the fuel temperature recorded at Ch A of the instrumented fuel element (IFE) and control rod positions with the fuel temperature reactivity coefficient as the targeted output. The ANFIS model was validated using testing data that consisted of fuel temperature measured at Ch B and results on -0.00571 $\$°C^{-1}$ (-3.997 pcm$°C^{-1}$) of actual output and -0.00592 $\$°C^{-1}$ (4.144 pcm$°C^{-1}$) of predicted output. Through the developed model, it showed that the ANFIS also can be used as a tool to estimate the fuel temperature reactivity coefficient with good prediction and accuracy shown from the calculation of mean absolute error (MAE) of 0.00349.

1. Introduction

Temperature reactivity coefficient is one of the important operational and safety parameters in nuclear reactors for monitoring of normal operation and transient safety analysis in research as well as in power reactors [1]. Fuel temperature is one of the main operating parameters that affect the reactivity during reactor operation. When there are changes in the fuel temperature, the reactivity will also be affected. This affected reactivity is called the reactivity feedbacks and characterized as reactivity coefficients. The reactivity coefficient can be defined as the change of reactivity per unit change in some operating parameter of the reactor. In this work, the operating parameters chosen is fuel temperature data which measured by instrumented fuel element in Reactor TRIGA PUSPATI (RTP) core. One of the basic safety aspects for TRIGA reactor is the fuel temperature coefficient [2]. The reactors were inherently designed with negative temperature reactivity coefficient that automatically reset the reactor back to stability in any case of power excursion event.

In TRIGA reactor, the fuel element consists of uranium zirconium hydride (UZrH) which characteristically to have large negative fuel temperature coefficient. When the temperature of the hydride rises, the probability that a thermal neutron in the fuel element will gain energy from the excited state of an oscillating hydrogen atom in the lattice will also increase. As the neutrons gain energy from the ZrH, the thermal neutron spectrum in the fuel element shifts to a higher average energy which causes the neutron spectrum to harden [3]. Besides, increases in fuel temperature will affect the Doppler broadening due to the increased resonance capture reaction rate in U-238. In TRIGA Doppler effect contributes less than half to total fuel temperature reactivity effect [4].

The fuel temperature reactivity coefficient (FTC) is one of the important safety parameters as it implies the neutron population in the reactor core during fission process to ensure safe and efficient reactor operations. As such, development of the online monitoring system at IRP-R1 TRIGA reactor that monitors the parameters such as control rod worth, reactor temperature reactivity coefficient and the loss of reactivity during reactor operation were previously initiated [5-7]. The developed monitoring system depend on the equations that were used in the data acquisition and processing system installed at one of the instrumented fuel rods in the reactor core.

In Rabir [8] works, the measurement of the fuel temperature reactivity coefficient, α_T at RTP has been conducted by measuring the reactor power, fuel temperature, and the control rod positions. The results were approximately -6.925 pcm $^{o}C^{-1}$ of the average temperature reactivity coefficient which is 30% lower than the safety limit given in Safety Analysis Report (SAR). The method was done through offline mode by using Equation 1 where the reactivity change, $\Delta\rho$ was obtained from the calibrated control rod curve and the ΔT were measured using instrumented fuel elements. However, this method is not feasible for the online monitoring system of research reactors. Thus, in this study, the goal is to develop a new methodology to estimate the temperature reactivity coefficient by using adaptive neuro-fuzzy inference system (ANFIS) for the preliminary assessment of online monitoring system in research reactors.

$$\alpha_T = \frac{\Delta\rho}{\Delta T} \quad (1)$$

2. Materials and Method

2.1 ANFIS

In nuclear industries, the artificial intelligence (AI) has been used widely in the monitoring system, fault detection system, nuclear reactor safety analysis, and others [9-12]. ANFIS is one of the soft computing techniques that combine the fuzzy system and neural network method which can be effectively used due to the easy interpretability of fuzzy logic, as well as the superior learning ability and adaptive capability of neural networks [13]. ANFIS can serve as a basis for constructing a set of fuzzy if-then rules with appropriate membership functions, μ_f to generate the stipulated input-output pairs. The integrated technique combines both methods' advantages and also eliminates their disadvantages which then are capable to be used as a new flexible tool to predict fuel temperature reactivity coefficient.

As studied by Salman *et. al* in [9], the ANFIS method has proven to be an excellent function approximation tools to predict the critical heat flux with high accuracy. Besides, works done by Antio *et. al* in [10] showed that the ANFIS has also been used to predict the power peaking factors accurately and in real time. As studied by Man Gyun Na *et. al* in [11], the prediction of power peaking factor has also been estimated using the fuzzy neural network which is suitable for local power density monitoring. Similarly, this study was intended to utilize ANFIS that can predict the fuel temperature reactivity coefficient at RTP using online fuel temperature and control rod positions measurements.

2.2 ANFIS Architecture

ANFIS or Sugeno-type fuzzy inference system (FIS) is chosen in this study as it is simple in computation and easy to be combined with optimizing and self-adapting methods [13]. The Sugeno-type FIS has been proposed by Tagaki and Sugeno, and by Sugeno and Kang [15]. Initially, Tagaki and Sugeno proposed to use fuzzy IF-THEN rules as followed:

$$L^{(l)}: \text{IF } x_1 \text{ is } F_1^l \text{ and} \dots \text{and } x_n \text{ is } F_n^l, \quad \text{THEN } y^l = c_0^l + c_1^l x^l + \dots + c_n^l x_n \quad (2)$$

where F_l^l are fuzzy, c_i are real-valued parameters, y^l is the system output due to rule $L^{(l)}$, and $l = 1, 2, \ldots$, M. ANFIS architecture consists of five layers as shown in Figure 1 below. A Circle and square in Figure 1 reflect the adaptive capabilities in which the circle indicates fixed node while square specifies the adaptive node that can change the parameters during training.

In Figure 1, there are two inputs labelled as x and y with the targeted output of z. Each node in the first layer generates a membership grade while nodes in the second layer calculate the firing strength of the rule. Next, nodes in the third layer calculate the ratio of the 1 and 2 rule's firing strength to the total of all firing strength. In the fourth layer, each node is an adaptive node which maps to the output membership functions. The node in the fifth layer gives the overall output [13].

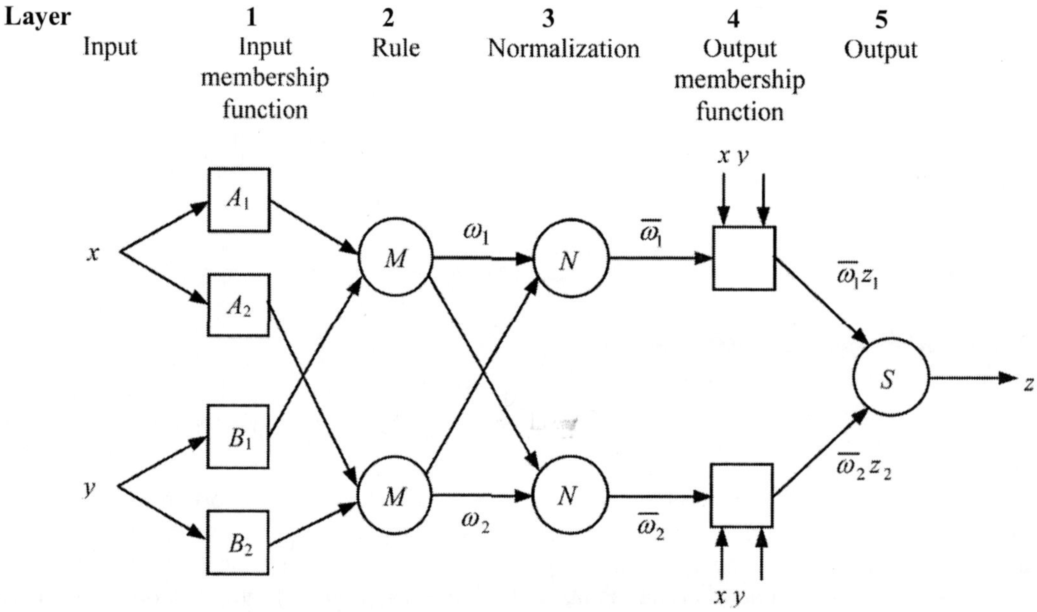

Figure 1. ANFIS architecture [10].

2.3 Methodology

The data input for the training and checking procedures are the four control rod positions and fuel temperature measured from the instrumented fuel elements of Channel A and Channel B. The targeted output is calculated using Equation (1). The reactivity change is determined from the calibrated control rod curve by considering each critical rod position, before and after each step. The error of the reading is estimated in ±0.1 cents, mainly due to the uncertainty in rod position [8].

The development of the network is constructed using Neuro-Fuzzy Designer toolbox in MATLAB software. The training and checking procedure for the model are shown in the flowchart in Figure 2. In this study, there are two types of membership function that are chosen in each FIS model which are *gaussmf* in Model 1 and *gauss2mf* in Model 2 as stated in Table 1 below. Figure 3 shows each of the membership function shapes for all inputs and Figure 4 displays the final membership function after the training session.

The optimization method used in this training procedure is hybrid which consists of two parts of the gradient method and least square method. The gradient method is applied to the calculation of input membership function parameters whereas the least square method is applied to the calculation of output function parameters [2]. The checking data was used to validate the FIS model developed during the training procedure and can be loaded before starting the training session. The purpose of having checking data associated with training data is to represent the point of overfitting in the developed model. The overfitting can occur during the training procedure, when the checking error should decrease but somehow at a certain point, increase. Then, the modelled FIS can be validated when the root means square error for checking data decreases as the training begins. Table 1 shows the details used for training and validating the FIS model. The least checking error of the model was later used for estimating the fuel temperature reactivity coefficient.

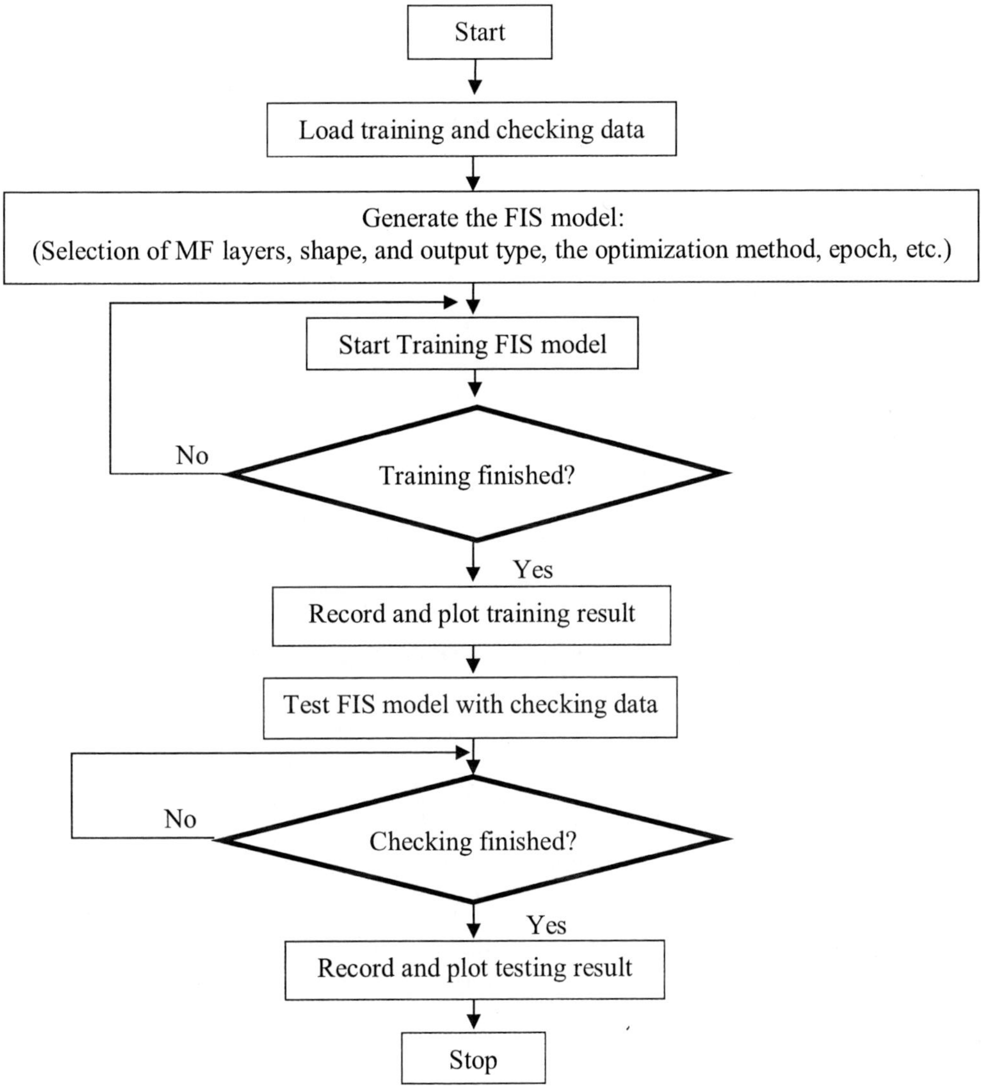

Figure 2. Flowchart for training and checking session of ANFIS in Neuro-Fuzzy Designer GUI.

IOP Conf. Series: Materials Science and Engineering **555** (2019) 012002 doi:10.1088/1757-899X/555/1/012002

Table 1. ANFIS input details before training sessions.

Model		1	2
No. of μ_f		2 2 2 2 2	2 2 2 2 2
Shape of μ_f	Input 1	*gaussmf*	*gauss2mf*
	Input 2	*gaussmf*	*gauss2mf*
	Input 3	*gaussmf*	*gauss2mf*
	Input 4	*gaussmf*	*gauss2mf*
	Input 5	*gaussmf*	*gauss2mf*
Type of μ_f Output		Linear	Linear
Training method		Hybrid	Hybrid
Epochs		500	500

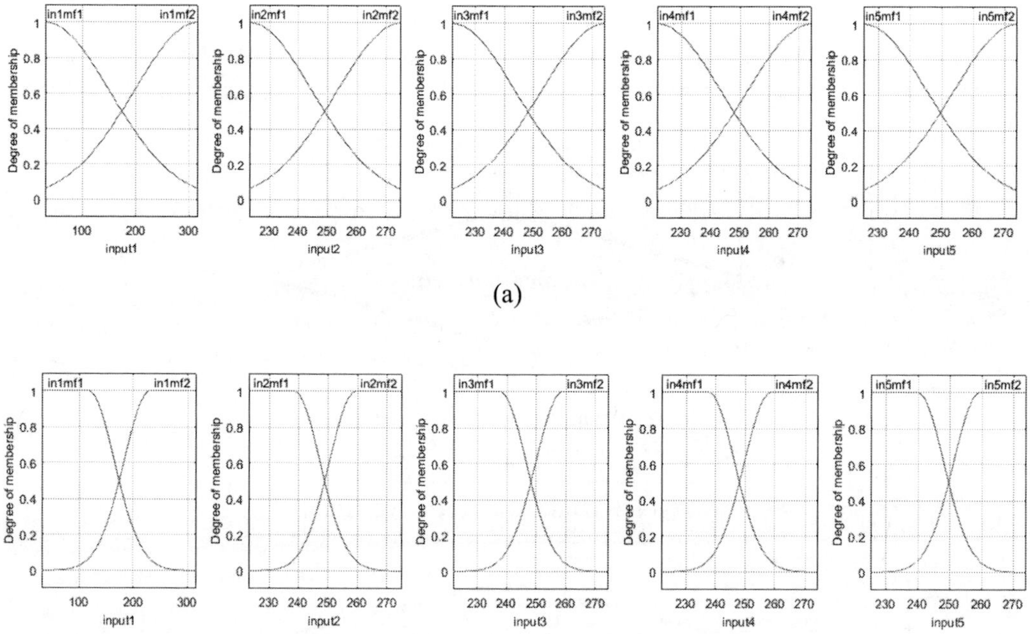

(a)

(b)

Figure 3. FIS model with different membership functions before training session: (a) *gaussmf* shape (b) *gauss2mf* shape.

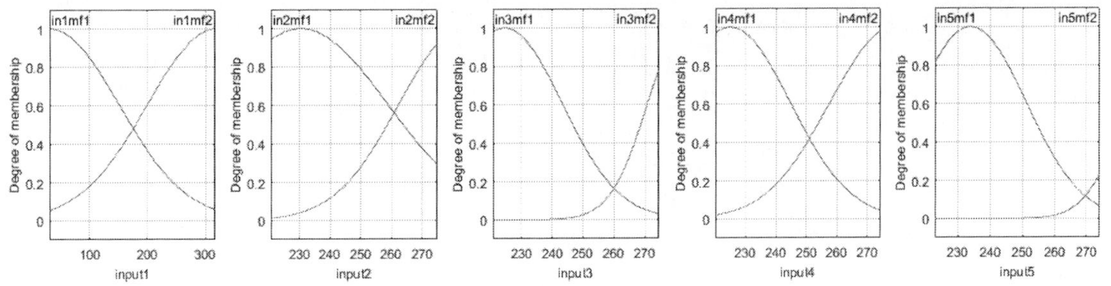

(a)

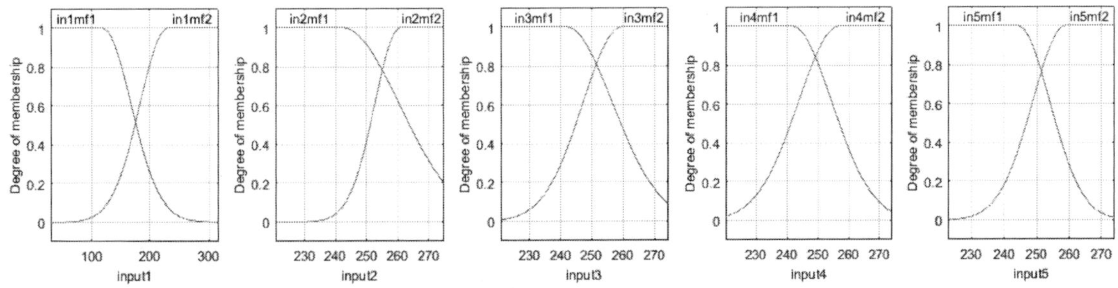

(b)

Figure 4. FIS model with different membership functions after the training session: (a) *gaussmf* shape (b) *gauss2mf* shape

3. Results

Table 2 lists the information from the ANFIS training procedures which show performance of the model on predicting the outputs. Model 2 shows the overfitting characteristic which makes the selection of membership functions not the best choice for modelling the training data. The overfitting occurs due to the sudden increases in checking error during the training session. Figure 5 shows each model's training error and checking errors. It is clear that the overfitting occurs in Model 2 while in Model 1, the training session does not exhibit any overfitting characteristics.

Table 2. ANFIS information during training session.

Model	1	2
No. of nodes	92	92
No. of linear parameters	192	192
No. of nonlinear parameters	20	40
Total no. of parameters	212	232
No. of Fuzzy rules	32	32
RSME Training Error	0.0031	0.0030
RSME Checking Error	0.0066	0.0318
Overfitting	No	Yes

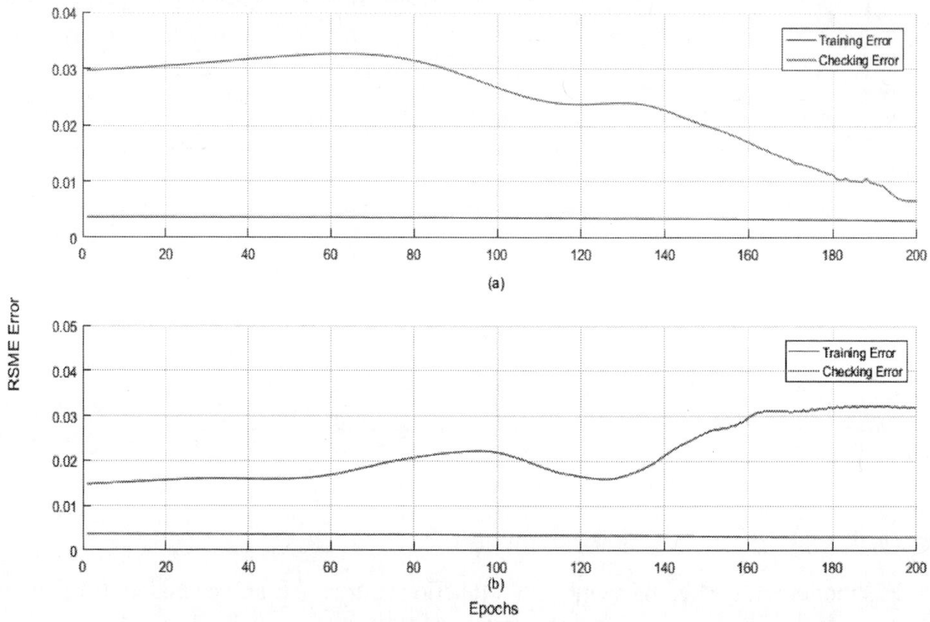

Figure 5. Training and checking error after training each Model; (a) Model 1, (b) Model 2.

Therefore, only Model 1 was used to test the real data as shown in Figure 6. The test data were obtained from fuel temperature measurement at Channel B. To evaluate the performance and the quality of the model, the mean absolute error (MAE) is calculated by using Equation 2 [16]. The MAE for Model 1 is 0.00349 which implies the model can predict the output accurately. In terms of fuel temperature reactivity coefficient, the average results from the actual output and predicted output of the testing data are -0.00571 $\$^{o}C^{-1}$ (-3.997 pcm $^{o}C^{-1}$) and -0.00592 $\$^{o}C^{-1}$ (4.144 pcm $^{o}C^{-1}$), both are less than the values reported in the RTP Safety Analysis Report of 0.014$\$^{o}C^{-1}$ (-10 pcm $^{o}C^{-1}$) [17].

$$MAE = \frac{1}{n}\sum_{i=1}^{n}|y_i - f_i| \quad (2)$$

where y_i and f are the actual outputs and predicted the output.

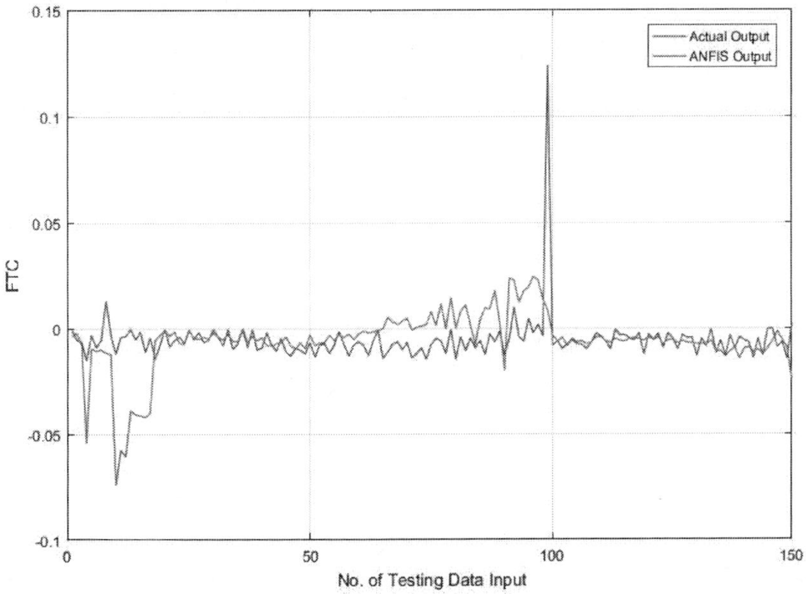

Figure 6. Testing Data output to validate the model performance and quality.

4. Conclusion

Estimation of the fuel temperature reactivity coefficient at RTP has been demonstrated using ANFIS techniques. The model developed by ANFIS are capable in predicting the output which results on -0.00571 $\$°C^{-1}$ (-3.997 pcm $°C^{-1}$) and -0.00592 $\$°C^{-1}$ (4.144 pcm $°C^{-1}$) of actual output and predicted output which are less than the values reported in the SAR. The performance and quality of the model are also good with 0.00349 of MAE. Thus, ANFIS can be one of the alternative methods to calculate the fuel temperature reactivity coefficient in the future. This concept also can be applied to develop real-time monitoring system.

5. References

[1] Phan T T G and Do Q B 2015 Evaluation of the reactivity feedback coefficients of the Da Lat nuclear research reactor using highly enriched uranium fuels and low enriched uranium fuels *11th Nat. Conf. on Nuclear Sci. and Tech.* (Danang, Vietnam: August 06-07, 2015).

[2] Nabeshima K, Suzudo T, Suzuki K 1998 Real-time nuclear power plant monitoring with neural network *J. Nuclear Sci. Tech.* p 93-100.

[3] https://www.nuclear-power.net/nuclear-power/reactor-physics/nuclear-fission-chain-reaction/reactivity-coefficients-reactivity-feedbacks/ (4/8/18), p 0035.

[4] http://www.rcp.ijs.si/ric/safety_parameters-a.html (4/8/18) p 0045.

[5] Mesquita A Z and Gomes do Prado Souza R M 2007 On-line monitoring of the reactivity and control rod worth at the IRP-R1 TRIGA Reactor *Int. Nuclear Atlantic Conf. INAC-2007* (Santos, SP, Brazil: Sept 20- Oct 5 ISBN: 978-8599141-02-1).

[6] Mesquita A Z and Gomes do Prado Souza R M 2008 On-line monitoring of the reactivity temperature coefficient of the IRP-R1 TRIGA nuclear research reactor *12th Brazilian Cong. of Thermal Eng. and Sci.* (Belo Horizont:, MG Nov 10-14, 2008).

[7] Mesquita A Z and Gomes do Prado Souza R 2010 On-line monitoring of the IRP-R1 TRIGA reactor neutronic parameters *Progress in Nuclear Energy* **52** p 292-297.

[8] Rabir M H 2013 Measurement of the power and temperature reactivity coefficient of the RTP TRIGA reactor *J. of Nuclear Eng. and Design* **265** p 269-271.

[9] Zaferanlouei S, Rostamifard D, Setayeshi S 2010 Prediction of critical heat flux using ANFIS. *Ann. of Nuclear Energy* **37** p 813–821.

[10] Guimarães A C F, Lapa Celso M F 2007 Power peak factor estimation using adaptive neural fuzzy inference system *Adv. in Comp. Sci. and Eng.*

[11] Man G N, Dong W J, Sun H S 2004 Estimation of the nuclear power peaking factor using in-core sensor signals *J. of the Korean Nuclear Society* **36(5)** p 420-429.

[12] Housiadas C, Antonopoulos-Domis M 2001 Estimation of the moderator temperature coefficient of reactivity via noise analysis using closed-loop transfer functions *Ann. of Nuclear Energy* **28** p 983-991.

[13] Guo Y, Gong C, Zeng H Y 2010 Tthe application of artificial neural network in nuclear energy *Proc.of the Ninth Int. Conf.on Machine Learning and Cybernetics* (Qingdao: IEEE, 11-14 July 2010).

[14] Jhy-Shing and Jang R 1993 ANFIS: Adaptive neuro-fuzzy inference system *IEEE Trans. on Sys., Man and Cybernetics* **23(3).**

[15] Walia N, Singh H, Sharma A 2015 ANFIS: Adaptive-Fuzzy Inference System- A survey *Int. J. of Comp. App.* **123(13)** p 0975-8887.

[16] Mathur N, Glesk I, Buis A 2016 Comparison of adaptive neuro-fuzzy inference system (ANFIS) and gaussian processes for machine learning (GPML) algorithms for the prediction of skin temperature in lower limb prostheses Med. Eng. and Phys. p 1-7.

[17] Malaysian Nuclear Agency 2008 *Reaktor TRIGA PUSPATI Safety Analysis Report.*

Acknowledgment

The author would like to thank all research officers from Agensi Nuklear Malaysia (ANM) and from Kementerian Pelajaran Tinggi Malaysia (GUP Tier 1. Q. J130000.2546.18H90) for their supports and helps in completing this work.

Performance analysis of random sampling algorithms on GNU Octave

Mohd Syahmi bin Yazid[1], Matthieu Lemaire[2], Mohsin bin Mohd Sies[1], Deokjung Lee[2]

[1]School of Chemical and Energy Engineering, Faculty of Engineering, Universiti Teknologi Malaysia, 81310 Johor Bahru, Johor, Malaysia.
[2]Department of Nuclear Engineering, Ulsan National Institute of Science and Technology, 50 UNIST-gil, Ulsan 44919, Republic of Korea.

deokjung@unist.ac.kr

Abstract. This paper presents the performance analysis of two random sampling algorithms, the inverse-transform method and the Vose aliasing method, on GNU Octave. The Monte Carlo code MCS developed by UNIST uses random sampling methods to simulate the physics of neutron and photon transport [1]. The goal is to optimize the sampling time of MCS for cases when the probability density function is a constant function throughout the simulation. For this purpose, the runtime of the inverse-transform method and Vose aliasing method are compared for increasing input size with scripts developed on GNU Octave. To compare the execution time, the initialization and generation time of both methods are determined and discussed.

1. Introduction

The random sampling of physical parameters modelled as random variables is an essential component of Monte Carlo simulations [4]. Monte Carlo simulations are able to simulate problems that are stochastic (probabilistic) by nature, such as the transport of neutrons and photons in matter [6]. Random variables can be characterized by so-called probability distribution functions (PDFs) and there are several algorithms available in the literature to perform random sampling from a PDF. The runtime of algorithms can become the major bottleneck that limit the performances of computer simulation and selecting faster algorithms is therefore key for efficient simulations yielding results in reasonable times [3]. In this paper, an analysis of the runtime of two random sampling algorithms for discrete distributions, the inverse-transform method and the Vose aliasing method, is conducted. Both methods are implemented in the mathematical freeware GNU Octave [2] and the initialization time and generation times of the two methods is compared and discussed.

The paper is organized as follows. In section 2, the theory of random sampling, the inverse-transform method (IVM) and the Vose aliasing method (VAM) are introduced. The runtime comparison results are analysed and discussed in section 3 and conclusions are drawn out in section 4.

2. Random sampling for discrete distribution

Random sampling are techniques to generate random values of a variable x distributed in the interval (x_{min}; x_{max}) according to a given PDF, p(x) [4]. In discrete distributions, the PDF p(x) is quantized: this case corresponds for instance to throwing a dice or tossing a coin. Technical vocabulary associated to discrete distributions and probability terms are defined in g over the simulated histories.

Table *1*. From a discrete PDF, a discrete cumulative distribution function (CDF) or a probability-alias table can be generated. This CDF / this probability-alias table can be used in conjunction with a random number generator (RNG) to sample random variables according to the initial PDF. The RNG commonly samples number uniformly in the interval (0;1). An illustration of the random sampling process is shown in Figure 1. If the number of generated random variables is large enough, quantitative information on the process being simulated (dice, neutron transport process, etc.) may be obtained by simply averaging over the simulated histories.

Table 1. Probability terms used for discrete distributions.

Terms	Notation	Definition/description
Random variable	x	Variable whose value is determine by chance
Sample space	N	Total number of possible x-values
Probability distribution functions, PDF	*PDF(x) = n/N*	Function which allocates probabilities for certain event to occur where: • n = number of values of x that fall within interval ($x_{min} \leq x \leq x_{max}$) • N = sample space
Cumulative distribution function, CDF	$CDF(x) = \sum_{i=1}^{N} p(x)$	• Shows cumulative summation of PDF for each x-value • Sum of all CDF equals 1 • CDF is non decreasing function
Random number generator, rand	$Rand = \xi$	rand is an Octave keyword to generates a random number, ξ uniformly between 0 and 1. ($0 \leq \xi < 1$).

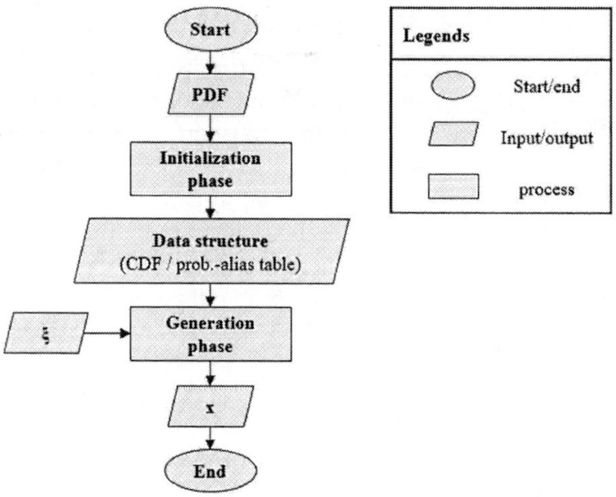

Figure 1. Random sampling for discrete distributions.

2.1 Inverse-transform method (IVM)

Inverse transform method is the conventional algorithm used in random sampling. This practical method is useful for generating random values of x using a generator of random numbers uniformly distributed in (0,1). It exploits the fact that the CDF, because it is the cumulative summation of PDF for each x-value, is a non-decreasing function that can be inversed.

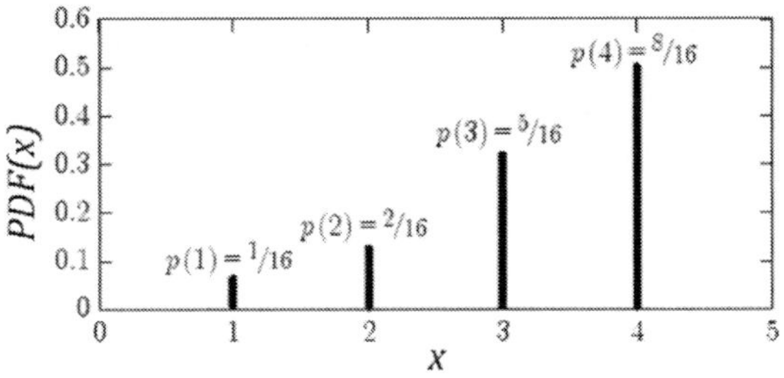

(a) **(b)**

Figure 2. (a) Example of PDF and, (b) CDF for sampling with the inverse-transform method.

The output of the uniform RNG $0 < \xi < 1$ is related to CDF by Equation 1 and Equation 2 consists in inversing the CDF (hence the name inverse-transformation method). The random variable, x can therefore be sampled according to Equation 2. The sampling process is illustrated in (b)

Figure *2* [4]. The initiation phase of IVM consists in creating a CDF by summing up the PDF cumulatively. The generation phase of IVM consists in solving Equation 2 for a random number ξ sampled uniformly in (0,1) to determine the random variable, x.

$$\xi = \mathrm{CDF}(x) \tag{1}$$

$$\mathrm{CDF}(\xi)^{-1} = x \tag{2}$$

2.2 Vose aliasing method (VAM)

The Vose aliasing method is another approach to generate a random variable from a RNG outputting numbers uniformly between 0 and 1. The VAM relies on the use of a so-called probability-*alias* table instead of a CDF. The initialization phase of VAM consists in generating the probability-*alias* table from the specified PDF. In the probability-*alias* table, the row *prob* stores the probability within the column that the original PDF will be chosen whereas the row alias stores the index of the cut PDF (if any) [5]. This index acts as a pointer referring to the original PDF (hence the name alias). The table is characterized by a length of N and the height is 1 (100% probability) as displayed in Figure 3.

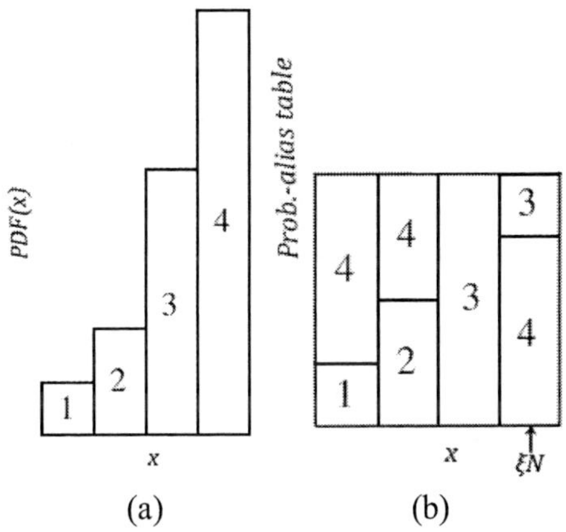

(a) (b)

Figure 3. (a) Example of PDF and (b) *prob.-alias* table for sampling with the Vose aliasing method.

In the generation phase, a random number, ξ_1 is generated by the RNG to select a column of the probability-alias table uniformly at random. The select column is called column i. Next, another random number, ξ_2 is generated by the RNG. If $\xi_2 \leq prob[i]$, the random variable, x in *prob*[i] is returned, else the random variable, x in *alias*[i] is returned. In this process, 2 random numbers from the RNG are required to sample a single random variable. A trick can be used to only one random number instead of two: it is indeed possible to replace ξ_2 by ξ_1. N – floor (ξ_1. N) and still obtain a fair sampling of the random variable.

3. Results

The data collected from the validation test is summarized in Table 2 and illustrated in Figure 4. The difference between the normalized PDF, the IVM frequency and VAM frequency is very small, thus showing that both IVM and VAM methods were properly implemented in GNU Octave and are valid for further random sampling usage.

Table 2. Validation test.

Dice value	Normalized PDF	1 million samples in total			
		IVM count	IVM frequency	VAM count	VAM frequency
1	1/16 = 0.0625	6.26×10^4	0.0626	6.23×10^4	0.0623
2	1/16 = 0.0625	6.23×10^4	0.0623	6.24×10^4	0.0624
3	2/16 = 0.125	1.24×10^5	0.124	1.25×10^5	0.125
4	3/16 = 0.1875	1.87×10^5	0.187	1.87×10^5	0.187
5	4/16 = 0.25	2.49×10^5	0.249	2.50×10^5	0.25
6	5/16 = 0.3125	3.12×10^5	0.312	3.12×10^5	0.312

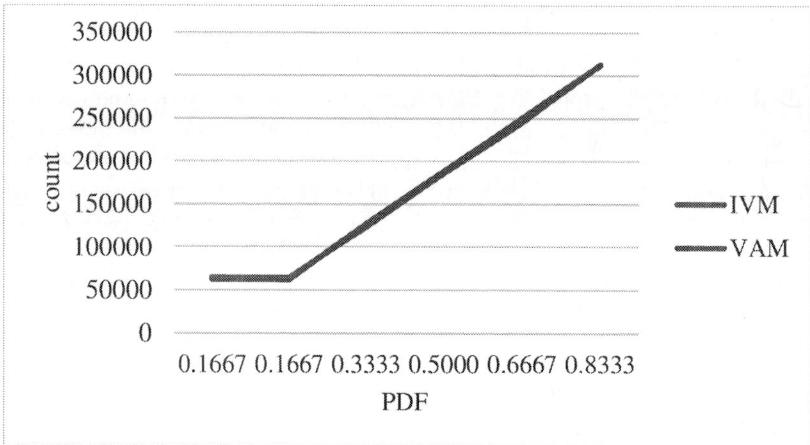

Figure 4. Validation test of implementation of random sampling methods.

The runtime data to generate one million samples from increasing PDF size is tabulated in Table 3. Figure 5 depicts the initialization time as a function of the PDF size. For both methods, the initialization time increases linearly as the PDF size increases. However, the slope for VAM is much steeper than for IVM due to a more complex initialization phase. Notice that the initialization time of both methods (runtime of one initialization phase) is very small compared with the generation time of one million samples, which makes it not significant to the total runtime.

Table 3. Runtimes of sampling methods for increasing PDF size.

		PDF size, N				
		20	40	60	80	100
Initialization time [s]	IVM	9.6×10^{-5}	2.8×10^{-5}	3.2×10^{-5}	6.2×10^{-5}	1.3×10^{-4}
	VAM	2.5×10^{-3}	3.2×10^{-3}	4.9×10^{-3}	8.8×10^{-3}	8.5×10^{-3}
Generation time [s]	IVM	209.5	348.0	478.3	616.7	746.6
	VAM	64.0	65.7	63.5	65.0	64.9

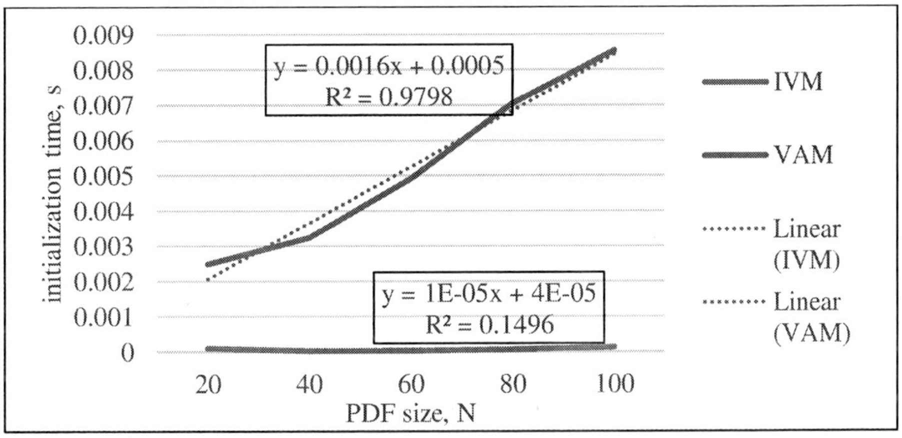

Figure 5. Initialization time [s] versus PDF size, N.

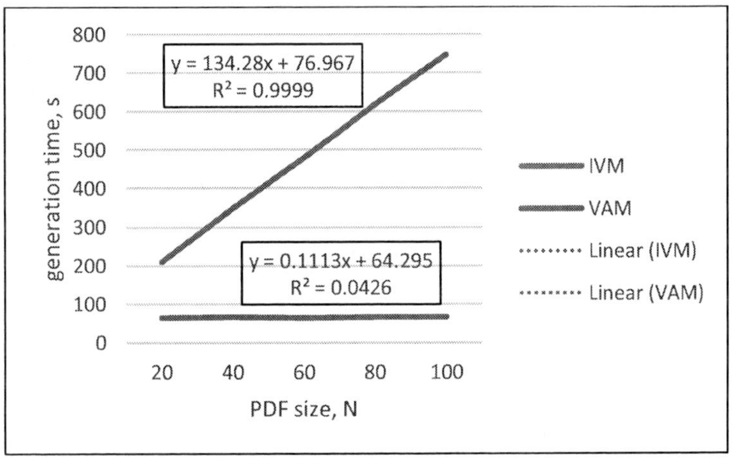

Figure 6. Generation time [s] versus PDF size, N.

In Figure 6, a proportional increase of the generation time as a function of the PDF size is observed. This relationship occurs because in the generation phase of IVM, a sequential search of the CDF is performed to output a sample, whose runtime increases linearly with the PDF size. Meanwhile,

VAM shows an almost constant generation time close to ± 64.5s regardless of the PDF size. This speed advantage is due to the smart data structure of the prob.-alias tables. Therefore, as long as the initialization phase is only performed once, this analysis shows that the VAM has a better runtime performance than the IVM.

4. Conclusions

The IVM and VAM have been implemented in GNU Octave and the complexity analysis of both algorithms have been performed. VAM is shown to have a better generation runtime performance compared to IVM but a worse initialization runtime. For random variables with constant PDFs, the initialization is only called once: VAM is therefore a good alternative to IVM to optimize the sampling time of random variables characterized by constant PDF. An example of constant PDFs can be for instance the electron shell transition probabilities in Monte Carlo codes simulating the transport of photons and the creation of fluorescence photons through atomic relaxation processes. Implementing the VAM instead of IVM for such a case will result in a more complex initialization phase and coding but a net gain of computing time. VAM instead of IVM may be used in other circumstances as well as long as the PDF governing the random variables is constant and the initialization phase is only called once.

5. References

[1] Lee D, lemaire M, 2017 MCS Photon Physics Retrieved from CORE | UNIST:http://reactorcore.unist.ac.kr/wpcontent/uploads/sites/229/2016/03/20180131_MCS_photon_transport.pdf

[2] Eaton J, 2018 About and History of GNU Octave Retrieved from GNU Octave: https://www.gnu.org/software/octave/about.html

[3] Sedgewick R, 2013 *An Introduction To The Analysis of Algorithms, Second Edition* . Pearson Education, Inc.

[4] Salvat F 2014 *PENELOPE-2014: A Code System for Monte Carlo Simulation of Electron and Photon Transport* (Barcelona, Spain: Organisation for Economic Co-operation and Development (OECD), Nuclear Energy Agency (NEA)).

[5] Schwarz K 2011 Darts, Dice, and Coins: Sampling from a Discrete Distribution Retrieved from http://www.keithschwarz.com/darts-dice-coins/

[6] Vujic J L 2008 Monte Carlo Sampling Algorithms (Berkeley: Nuclear Engineering Department, University of California).

Acknowledgement

This project was conducted as a two-month undergraduate internship under the supervision of the UNIST CORE staff members who provided relevant advice all the way to complete this paper. The publication of this paper is financially funded by the Malaysian Nuclear Society (MNS). Their contributions are highly appreciated.

IOP Conf. Series: Materials Science and Engineering **555** (2019) 012004 doi:10.1088/1757-899X/555/1/012004

Applicability of one-body model to predict thermal distribution in spent nuclear fuel pool

Husaini Roslan[1,a)], Nahrul Khair Alang Md Rashid[1], Muhammad Arif Sazali[1]

Department of Energy Engineering, Faculty of Chemical and Energy Engineering, Universiti Teknologi Malaysia, 81300 Johor Bahru, Johor, Malaysia.

[a]husainiroslan93@gmail.com

Abstract. An assumption system with a one-body model of spent nuclear fuel (SNF) assemblies was developed to predict the thermal distribution in the spent fuel pool (SFP) in the absence of external cooling system. It was based on the three-dimensional (3D) thermal hydraulic behaviour computed using the computational fluid dynamic (CFD) software, Fluent 18.0. This study aims to determine the applicability of the one-body model (average decay heat) by comparing it with the individuals-body model (variation of decay heat) computation results. The thermal distribution for both models were compared and evaluated, neglecting other effects such as position of the SNF assemblies, decay heat, and axial heat flux profile. It was found that the one-body model was applicable to predict the thermal distribution of the SFP during loss of active cooling system with error less than 1°C.

1. Introduction

Spent fuel pool (SFP) is a type of storage to cool spent nuclear fuel (SNF) unloaded from nuclear reactors. The SNF is loaded into the pool immediately to reduce the radioactive decay heat and radiation level [6]. SFP usually equipped with storage rack assemblies to hold the SNF rods. To cool the SNF and to ensure enough decay heat removal, SFP is generally designed with external active cooling system. However, this cooling system will stop working if there is a loss of electrical supply. As the cooling capability of the SFP is unavailable, the temperature of SFP water and SNF will increase continuously [8]. Therefore, it is important to evaluate the thermal behavior of the SFP during a loss of cooling accident.

The accident in Fukushima Daiichi Nuclear Power Plant in March 2011 had proven the importance of consistent monitoring on the thermal behavior of SFP. There is also an increase in awareness to improve the safety level of SFP when there is a loss of active cooling system [10]. Many researches have been done to investigate the thermal behavior of the SFP and to improve the safety level of the SFP during loss of cooling accident. Yanagi *et al.* [1-4] have developed a system to predict the thermal distribution in the SFP during loss of cooling accident. They carried out a series of studies comprising of prediction on decay heat, calculation of heat flux to the air, and three-dimensional (3D) hydraulic model of the SFP using computational fluid dynamic (CFD) software Fluent 18.0.

Since the CFD solving the conservation equation has become more efficient and robust due to the tremendous advancements of both software and hardware technologies in the past decade, it has becoming a useful tool for engineering design and analysis [6]. CFD method also has become one of the most popular methods to predict the thermal behavior of the SFP. However, one of its disadvantages is that 3D computation requires a long computing time. Many researches had been

IOP Conf. Series: Materials Science and Engineering **555** (2019) 012004 doi:10.1088/1757-899X/555/1/012004

conducted to investigate the applicable model to reduce the computing time and modelling complexity. Yanagi *et al.* [1] had developed one-region model to predict thermal behavior of the SFP. They also investigated the effect of computing parameters such as the mesh size and the timestep size. Based on the previous works, the authors had confirmed the applicability of the one region model to predict thermal behavior of the SFP. They also found that the mesh size and the timestep size had low effect on the result of the computation. Therefore, other computation parameters could be optimized to reduce the computing time.

The current study aims to investigate the applicability of the one-body model of the SNF assemblies on predicting the thermal distribution in the SFP during loss of cooling accident using computational fluid dynamic software, Fluent 18.0. This study compares the measured point temperature from both one-body assemblies model (average decay heat) and individual-body (variation of decay heat) assemblies model.

2. Materials and Method

2.1. Computational Model Development

The conceptual design of SFP established by Yanagi *et al.* [4] is used to develop the computational model of SFP which is shown in Figure 1(a). The dimensions of the SFP are 15 m (W) × 9 m (L) × 12 m (D). The elevation of the SNF is 0.5 m from the bottom of the SFP. The distance between each SNF rack (center to center) are 1.9 m and 2.8 m for x and z direction respectively. The water surface level is 7 m above the top of the SNF assemblies which functions as gamma radiation shielding [3]. No external active cooling system is involved in this study.

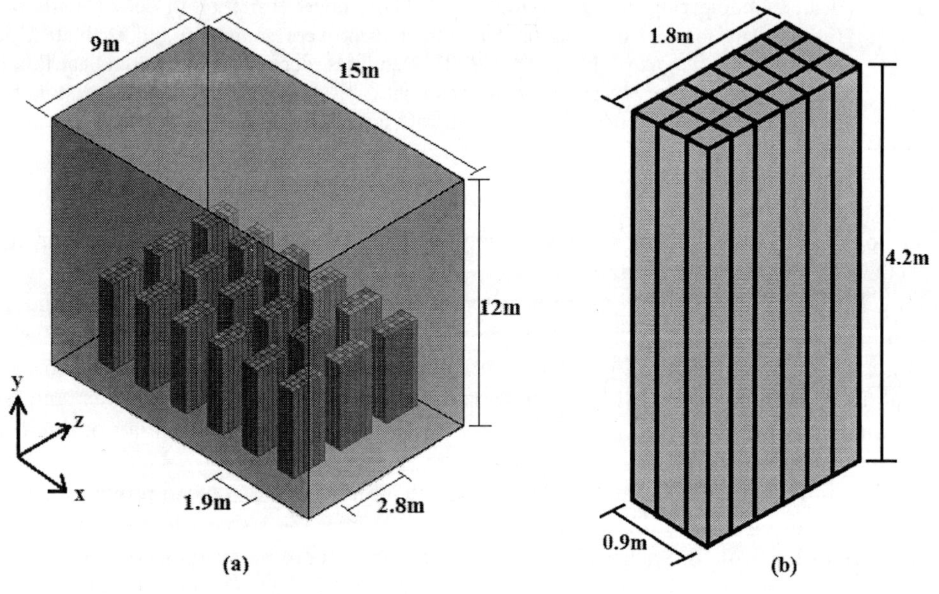

Figure 1. Conceptual design of SFP [4].

There are 18 SNF racks in the SFP in which each rack contains 18 cells and each cell stores 17x17 SNF rod assemblies. The dimension of a single SNF rack are 1.8 m (W) × 0.9 m (L) × 4.2 m (D) as shown in Figure 1(b). To reduce the computation complexity and errors during modelling of the real SFP, some geometry simplification usually was made. Based on previous studies, all SFP models were described as individual-body model in which each SNF cells was treated as an individual body source with respective decay heat value, regardless the number of the SNF rod. In this study, the effect of decay heat distribution in the SFP was evaluated by using the one-body model, in which all the SNF cells in each rack were treated as one body source with average decay heat value. Figure 1 shows the computational design of the SFP loaded with individuals-body assemblies, while Figure 2 (a) consists of one-body SNF assemblies. Figure 2(b) shows the side view of the computational design of SFP loaded with one-body model. For the individuals-body model, there were 18 assemblies of SNF rack loaded with 18 SNF cells. The dimensions of each SNF cells are 0.3 m (W) × 0.3 m (L) × 4.2 m (D) (refer Figure 1(b)). The modeled SFP were also assumed to have SNF assembly in the form of SNF cell box placed in the rack where there are no holes or gap in the cell rack. The value of decay heat generated from each SNF cells box was varied from 5MW to 15MW. For the one-body model, there was also 18 assemblies of SNF racks, but each SNF racks was treated as one body source with averaged decay heat value which is 10MW.

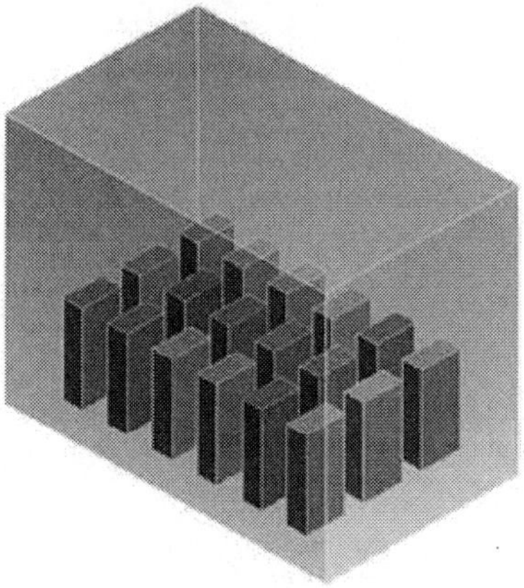

(a)

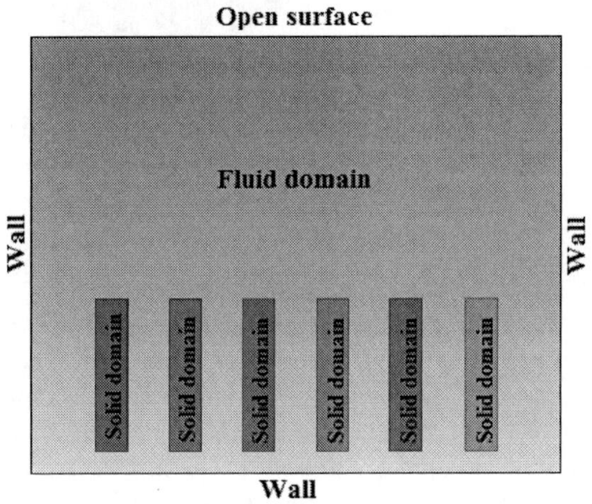

(b)

Figure 2. Computational design of SFP loaded with one-body assemblies.

Figure 3 and Figure 4 show the computational grid model and cross-sectional view of the grid model respectively. They were applied to investigate the thermal hydraulic behavior of the SFP. Uniform structured mesh was used for all body involved in this simulation. The maximum size of the mesh cells was about 0.3 m. Different name selections such as fluid domain, solid domain, wall domain, and open domain were assigned to each body and surface as shown in Figure 2(b). The total number of computation cells was set to about 72,000 cells with 0.2 m mesh size near the wall to captured relevant turbulence flow and 0.7 orthogonal quality.

IOP Conf. Series: Materials Science and Engineering **555** (2019) 012004 doi:10.1088/1757-899X/555/1/012004

Figure 3. Computation grid model.

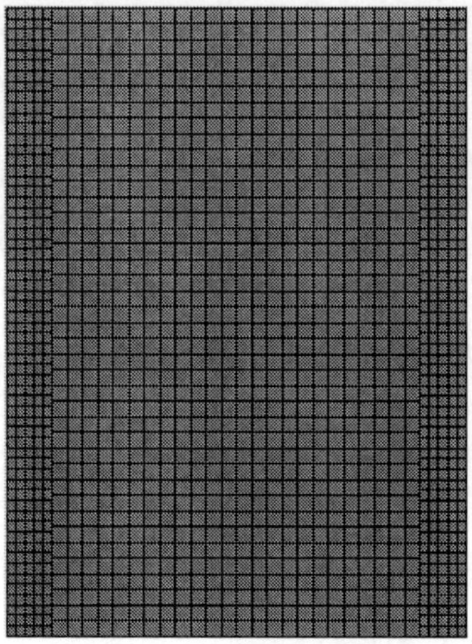

Figure 4. Cross sectional view of computational grid model.

2.2. Computation condition and solution

This research deals with thermal and fluid flow problems. Therefore, the boundary condition had been set up based on the appropriate parameters. The computation model approach and boundary condition used in this simulation were summarized in Table 1. To model the fluid flow field in the SFP, standard k-epsilon turbulence model was used with scalable wall function as the model walls treatment. This turbulence model was used because there were regions in the SFP where Rayleigh number could be high [5]. The wall of the pools was set with non-slip condition and standard wall function in which zero heat flux passes through. The heat transfer between the water and concrete wall was neglected since the heat loss to concrete is small because of its low conductivity and only 10% of the heat is loss to the air [2]. For conjugate heat transfer part between solid (SNF) and fluid (water), both the interfaces were set up as coupled wall. To calculate the natural convection force, and others temperature dependent parameters, the polynomial function based on standard thermodynamic properties of saturated water was used. The transient simulations were done for both models. Time step size was set up at 10 seconds with more than five hours of simulation real time.

Table 1. Computation parameters and description

Computation Parameters	Descriptions
1. Turbulent model	Standard k-epsilon model
2. Convective term	First order upwind
3. Bodies condition	Pool: fluid medium, SNF: Source term in solid state
4. Boundary condition	Top surface: Ambient pressure (1atm) and temperature (300K) Pool bottom and side wall: Stationary with no slip condition Pool and SNF interface: Coupled wall
5. Solution Discretization Method	Energy: Second order upwind Momentum: Second order upwind Turbulent kinetic energy: First order upwind Turbulent Dissipation rate: First order upwind
6. Calculation	Time step size: 10s Number of timesteps: 10 000 Recursion Termination: Temperature of water reach 373K

In terms of decay heat generation, all rack assemblies and individual cells assembly were set up as volumetric heat source. The standard solid zircaloy properties were used to define the properties of the heat source. According to Yanagi *et al.* [2], decay heat has no significant effect on the relationship between the average temperature and the surface temperature, but it affects the point temperature and the computation time. The smaller the decay heat, the higher the computing time. In addition, the maximum difference of water temperature for different SNF arrangement was about 4°C when there is no external active cooling system [9]. Therefore, simulations for one-body model were conducted in which average of 10 MW decay heat were applied to all SNF racks regardless the arrangement of the SNF cells. For individual-body model, the decay heat varies from 5MW to 15MW randomly. Each rack contains nine cells which have decay heat value below 5MW, and the other cells have decay heat values below 15MW. The heat source was set at the center of each SNF cells and racks for individual-body and one-body model respectively. This was to determine the temperature and flow field in the SFP caused by natural convection process due to decay heat transfer. All SFP conditions used in this computation were based on pressurized water reactor SFP, having maximum burnup value of 55GWd/t. The value of 10MW corresponds to the decay heat value of an SNF that is unloaded one week after the reactor shutdown. All SNF were assumed to have the same SNF decay heat profiles and arrangements. The simulation was terminated after five hours of real time just before the temperature at one of the points reached a maximum value of about 373K.

3. Results and Discussion

The temperature distribution and three different point temperatures for individual-body model and one-body model are shown in Figure 5 and Figure 6 respectively. These three points were selected based on the axial distance from the SNF (Point 1: 0.5 m, Point 2: 1.5 m, and Point 3: 2.5 m). Both Figures show the temperature at the points of the SFP after five hours simulation time. The time taken for individual-body and one-body models to achieve the result was about six days and three days respectively.

Based on the results, the temperature at the bottom part of the SFP was higher than that of the upper part since the SNF was located at the bottom of the SFP. Although the temperature at the nearby SNF in the SFP had increased to almost 100 °C, the surface of the water inside the SFP was still at its initial temperature. The results were consistent with the analysis result from Chan et. al [7], which indicates that the upper part temperature of SFP was found to increase more slowly than the bottom part of the SFP. This suggests that the rate of heat transfer due to natural convection process is low. Thus, it is unable to transfer the decay heat from the SNF to the surface of the water efficiently. It was also observed that the temperature between the SNF assemblies was higher than in other locations in the SFP.

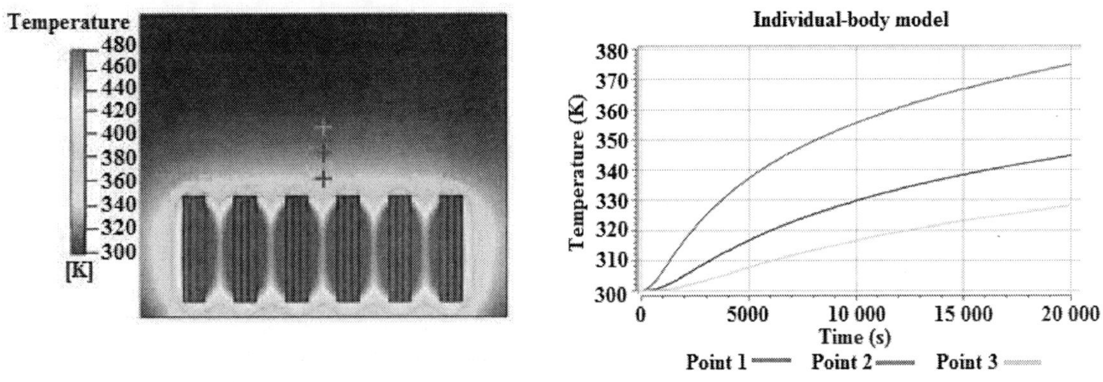

Figure 5. Temperature distribution and three different point temperature of individual-body model.

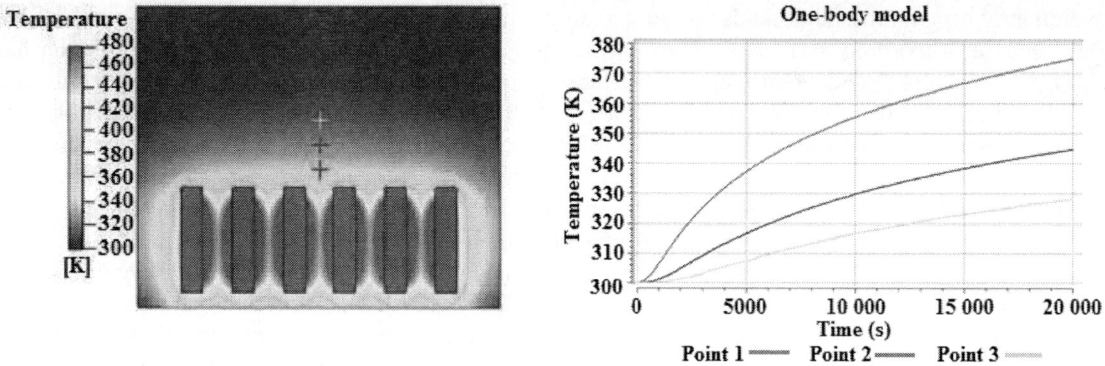

Figure 6. Temperature distribution and three different point temperature of one-body model.

3.1. Comparison of individuals-body and one-body models simulation results

Figure 7 shows the temperature differences between the individual-body and the one-body models at three different points in five hours of simulation time. The temperature differences between these two models were very low at less than 1°C and they increased at a very low rate as simulation time passed. The temperature differences also increased at a very low rate when the distance of the measuring point from the SNF increased. These results show that using the one-body model only had a very small effect on the computation of SFP temperature distribution. This is due to the conduction process that occurs between each SNF cells which transfer the heat to achieve equilibrium condition and standardize the temperature of all the SNF assembly. This results also might be because of heat transfer due to natural convection process in the SFP which has certain limit regardless the amount of the decay heat introduced. Besides that, the small difference in temperature occurs between these two models is due to the effect of the nearest SNF cell which has generated different amount of heat. The real simulation time taken for individuals-body and one-body model was six days and three days respectively. The latter only took three days less to obtain a comparable SFP temperature distribution with 1°C difference than the individuals-body model.

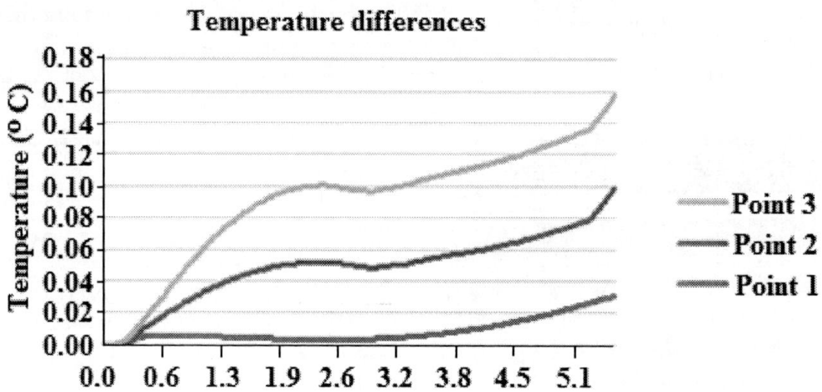

Figure 7. Temperature difference between individual-body and one-body models at three different point location.

IOP Conf. Series: Materials Science and Engineering **555** (2019) 012004 doi:10.1088/1757-899X/555/1/012004

4. Conclusions

In this study, a 3D thermal behavior of SFP during the loss of external cooling system was computed using Ansys Fluent 18.0 CFD software to investigate the applicability of the one-body model to predict the thermal distribution in the SFP. The investigation was done by comparing the one-body model thermal distribution with the individual-body model thermal distribution. The following conclusions were obtained:

- There was a high temperature difference (about 15 ºC) between the bottom part (near the SNF) and the upper part of the SFP. The temperature at the surface of the water was still at its initial temperature even when certain parts in the SFP had reached 100 ºC. This shows that natural convection process has a low rate of heat transfer and therefore, it is unable to transfer the decay heat from the SNF to the surface of the water efficiently.
- Compared to the thermal distribution of the individual-body model, the thermal distribution obtained by using one-body model had no significant difference. Based on the three-point measuring temperature, the temperature difference was low at less than 1 ºC. The one-body model simulation was completed in three days, while the other took six days. Therefore, using one-body model seems to be favorable in predicting thermal distribution in nuclear safety analysis.

5. References

[1] Yanagi C, Murase M, Yoshida Y, Utanohara Y 2016 Effect of decay heat distribution on water temperature in a spent fuel pit and prediction errors with a one-region model *J. Nuclear Eng. and Rad. Sci* **2(3)** p 031001(1)-031001(10).

[2] Yanagi C, Murase M, Yoshida Y, Utanohara Y, Iwaki T, Nagae T 2012 Numerical simulation of water in a spent fuel pit during the shutdown of its cooling system *J. Power Energy Sys.* **6(3)** p 423-430.

[3] Yanagi C, Murase M, Yoshida Y, Kusunoki T 2015 Prediction of temperature and water level in a spent fuel pit during loss of all AC power supplies *J. Nuclear Sci. and Tech.* **52(2)** p 193-203.

[4] Yanagi C, Murase M, Yoshida Y, Iwaki T, Nagae T 2012 Evaluation of heat loss and water temperature inn a spent fuel pit *J. Power Energy Sy.*, **6(2)**, p51-62

[5] Hasan R, Tudor J, Ramadan A 2015 Modelling of flow and heat transfer in spe.nt fuel cooling ponds *Int. Congr. on Adv. in Nuc. Power Plant* p 15239

[6] Tzu C H, Vijay K D, Bau S P, Yen S C, Fengjee P T 2013 The development of a three-dimensional transient CFD model for predicting cooling ability of spent fuel pools *Appl. Thermal Eng.* p 496504.

[7] Chan H P, Arim L, Jung M L, Joo H M 2013 Estimation of the spent fuel pool water temperature at a loss of pool cooling accident *J. Math.l, Comp., Phys., Eelec. and Comp. eng.* **7(7)** p 1183-1186.

[8] Chen S R, Lin W C, Ferng Y M, Chieng C C, Pei B S 2014CFD simulating the transient thermal-hydraulic characteristics in a 17x17 bundle for a spent fuel pool under the loss of external cooling system accident *Ann.s of Nuc. Energy* **73** p 241-249.

[9] Hung T C, Dhir V K, Pei B S, Chen Y S, Tsai E P 2013The development of a three dimensional transient CFD model for predicting cooling ability of spent fuel pools *Appl. Thermal Eng.* **50(1)** p 496-504

[10] Xiaoli W, Wei L, Yapei Z, Wenxi T, Guanghui S, Suizheng Q 2014Analysis of the loss of pool cooling accident in a PWR spent fuel pool with MAAP5 *Ann. of Nuc. Energy p* 198-213

IOP Conf. Series: Materials Science and Engineering **555** (2019) 012005 doi:10.1088/1757-899X/555/1/012005

A new framework for optimisation of Pressurised Water Reactor design as a trigeneration system

K Jamaluddin[1, 2], S R Wan Alwi[1, 2.a)], K Hamzah[2], Z Abdul Manan[1, 2] and J J Klemeš[3]

[1]Process Systems Engineering Centre (PROSPECT), Research Institute of Sustainable Environment, Universiti Teknologi Malaysia, 81310 UTM Johor Bahru, Malaysia

[2]School of Chemical and Energy Engineering, Faculty of Engineering, Universiti Teknologi Malaysia, 81310 UTM Johor Bahru, Malaysia

[3]Sustainable Process Integration Laboratory–SPIL, NETME Centre, Faculty of Mechanical Engineering, Brno University of Technology–VUT BRNO, Technická 2896/2, 616 69 Brno, Czech Republic

syarifah@utm.my

Abstract. Conventional nuclear power reactors convert between 30 – 35 % only of the total energy input into electricity while the remaining was wasted. The waste heat is sometimes used for desalination processes while the remaining heat is transferred to a cooling media or lost to the surrounding. Therefore, heat from nuclear reactor can be used to produce heat and power, as well as for cooling in a trigeneration system. This paper presents the Trigeneration System Cascade Analysis (TriGenSCA) for an optimal Pressurised Water Reactor (PWR) design. The TriGenSCA framework allows engineers to determine an optimum utility generation system size and estimate the required amount of external utilities. The analysis includes data extraction, cascade analysis for size estimation and calculation of the new trigeneration system size. The technique also enables users to determine accurate results for energy minimisation based on demand fluctuations. Application of the framework on a case study presented on this paper demonstrates the trigeneration PWR system successfully saved energy of 328GWh/y (97 %).

1. Introduction

Trigeneration is a system of technology that produces power, heating and cooling from a single source of fuel. Trigeneration has been one of the most effective use of energy available with a history of more than 100 years [1]. The trigeneration was mainly utilized in a large-scale centralized power plants and industrial applications. Conventionally, production of electricity and heat are by purchasing electricity from the local grid and heat is generated through burning fuel in a boiler. Trigeneration system, however, reuse by-product heat which can be as much as 60 – 80 % of total primary energy for different applications. Most nuclear power plants are almost entirely comprised of PWR [2]. Khamis *et al.* [3] stated installing a trigeneration system within a nuclear power plant ecosystem allows investors to achieve return on investment within five to six years. Pinch Analysis (PA) is a well-established methodology for designing an efficient resource utilization system. Manan *et al.* [4] first introduced the cascade analysis technique to determine the minimum water targets for water networks.

The techniques have been extended to determine the minimum targets for CO_2 [5], resource supply chain and biomass energy generation and storage capacity [6] as well as energy production in integrated energy system such as trigeneration system [7]. Jamaluddin *et al.* [7] developed a novel systematic methodology based on Pinch Analysis for trigeneration system to determine the minimum target for outsourced power, heating and cooling during the first and continuous 24 h operations, and the maximum storage capacity, simultaneously. Varbanov *et al.* [8] introduced Time Slice (TSL) as set of time intervals in integrated energy system to minimize heat waste and carbon footprint of the considered sites. The development of trigeneration system based on Pinch Analysis, however, still open for further development. The objective of this paper is to design an optimum PWR with a trigeneration system and to estimate the amount of external utilities required. The work presented in this paper can help engineers to determine the appropriate size of PWR with a trigeneration system for the load demand, thereby reducing operation cost of the power plant.

2. Methodology

Initially, PWR as a trigeneration system includes a steam generator to generate very high-pressure steam (VHPS) at varying pressures: from high-pressure steam (HPS) to low-pressure steam (LPS). A condenser is used to produce hot water (HW) whereas an absorption chiller is used to produce chilled water (ChW). Figure 1 shows the conceptual design of a trigeneration PWR system. The usage of electricity, heating and cooling as upstream and downstream of the systems is provided. All parameters for simulation are based on PWR NPP in Three Miles Islands [9]. The demand for heating and cooling are taken from [10] for heating and cooling and [11] for electricity. The minimum temperature difference between utility and process as an energy trade-off is set for both sides. As stated by [9], the minimum temperature difference of PWR NPP in Three Miles Island is 27.8 °C. The minimum temperature difference of Plants A, B, C and D are 12 °C [10]. Methodology of designing an optimal design trigeneration PWR system is shown in Figure 2.

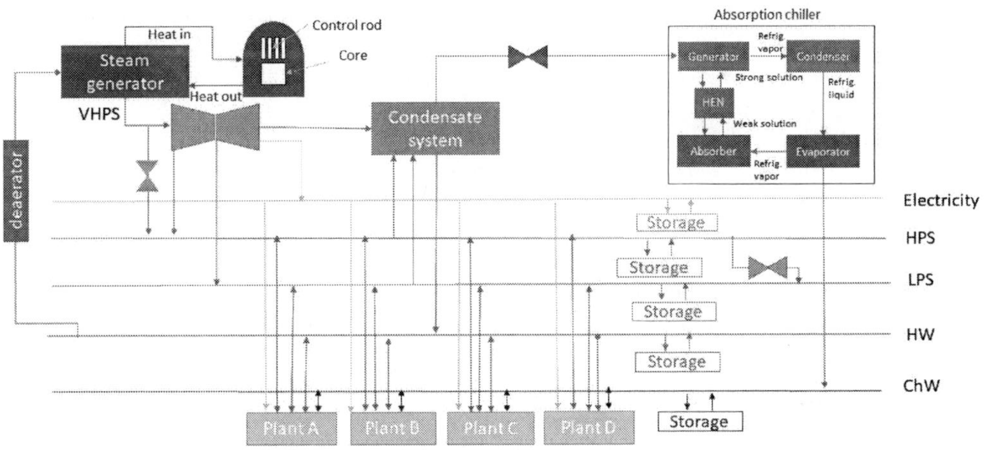

Figure 1. Conceptual design of trigeneration PWR system

2.1. Data Extraction

In the first step, data need to be extracted at demand side are the power rating and time for electricity demand requirement, and flow rate, heat capacity, temperature and time for heating and cooling demand requirements from each industrial sites served by the centralized PWR power supply system. In this paper, the methodology is illustrated using an illustrative case study. PWR as a source side for power, heating and cooling can be obtained from [9]. Energy storage systems for power, heating and cooling are also taken into consideration in the total utility sites system. The energy losses due to charging and discharging of power, heating and cooling are considered in the energy storage systems. Energy losses due to the conversion of the power of AC-DC and DC-AC is also considered. Sodium sulphide (NaS) is used for the power storage system. NaS battery is highly efficient and can accommodate a large storage capacity of up to 245 MWh. Meanwhile, zinc carbonate ($ZnCO_3$) is used for HPS and LPS storage systems. HW and ChW use Iron (II) hydroxide ($Fe(OH)_2$) as energy storage systems. $ZnCO_3$ and $Fe(OH)_2$ are phase change materials (PCM) which enables them to store important quantities of heat over narrow temperature ranges, without a large volume changes.

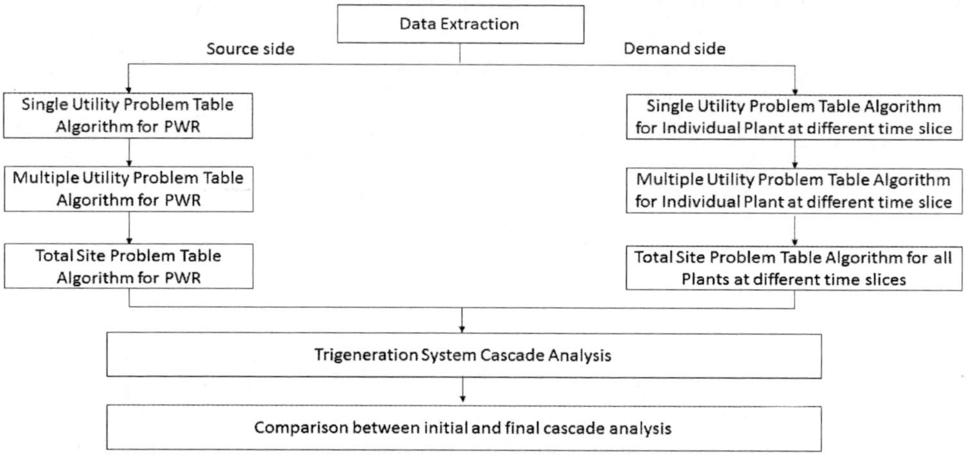

Figure 2. Methodology for determining the optimal trigeneration system

2.2. Single Site Utility Problem Table Algorithm

Single Site Utility Problem Table Algorithm (PTA) is proposed by [12] as the second step to determine the Pinch Temperature as well as to conclude values of minimum external heat and cooling needed in the systems. There are four-time slices for industrial Plants A, B, C and D can be determined which are 0 to 6 h, 6 to 17 h, 17 and 20 h and 20 to 24 h, and every time slice is performed through Single Site Utility PTA. On the other hand, the source side of PWR only has one-time slice which is 0 to 24 h since the PWR generates constant power, heating and cooling for 24 h of operation. The methodology of Single Site Utility PTA can be referred to [12].

2.3. Multiple Site Utility Problem Table Algorithm

Liew *et al.* [13] have extended the PTA to identify pockets and target the exact amounts of utilities needed at specific utility temperature interval. Pinch Temperature from Single Site Utility PTA is needed in this method to determine heating and cooling regions. The method is called the Multiple Utility Problem Table Algorithm. The method is conducted as a third step for all four-time slices for industrial Plants A, B, C and D and one time slice for PWR. Further detail of the methodology can be referred in [13].

2.4. Total Site Utility Problem Table Algorithm

Total Site Utility Problem Table Algorithm (TS PTA) proposed by [13] is used as the fourth step to summarize the different utilities from the highest to the lowest temperature. The TS PTA is conducted at different time slices and total external utility requirements can be obtained on different temperatures at every time slices. The construction of TS PTA can be followed as presented by [13].

2.5. Trigeneration System Cascade Analysis

Trigeneration System Cascade Analysis (TriGenSCA) is a numerical technique of Pinch Analysis methodology to determine the optimal storage size and amount of energy needed from the PWR power plant with trigeneration to supply the energy needed in terms of electricity, heating and cooling by the industrial plants. This methodology is modified from the work of [6] which proposed Electric System Cascade Analysis (ESCA) which only cascade power as the utility to the load demand. There are four main steps that can be used in the TriGenSCA which are data extraction, cascade analysis for sizing

IOP Conf. Series: Materials Science and Engineering **555** (2019) 012005 doi:10.1088/1757-899X/555/1/012005

estimation, calculation of the new size of trigeneration system, calculation of new sizing of trigeneration system and percentage change of the new and previous sizing of trigeneration system. Iteration method is involved in this method and will be stopped when the percentage change of new and previous sizing of trigeneration system is equal to or less than 0.05%.

In the first step, results from TS PTA for source and demand sides for heating and cooling is used as data in TriGenSCA. Power data for demands side can be obtained directly from [11]. Power data for source side is obtained from [9]. The second step for constructing TriGenSCA is the construction of cascading analysis for sizing estimation is modified from ESCA in [6] work. The detail construction of TriGenSCA can be referred to [7] and can be shown in Table 1. Time interval of 1 h is used to obtain the initial and final energy of the trigeneration system. The initial and final energy of the trigeneration system is used to calculate the new size of the trigeneration system. The calculation of the new size of the trigeneration system is in the third step to determine the optimal energy generation. Equation (1) shows the calculation of the new size of trigeneration system. The calculation of the new size of trigeneration system is iterated until differences between the new and previous size of trigeneration system equal to or less than 0.05%. As stated by [6], a value of 0.05% is set as a tolerance to make sure the accuracy of the results. Percentage change between the previous and new size of trigeneration system need to be iterated until the value is equal to or less than 0.05%. Equation (2) shows that the percentage change between the new and previous sizing of trigeneration system.

$$S_{B(new)} = \frac{E \ldots -}{T} \qquad (1)$$

where;

$S_{B(new)}$ = New size of trigeneration system;

S_B = Previous size of trigeneration system;

$E_{t(final)}$ = Final energy content at t = 24 h;

$E_{t(initial)}$ = Initial energy content at t = 0 h;

T = Total time duration

$$P = \frac{S}{S_B} \times \qquad (2)$$

where;

P = Percentage change;

$S_{B(new)}$ = New size of the trigeneration system;

S_B = Previous size of the trigeneration system

Table 1a. Construction of TriGenSCA before iteration

Time (h)	Demand (MW)					Energy generation (MW)				
	Power	HPS	LPS	HW	ChW	Power	HPS	LPS	HW	ChW
1	52.25	1.148	78.61	24	44.24	60	0.012	64.89	97.39	0
.	.	.	.	.	.	.	.	.	.	.
.	.	.	.	.	.	.	.	.	.	.
24	52.25	1.148	78.61	24	44.24	60	0.012	64.89	97.39	0

Table 1b. Construction of TriGenSCA before iteration.

Time (h)	Net energy requirement (MW)					New net energy requirement (MWh)				
	Power	HPS	LPS	HW	ChW	Power	HPS	LPS	HW	ChW
1	7.75	-1.14	-13.72	73.39	-44.24	7.75	0	-14.86	15.88	0
.	.	.	.	.	.	.	.	.	.	.
.	.	.	.	.	.	.	.	.	.	.
24	-35.67	-1.14	-17.72	73.39	-44.24	-35.67	0	-18.86	15.88	0

Table 1c. Construction of TriGenSCA before iteration.

Time (h)	Charging energy (MWh)					Discharging energy (MWh)				
	Power	HPS	LPS	HW	ChW	Power	HPS	LPS	HW	ChW
1	5.58	0	0	12.70	0	0	0	-25.61	0	0
.	.	.	.	.	.	.	.	.	.	.
.	.	.	.	.	.	.	.	.	.	.
24	0	0	0	12.70	0	-49.54	0	-32.51	0	0

Table 1d. Construction of TriGenSCA before iteration.

Time (h)	Cumulative energy (MWh)					New cumulative energy (MWh)				
	Power	HPS	LPS	HW	ChW	Power	HPS	LPS	HW	ChW
	0	0	0	0	0	811.94	0	153.69	0	0
1	.	.	.	.	.	.	.	.	Initial energy content	
.	.	.	.	.	.	.	.	.	.	.
.	.	.	.	.	.	.	.	.	.	.
24	.	.	.	.	.	.	.	.	.	.
	-811.94	0.13	-91.82	114.32	0	0	0.13	61.87	114.32	0
									Final energy content	

2.6. Comparison between Initial and Final Cascade Analysis

The results of initial and final energy contents using TriGenSCA can be compared to determine the total energy that is deficit/excess. A positive value of total energy represents energy in excess whereas a negative value of total energy represents energy in deficit. Figure 3 shows the comparison between the initial and final cascade of PWR as trigeneration system using TriGenSCA. Equation (3) shows the total energy that is required/excess using TriGenSCA. Based on the calculation, the energy that is required from external at initial cascade analysis is 922.15 MWh whereas 22.62 MWh of energy are in excess at final analysis.

$$E_T = \sum \qquad \sum \tag{3}$$

where;

$E_T =$ Total energy required (-)/excess (+);

$\sum$ Summation of final energy content;

$\sum$ Summation of initial energy content

3. Conclusions

A new methodology based on the optimisation technique to obtain the optimal size of trigeneration system called the Trigeneration System Cascade Analysis (TriGenSCA) has been developed. The work is an extension of the methodology by [6] which only considered power whereas this method considers power, heating and cooling optimal energy, simultaneously. The development of TriGenSCA can save energy up to 899.53 MWh/d (922.15 MWh - 22.62 MWh = 899.53 MWh) or 328 GWh/y. Work is in progress to extend the methodology to consider cost and maximum energy storage capacity. Development of this methodology can help designers, engineers and energy manager to minimise the total energy usage and operation cost.

4. References

[1] Wu D W and Wang R Z 2006 Combined cooling, heating and power: A review *Progress in Energy and Combustion Sci.* **32** p 459 – 495.

[2] Shaw S 2017 Advantages of pressurized water reactors (PWR) *Physics 241.* From <large.stanford.edu/courses/2017/ph241/shaw-s1/> accessed on 29 July 2018.

[3] Khamis I, Koshy T and Kavvadias K C 2013 Opportunity for cogeneration in nuclear power plants *The 2013 World Congr. on Advances in Nano, Biomechanics, Robotics, and Energy Research. (*Seoul, Korea: 25-28 August).

[4] Manan Z A, Foo C Y and Tan Y L 2004 Targeting the minimum water flowrate using water cascade analysis technique *AIChE* **50(12)** p 3169 – 3183.

[5] Manan Z A, Wan Alwi S R, Sadiq M M and Varbanov P 2014 Generic carbon cascade analysis technique for carbon emission management. *Appl. Thermal Eng.* **70 (2)** p 1141 – 1147.

[6] Ho W S, Hashim H, Hassim M H, Muis Z A and Shamsuddin N L M 2012 Design of distributed energy system through Electric System Cascade Analysis (ESCA) *Applied Energy* **99** p 309-315.

[7] Jamaluddin K, Wan Alwi S R, Manan Z A and Klemeš J J 2018 Pinch analysis methodology for trigeneration and energy storage system design. *Chem. Eng. Trans.* **70** p 1885-1890.

[8] Varbanov P S and Klemeš J J 2011 Integration and management of renewables into total sites with variable supply and demand *Computers and Chem. Eng.* **35** p 1815-1826.

[9] Barnes S 2013 Using Pinch Analysis to optimize the Heat Exchanger Network of a regenerative Rankine Cycle for an existing modern Nuclear Power Plant (Rensselaer Polytechnic Institute, Hartford, Connecticut, USA: MSc Dissertation).

[10] Liew P Y, Wan Alwi S R, Klemeš J J, Varbanov P S and Abdul Manan Z 2014 Algorithmic targeting for total site heat integration with variable energy supply/demand *Appl.Thermal Eng.* **70** p 1073 – 1083.

[11] Esmailnajad S and Sundquist J 2014 *Demand side management in Swedish industry* (Chalmers University of Technology. Goteborg, Sweden: MSc. Dissertation).

[12] Linnhoff B and Flower J R 1978 Synthesis of heat exchanger networks: 1. Systematic generation of energy optimal networks *AIChE J.* **24(4)** p 633-642.

[13] Liew P Y, Wan Alwi S R, Varbanov P S, Abdul Manan Z and Klemeš J J 2012 A numerical technique for total site sensitivity analysis *App.Thermal Eng.* **40** p 397-408.

Acknowledgments

The authors would like to thank Universiti Teknologi Malaysia and the Ministry of Education of Malaysia for the financial support provided under Skim Latihan Akademik Muda (SLAM) and the Research University Fund under Vote Numbers Q.J130000.2509.12H88 and Q.J130000.2546.18H90. The authors also gratefully acknowledge the EC project fund No. CZ.02.1.01/0.0/0.0/15 003/0000456, provided to the Sustainable Process Integration Laboratory-SPIL, which is funded as the Operational Programme Research, Development and Education of the Czech Ministry of Education, Youth and Sports by EU European Structural and Investment Funds, Operational Programme Research, Development and Education under the collaboration agreement with Universiti Teknologi Malaysia.

IOP Conf. Series: Materials Science and Engineering **555** (2019) 012006 doi:10.1088/1757-899X/555/1/012006

Preliminary quantitative feasibility analysis of proposed Thorium-based nuclear reactor

Anas Muhamad Pauzi[1,a)], Azril Wasim Abdul Wahid[1], Juniza Md Saad[1]

[1]College of Engineering, Universiti Tenaga Nasional, Jalan IKRAM-UNITEN, 43000 Kajang, Selangor, Malaysia

[a]anas@uniten.edu.my

Abstract. In the beginning of the 21st century, several research institutions and also private companies had proposed various design of thorium-based nuclear reactor, ranging from solid fuel to molten fuel, fast and thermal neutron spectrum and also various path of waste management. This paper studies 10 of the proposed reactor designs by 10 different organizations, three key aspects analysed quantitatively namely price per kilowatt, safety features and spent fuel managements. Corresponding factors contributing to the key aspects mentioned above were gathered, weighted based on evidence available and analysed using decision matrix. Based on the information collected, preliminary ranking were constructed based on trends between various factors.

1. Introduction

1.1. Thorium reactor developer of the 21st century

Thorium fuel reactor was first developed by the Oak Ridge National Laboratory (ORNL) in 1950s, under the Aircraft Reactor Experimental (ARE) by Alvin Weinberg, then director of ORNL. Thorium fuel was initially proposed in 1960s using the Molten Salt Reactor (MSR) and had proved its reliability for three years until it was shut down due to piping problems. Since then, no thorium fueled reactor was in operation and researches on the thorium fuel was almost non-existence [1]. Van Gosen in 2016 from the US Geological Survey in US had compiled a total of 17 experiments on thorium fuel in reactor from Canada, Germany, India, Japan, the Netherlands, Norway, the Russian Federation, Sweden, Switzerland, the United Kingdom, and the United States [2][3].

Beginning the 21st century, thorium fuel reactor had again received wide interest as a promising technology listed in the Generation IV International Forum (GIF) to meet future requirements for nuclear power plant. Institutions and organization from all around the world, both governments and private companies are proposing various reactor designs utilizing thorium as source of reactor fuel [4]. Table 1 below lists most of the organization proposing for future thorium fueled nuclear reactor.

IOP Conf. Series: Materials Science and Engineering **555** (2019) 012006 doi:10.1088/1757-899X/555/1/012006

Table 1: Organization proposing Thorium-based fuel nuclear reactor design [5].

#	Name of organization	Headquarter	Established	Reactor Name
1	TerraPower [6]	USA	2006	Travelling Wave Reactor/ Molten Chloride Fast Reactor
2	FliBe Energy [7]	USA	2011	Liquid Fluoride Thorium Reactor
3	Thorium Tech Solution Inc [8]	Japan	2011	FUJI Thorium Molten Salt Reactor
4	Transatomic Power [9]	USA	2011	Waste Annihilating Molten Salt Reactor
5	Terrestrial Energy [10]	USA, Canada, UK	2013	Integrated Molten Salt Reactor
6	Copenhagen Atomics [11]	Denmark	2014	CA Waste Burner
7	Moltex Energy [12]	UK	2014	Stable Salt Reactor
8	Elysium Industries [13]	USA	2015	Molten Chloride Salt Fast Reactor
9	Seaborg Technologies [14]	Denmark	2015	Seaborg Wasteburner Molten Salt Reactor
10	ThorCon International [15]	USA, Indonesia	2016	Molten Salt Reactor

From Table 1 above, all of the organizations proposing future thorium-based nuclear reactor are private companies and mostly are based in the United States. None of the major or current nuclear power plant manufacturer had yet to propose any conceptual design for the thorium fueled nuclear reactor.

1.2. Nuclear Reactor Technology Assessment

Quantitative tools for nuclear reactor technology assessment were developed all around the world as many nations are starting to reintroduce new build nuclear reactor after a long pause since Chernobyl and also new countries are starting to embark in nuclear power to meet the demand of clean energy. The IAEA had introduced the Reactor Technology Assessment (RTA) tool for basic guideline [16]. Harper Mark in 2015 had compiled key features for technology assessment in as such he sorted out features that have high priorities, where the high priority criterion could be grouped into three main areas which are economics, safety and waste management [17]. These three groups are also part of the goals for future reactor system as listed in the Generation IV International Forum (GIF) [18]. Meanwhile, proliferation resistance and physical protection were considered requirement for all future reactor systems, hence, not a differentiator for assessing feasibility [17]. Economics, safety and waste management were chosen as the top priority for this assessment, where information on these 3 objectives were gathered based on journal papers published by the listed organizations or information made available officially on their websites.

2. Methodology

An overview of the flow chart to achieve this assessment is shown in Figure 1. This study requires a formulation of quantitative values for each criteria and an intelligent estimation based on available information. The process would be iterative where additional information would be included and thus affect the weightage values assigned. The output of this process is to automatically generate feasibilities ranking for future thorium reactor design provided basic input information.

IOP Conf. Series: Materials Science and Engineering **555** (2019) 012006 doi:10.1088/1757-899X/555/1/012006

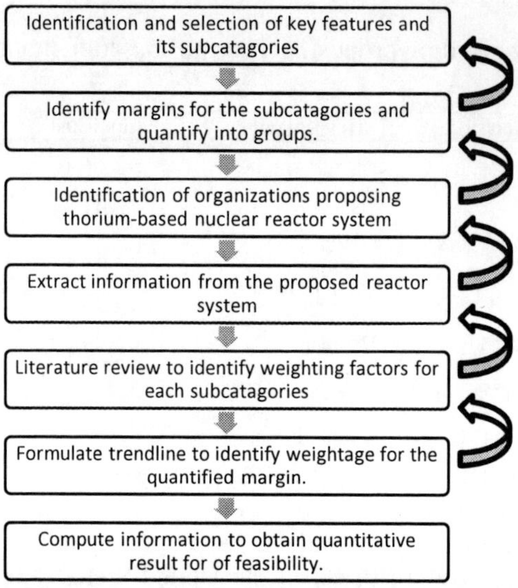

Figure 1. Methodology for quantitative feasibility study for future thorium based nuclear system.

2.1. Price of electricity

Assessing the expected price per kilowatt for a technology at the conceptual early stage of its readiness level would be very speculative as it is prone to high level of uncertainties [19]. With available information, a preliminary ranking was formed.

The total cost per kilowatt are broken down into categories which are reactor size, fuel type, fuel state, neutron spectrum and type of working fluid. The weightage of each categories were initially assigned equally.

2.1.1. Price: Reactor Size

Smaller reactor size are less economical compared to larger unit due to economy of scale, inefficient utilization of power plant infrastructure, transmission grid and management, hence, most design of small reactor would suggest to be constructed in multiple unit [20]. A simple analysis on the announced construction cost and its corresponding power station output is gathered in Table 2.

Table 2. Construction cost and power of new nuclear power plant [21][22].

Nuclear Power Station	Country	Construction Cost, C (Billion USD)	Capacity, P (MW)	Cost, C/Capacity, P (Bil. USD/MW)
Barakah	United Arab Emirates (UAE)	24.4	5380	4.535
Akkuyyu	Turkey	20	4800	4.167
Shidao Bay	China	15.7	3400	4.618
Hinkley Point C	United Kingdom	20.3	3260	6.227
Ruppur	Bangladesh	13.5	2160	6.250

Based on information from Table 2, a trend line was developed to analyse the weightage for price per power produced based on the construction cost as in Equation 1. At a correlation of 80%, the trend showed linear line, where it would be economical to have larger amount of power output in a power station, either by multiple units or larger reactor size.

$$\text{Weightage-Plant Size, } W_S = \text{Construction cost per Power, } C/P_R = -0.0006P_i + 7.5224 \qquad (1)$$

2.1.2. Price: Neutron Spectrum.

Further analysis on the construction cost, the difference between the thorium reactor operating in fast and thermal neutron spectrum were compiled and weightage based on neutron spectrum, W_{NS} were assigned based on the average values as shown in **Table 3**. W_{NS} is inversely proportional to the average cost per megawatt power.

Table 3. Construction cost and power output for fast and thermal reactors [22].

Neutron Spectrum	Reactor name	Capital cost (Billion USD)	MW	Cost (mil) /MW	Country	Average Cost (Bil) /MW	Weightage, W_{NS}
Fast	BN-1200	6	1220	4.918	Japan	8.178	<0.5>
	BN-600 reactor	4.957	600	8.262	Russia		
	PFBR	5.677	500	11.354	India		
Thermal	Barakah APR-1400	24.4	5380	4.535	UAE	4.984	<1.0>
	Ruppor VVER 1200	13.5	2160	6.250	Bangladesh		
	Akuyyu VVER-1200	20	4800	4.167	Turkey		

2.1.3. Price: Fuel State and Fuel Type. There are two state of fuel used in a thorium fuelled reactor which are solid fuel and molten fuel. Molten fuel are technically easier to manufacture since fuel conversion, enrichment and fabrication process can be skipped and uranium hexafluoride could be directly mixed and used inside the reactor. The process mentioned consists of 57% of the total cost of fuel preparation [24][25]. Weightage on the price per kilowatt effect as proportion to the total cost as shown in Table 4.

Small number of thorium based reactor were designed to be able to operate without requiring the use of enriched uranium or minimal amount of enriched uranium during its initial start-up of the reactor. Hence, the weightage is assigned based on the ratio of the difference in the cost of natural uranium with and without enrichment process as shown in Table 4.

Table 4. Fuel state and cost fraction [24][25].

Processes	Cost fraction	Fuel state- <Weightage, W_{FS}>	Fuel type <Weightage, W_{FT} >	
Uranium mining	43%	Molten **<1.0>**	Natural Uranium **<1.0>**	Enriched Uranium **<0.65>**
Conversion	8%	Solid **<0.43>**		
Fuel fabrication	22%			
Enrichment	27%			

2.1.4. Price: Working Fluid. All proposed reactor design utilizes either molten salt or liquid metal as its working fluid, either as its coolant or molten with fuel mixture. Price of the working fluid were estimated by comparing bulk price for the elements contained within the working fluid from popular online chemical elements shopping which are *Chemical Book* and *Chemicool* [25-27].

Table **5** shows the cost of the elements in the work fluid and the corresponding weightage, W_{LT}. It shows that molten salt are more expensive than liquid metal. Table 5 shows the cost of the elements in the work fluid and the corresponding weightage, WLT. It shows that molten salt are more expensive than liquid metal.

Table 5. Type of working fluid and cost [26][27].

Working Fluid Type	Fluid name	Mixture	Composition (%)	USD/1g	USD/1g	Average USD/1g - <Weightage, W_{LT}>
Salt	FLiBe	LiF (98.5%)	66.7%	0.516	0.68	0.51 **<1.0>**
		BeF (99%)	33.3%	1		
	FLiNaK	LiF (98.5%)	46.5%	0.516	0.34	
		NaF (99%)	11.5%	0.0628		
		KF (99%)	42.0%	0.225		
Metal	Lead-Bismuth Eutectic (LBE)	Lead	44.5%	0.0245	0.23	0.24 **<0.5>**
		Bismuth	55.5%	0.39		
	Liquid metallic Sodium (Na)	Na-23	100.0%	0.25	0.25	

2.2. Reactor safety

Safety of the proposed thorium reactors focusses on the molten state of fuel, since most of the proposed reactor suggested molten state because of the significant safety advantages compared to solid state fuel reactors such as its low or atmospheric operating pressure, high boiling temperature, large negative temperature reaction coefficient, continuous removal of gaseous fission product that minimize chemical reprocessing and various inherit passive safety features [1][28].

Tritium, ^3H, commonly written as T which is a radionuclide of hydrogen, are produced when neutron interact with Lithium-Beryllium Fluoride (FLiBe) forming Tritium Fluoride, TF. This causes corrosion and embrittlement on the structural metal. In case of an accident of core exposure, Tritium may diffuse to the environment and dilute in air to form *Tritiated water* (HTO) which could easily be inhaled and cause serious internal exposure [28]. Comparing with PWR, the tritium production rate of MSR is about 200 times higher; hence, tritium management becomes one of the most critical issues for using molten salt as a working fluid [29].

As an alternative to molten salt as working fluid, thorium reactors can apply liquid metal such as Lead-Bismuth Eutectic (LBE) which would have higher risk of corrosion on structural metal and may also face problem of solidification when the reactor operates at low temperature. Furthermore, LBE coolant reactors would require extra precautions of alpha contaminants due to Polonium-210 build-up with neutron flux [30].

Table 6 summarizes some of the safety criteria with the lower and upper weightage, W_S, with preliminary assumed as 0.5 and 1.0. Further analysis would provide more input on the risk and hazard mitigation related to the safety criteria.

Table 6. Safety criteria of Thorium-based reactor.

Criteria	Description	Weightage, W_S	
		Lower <0.5>	Upper <1.0>
Number of loops	Adding more shielding later	2 Loops	3 Loops
Working fluid	Both liquid metal and molten salt have their pros and cons, but overall molten salt shows better safety criteria	Liquid metal	Molten salt
Reduced number of moving parts	Some reactor design claim to operate fully via natural convection and does not require any online fuel management	No moving parts	Moving parts in fuel management and pump

2.3. *Spent fuel management*

The effective management of spent fuel is another important issue that must be taken into account when using Th-based fuel for nuclear reactor operations. In general, there are four main concerns in dealing with spent nuclear fuels from Th-based reactor namely; decay heat, spontaneous neutron emission rates and radiotoxicities (gamma-ray) per discharge assembly, as well as the volume of spent nuclear fuel [31]. It is advisable that the spent fuel is stored in waste-form package, or wet/dry storage for long period due to high amount of decay heat. Apart from that, stronger shielding measurement with the help of remote operation is necessary in order to contain the gamma ray production. In this study, in spite of the issues regarding spent fuel management, none of the working organizations seems to address them. This is possible because these organizations claimed that very small amount of spent fuel production is produced on Th-fueled compared to U-fueled reactor, hence only small repositories of spent fuel are required. Some even believed that no long term storage is even needed due to the short decay period, for example, SWaB reactor by Seaborg Technologies declared that their technology reduce the year of storage to 20-30 years [14]. The ThornCon Power on the other hand claimed that their technology does not require to separate the spent-fuel cooling and storage system. The decay heat is reduced to 0.25% of the original (80kW) before being pumped the old fuelsalt to shipping casks [15].

3. Conclusion

Overall, preliminary quantitative feasibility study were performed on proposed Thorium based nuclear reactor involving 3 main criteria, which are economics, safety and waste management. Ten companies were studied where some basic differences such as size, working fluid, neutron spectrum, number of loops were tabulated. Quantitative values were predicted with publically available evidence on the analysis of its advantages and disadvantages. More study on the safety need to be performed to provide reliable quantitative feasibly. Much of the challenges regarding spent fuel management are not currently addressed by all the proposed design.

4. References

[1] IAEA 2005 Thorium fuel cycle - Potential benefits and challenges (IAEA Tecdoc 1450) p 1–113.
[2] K Jones 2012 *Richard Martin: SuperFuel (Thorium, the Green Energy Source for the Future)* **76**.
[3] Van Gosen B S and Tulsidas H 2016 *Thorium as a nuclear fuel* (Elsevier Ltd).
[4] J. Serp *et al.*, “The molten salt reactor (MSR) in generation IV: Overview and perspectives,” *Prog. Nucl. Energy*, vol. 77, pp. 308–319, 2014.
[5] LinkedIn [Available: https://www.linkedin.com/ accessed: 20-Nov-2018].

[6] Hejzlar P *et al.* 2013 Terrapower, LLC traveling wave reactor development program overview *Nucl. Eng. Technol.* **45(6)**, p 731–744.
[7] Sorensen K 2016 Liquid-Fluoride Thorium Reactor Development Strategy *Thorium Energy for the World* (Springer) p 117–121.
[8] Furukawa K *et al.* 2008 A road map for the realization of global-scale thorium breeding fuel cycle by single molten-fluoride flow *Energy Convers. Manag.* **49(7)** p 1832–1848.
[9] Transatomic Power Corporation 2016 Technical White Paper v 2.1.
[10] LeBlanc D and Rodenburg 2017 C *Integral molten salt reactor* (Elsevier).
[11] Pedersen T J 2017 Copenhagen Atomics waste burner *Molten Salt Reactors and Thorium Energy*, (Elsevier) p 599–607.
[12] Scott I 2015 Static fuel molten salt reactors-simpler,cheaper and safer *Proc. of the Int. Thorium Energy Conf.: Gateway To Thorium Energy.*
[13] Elysium Industries 2018 Advanced nuclear technology to close the fuel cycle.
[14] Seaborg Technologies 2015 Seaborg wasteburner - molten salt reactor.
[15] Jorgensen L 2017 *ThorCon reactor* (Elsevier).
[16] International Atomic Energy Agency 2013 Nuclear reactor technology assessment for near term deployment (Vienna, Austria: IAEA).
[17] Harper M J 2012 Reactor technology assessment methodology: the assessment process and approach.
[18] OECH 2014 Nuclear Energy Agency Technology roadmap update for Generation IV nuclear energy systems.
[19] Mankins J. C 1995 Technology readiness level.
[20] Carelli M. D. *et al.* 2010 Economic features of integral, modular, small-to-medium size reactors *Prog. Nucl. Energy* **52(4)** p 403–414,.
[21] World Construction Network 2018 ENEC completes construction of unit 1 at $32bn Barakah nuclear plant in UAE [Available: https://www.worldconstructionnetwork.com/news/enec-completes-construction-of-unit-1-at-32bn-barakah-nuclear-plant-in-uae/].
[22] Power-technology.com 2018 Akkuyu nuclear power plant, Mersin [Available: https://www.power-technology.com/projects/akkuyu/].
[23] National Nuclear Laboratory 2012*Nuclear Fuel Cycle Science and Engineering* (Cambridge, UK: Woodland Publishing).
[24] Nuclear Energy Agency 2018 The economics of the nuclear fuel cycle **8** p 175.
[25] Chemical Book Inc *Chemical Book* [Available: https://www.chemicalbook.com/ accessed: 20-Nov-2018]
[26] Chemicool 2018 Chemicool periodic table [Available: https://www.chemicool.com accessed: 20-Nov-2018].
[27] Elsheikh B M 2013Safety assessment of molten salt reactors in comparison with light water reactors *J. Radiat. Res. Appl. Sci.* **6(2)** pp 63–70.
[28] Huang Y *et al.* 2011 *Control and monitoring of tritium in molten salt reactor* **34**.
[29] Wang C, Qin H, Zhang D, Tian W, Qiu S and Su G H 2018Numerical study of tritium transport characteristics in Thorium Molten Salt Reactor with Solid Fuel (TMSR-SF) *Nucl. Eng. Des.*, **335** p 391–399.
[30] Dobosz A, Plevachuk Y, Sklyarchuk V, Sokoliuk B, Tkach O and Gancarz T 2018 Liquid metals in high-temperature cooling systems: The e ff ect of bi additions for the physicochemical properties of eutectic Ga − Sn − Zn.
[31] Humphrey U E and Khandaker M U 2018 Viability of thorium-based nuclear fuel cycle for the next generation nuclear reactor: issues and prospects *Renew. Sustain. Energy Rev.* **97** p 259–275.

IOP Conf. Series: Materials Science and Engineering **555** (2019) 012007 doi:10.1088/1757-899X/555/1/012007

An overview of the development of a multigroup Monte Carlo code for TRIGA reactors

M R Omar[1], J A Karim[2] and T L Yoon[1, a)]

[1]School of Physics, Universiti Sains Malaysia, 11800 USM, Penang, Malaysia.
[2]Reactor Technology Center, Technical Support Division, Malaysian Nuclear Agency, 43000 Bangi, Selangor, Malaysia.

[a]tlyoon@usm.my

Abstract. This paper discusses the overview of the development of TRIMON, a next-generation core management code for TRIGA reactors. The newly developed code employs Monte Carlo method which uses diffusion theory type homogenized neutron cross section data, thus, speeding up the stochastic simulation via simplification of a complex core geometry. It has the capability of running three-dimensional core simulation up to 16 neutron energy groups. In general, the group neutron cross sections are pre-processed and homogenized via the Effective Diffusion Homogenization (EDH) method. The important features of TRIMON include core eigenvalue calculation, fuel burnup management, three-dimensional power and neutron flux distribution prediction. In this paper, the numerical results obtained using TRIMON were compared with the actual measured operational data. The performance comparison was also made between TRIMON and MCNP.

1. Introduction

In this work, the development of a new neutronics calculation code coupled with direct fuel burnup calculation for TRIGA reactors is reported. TRIMON is a stochastic code written in Fortran which integrates homogenized cross section data generated using diffusion theory with multi-group Monte Carlo method.

Majority state-of-the-art Monte Carlo codes are still inferior in core design and direct fuel burnup calculation despite their extremely powerful computational capabilities [1]. In this context, transferring the fuel burnup information of the previous core calculation to correct the neutron cross-section data for the use of current core calculation can be tedious in most general multi-purpose Monte Carlo codes. As a result, various strategies have been published by researchers on linking Monte Carlo codes to fuel depletion codes such as ORIGEN and CINDER [2]. However, these strategies are poorly outlined and require an intermediate linking code that is not accessible to all nuclear engineers.

Unfortunately, authors in [3] demonstrated that thousands of MCNP eigenvalue calculation cycles were required to converge at a good estimation of core effective multiplication factors, k_{eff}. Essentially, TRIMON addressed this issue by introducing diffusion-theory type homogenized neutron cross sections in the Monte Carlo method. Apparently, there is a lack of research on such initiative. Previous work [5] involves modifying scripts to feed in homogenized cross section data into MCNP, a continuous energy Monte Carlo code. Yet, the validity and the theoretical aspect of such trial are poorly understood.

As such, we have attempted to outline a systematic procedure to computationally utilize homogenized group cross section data in Monte Carlo method. The new code is also capable of modelling three-dimensional reactor core, thus enabling nuclear engineers to study the axial power peaking characteristics of the reactor core. In order to account the effects of fuel depletion on reactor behaviour, it is also integrated with core direct fuel burnup calculation.

TRIGA is a commercial research reactor manufactured by General Atomics, USA. To date, the reactor has been installed in 24 different countries. The reactor has been used for various applications such as radioisotopes production, non-destructive testing, and research on the properties of matter and for education and training. The reactor is a pool-type water reactor and the reactor core is loaded with hydride fuel-moderator element, specifically UZrH.

Reaktor TRIGA PUSPATI (RTP) is a 1 MWth TRIGA installed at Malaysian Nuclear Agency, Bangi Malaysia. RTP core is a cylindrical-shaped core containing 127 designated core locations to accommodate fuel elements and other non-fuel elements such as control rods and irradiation facilities. The reflector is made up of graphite and the reactor assembly is equipped with four boron carbide control rods. RTP is fuelled with 8.5%wt, 12%wt and 20%wt UZrH fuel with the enriched to 20% U-235. The fuel is clad by a 0.05cm thick SUS-304 can.

2. Theory

2.1 Unit cell homogenization

At the most fundamental level, a reactor core can be constructed using its basic lattice structures which are known as unit cells. The geometry of a unit cells is defined such that the entire reactor core can be build using repetitions of the unit cell. For instance, the unit cells forming a TRIGA core are illustrated in Fig. 1.

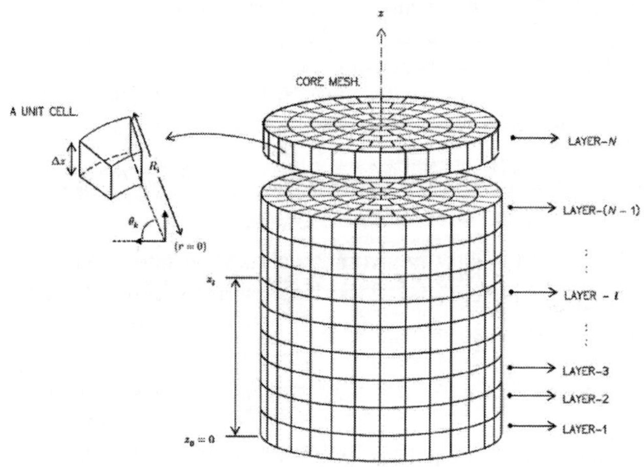

Figure 1. Geometry of a unit cell and TRIGA core mesh.

In TRIMON, each unit cell represents a homogenized reactor sub-region with constant homogenized group neutron cross sections, Σ. Ordinarily, when tracing the random walks of a neutron history, the current unit cell which contains the neutron is mapped so that the value of Σ for the cell can be retrieved and used for simulating the subsequent nuclear events.

IOP Conf. Series: Materials Science and Engineering **555** (2019) 012007 doi:10.1088/1757-899X/555/1/012007

2.2 Fuel burnup effect

In TRIMON, fuel burnup accumulation due to previous core operations is not neglected during cross section data pre-processing. The burnup, b, for each fuel cell in percent U^{235} is given by the general power series correlation,

$$b = \sum_{j=1}^{N} \beta_j p_{\text{cell}}^j (\Delta t)^j \tag{1}$$

where N is the order of the series, β_j are the coefficients of the series and Δt is the burnup increment in days. During neutron cross section data pre-processing, the fuel density and enrichment are corrected according to the prescribed value of b so that the number density of the fuel region within the unit cell can re-calculated.

2.3 Fuel temperature effect

Inevitably, fuel temperature gives a significant impact on homogenization calculation due to Doppler broadening effect of the neutron cross section at the resonance neutron energy region. For a typical TRIGA reactor, the temperature of a fuel meat in a unit cell is derived based on the empirical formula suggested by [1],

$$T_{\text{cell}} = T_{\text{cool}} + \sum_{n=1}^{N} a_n \alpha_r \alpha_z \left\{ \frac{m}{M} \left(1 - \frac{b}{100}\right) P \right\}^n \tag{2}$$

where T_{cool} is the temperature of coolant, N is the power series order, a_n is the coefficient of the power series, α_r is the radial power form factor at the fuel cell location, α_z is the axial power form factor at the fuel cell location, m is the mass of Uranium in the fuel rod, M is the total mass of uranium within the reactor core, b is the fuel rod burnup in percent, and P is the nominal core power.

2.4 Monte Carlo random walk overview

To begin with, neutrons are assumed to be moving in a straight line in the direction $\mathbf{\Omega}$ from an original position, $\mathbf{r}$. In fact, all neutrons are only produced by fission reactions. According to [6], other types of neutron production can be neglected, and their effects are imperceptible.

In general, a random walk step of a neutron is initiated at position $\mathbf{r}$ and ends at the position of the next collision event, $\mathbf{r}_c$. Next, the current unit cell, Λ, is mapped where the position $\mathbf{r}$ is located inside Λ. Thus, the distance to the next collision, d_c, can be predicted,

$$d_c = -\frac{\ln \xi}{\Sigma_t} \tag{3}$$

where Σ_t is the total homogenized neutron cross section of the unit cell Λ and ξ is a random number where $\xi \in [0,1)$.

Next, the distance to the nearest boundary of the current unit cell, d_b, along the neutron trajectory direction, $\mathbf{\Omega}$, is determined. If the distance to the next collision event is greater than the distance to the nearest cell boundary, i.e. $d_c > d_b$ then the neutron is transported to the boundary instead. Otherwise, the neutron is transported to the next collision point.

At the collision site, the neutron will proceed for a neutron–nuclear interaction. All interactions such as neutron scattering, absorption and fission reaction are processed according to their corresponding algorithms outlined by [7]. On the other hand, if the neutron is transported to the cell boundary, the distance to the next collision will be calculated using the total neutron cross section of the neighbouring unit cell. This process reiterates until the neutron is absorbed by a nucleus or escapes out of the reactor core. When a neutron is absorbed, the simulation stops and the next neutron from the fission bank are simulated.

In general, eigenvalue calculation is a transport simulation of neutrons studying the ability of a system, i.e. a nuclear reactor core, to sustain fission chain reactions. In reactor theory, the eigenvalue, k_{eff}, is interpreted as the ratio between the number of neutrons in successive generations, with the fission reaction regarded as the neutron birth event that distinguishes the generations of neutrons.

In general, TRIMON employs the method of successive generations (MSG) which was first introduced by [8]. Interestingly, MSG necessitates history tracking of a finite number of neutrons from one generation to another. Initially, an initial guess of k_{eff} and the fission site distribution, $\Psi(\mathbf{r})$, must be done. Here, $\Psi(\mathbf{r})$ contains the information where fission reaction is possible. As the eigenvalue calculation progresses, k_{eff} and $\Psi(\mathbf{r})$ will stochastically converge and equilibrate. It is important to ensure that the value of k_{eff} and $\Psi(\mathbf{r})$, converge and this can be ensured by skipping a few fission generation cycles before tally accumulation is made. In fact, $\Psi(\mathbf{r})$ converges slower than that of k_{eff} [9].

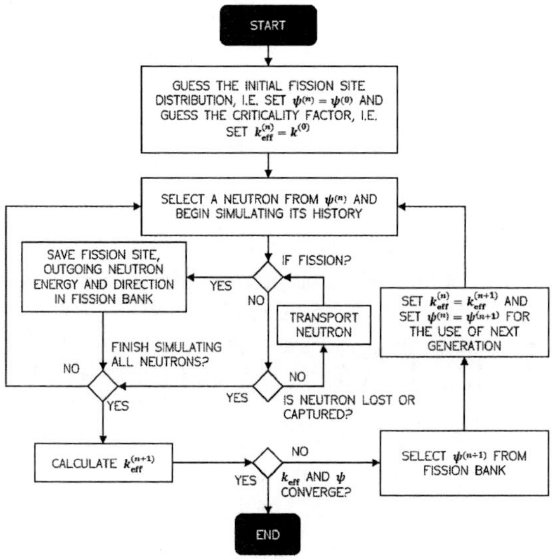

Figure 2. Method of successive generations (MSG) implemented in TRIMON.

3. Results and discussions

Throughout all eigenvalue calculation done in this work, the temperature of the coolant was set at 293K, the reactor core was xenon-free, and the power of the reactor is at negligible thermal power (<0.01kW). A total of 50000 neutron histories were simulated per fission cycle.

3.1 TRIMON validation with RTP operational data

The details of the operation histories of RTP for the first three operational cores are given in Table 1. The summary of the results of the comparison is given in Table 2. In the operational core benchmark problem, the mixed core configurations of Puspati TRIGA Reactor (RTP) which consists of 8.5%wt and 12%wt $UZrH_{1.6}$ fuels were simulated using TRIMON. The results obtained using TRIMON were compared with the actual operational data. Each core configuration is assigned to a specific core identification, i.e. Core-*N* for the Nth operational core. When a specific core configuration has reached its end of cycle and reconfigured, the new core configuration is assigned to a new identification, i.e. Core-(*N*+1).

Table 3 indicates a reduction of k_{eff} at the end of each operational core cycle (EOC). Then, the value of k_{eff} was restored every time the core is reconfigured at the beginning of the cycle (BOC).

During fuel reconfiguration, the existing high burnup fuels may be replaced with new fresh fuels, and some fuels loaded in the core may be reshuffled with the existing lower burnup fuels.

Table 1. First three operational core histories.

Configuration	Average Power (kW)	Total Burnup Given (MWd)	Accumulated Burnup (MWd)	Operating Hours (days)
Core-1	455.015	43.4	43.4	95.3
Core-2	600.580	45.8	89.2	76.4
Core-3	674.223	42.6	131.8	63.1

Table 2. Summary of eigenvalue calculations using TRIMON for Core-1 to Core-3. Measured k_{eff} obtained from the RTP operational log book is also given. Difference between calculated and measured, Δk_{eff}, is also given.

Core	k_{eff} TRIMON	k_{eff} Measured	Δk_{eff}
Core-1 BOC	1.05190 ± 0.00028	1.05312 ± 0.00038	0.00122 ± 0.00047
Core-1 EOC	1.03310 ± 0.00030	1.03095 ± 0.00037	0.00215 ± 0.00048
Core-2 BOC	1.05701 ± 0.00043	1.05377 ± 0.00039	0.00324 ± 0.00058
Core-2 EOC	1.03702 ± 0.00038	1.04004 ± 0.00038	0.00298 ± 0.00054
Core-3 BOC	1.05432 ± 0.00026	1.05323 ± 0.00039	0.00109 ± 0.00047
Core-3 EOC	1.04258 ± 0.00039	1.04193 ± 0.00038	0.00065 ± 0.00055

BOC: Beginning of operational core cycle.
EOC: End of operational core cycle.

3.2 TRIMON performance benchmark with MCNP

In this benchmark problem, the eigenvalue calculations of RTP first operational core (Core-1) was performed using TRIMON and MCNP. This section reports the evaluation of the gain brought by TRIMON compared to the local information lost due to the use of homogenized neutron cross section data. Figure 3 shows TRIMON imposes a slightly greater relative error of k_{eff} compared to MCNP. However, TRIMON exhibits better simulation time where the computational rate is at 15 fission cycles per minute compared to MCNP which is at 5 fission cycles per min.

In this work, the initiative to integrate homogenized group data in the Monte Carlo method is merely to reduce the complexity of the core model and to increase the performance of the Monte Carlo simulation. Notably, this approach reduces the spatial variation of neutron cross sections within the reactor core model. However, this attempt may introduce local information lost due to spatial averaging, thus, increasing the relative error of k_{eff}. Interestingly, the presented result shows that the performance gain brought by TRIMON is greater than the local information lost. Thus, the efficiency of TRIMON and MCNP is expressed using the figure of merit (FOM) as given in Table 3. Note also that a higher FOM value indicates better computational efficiency.

Table 3. TRIMON and MCNP performance result for Core-1. Calculation was done using 50000 neutrons per fission cycle with 200 fission cycles.

Parameters	TRIMON	MCNP
Simulation Time, T (min)	13.7	45.6
Converged k_{eff}	1.05154	1.05971
Δk_{eff}	0.00817 ± 0.00057	
Relative Error, R	0.00044	0.00032
Figure of Merit, $1/R^2T$	517744	154784

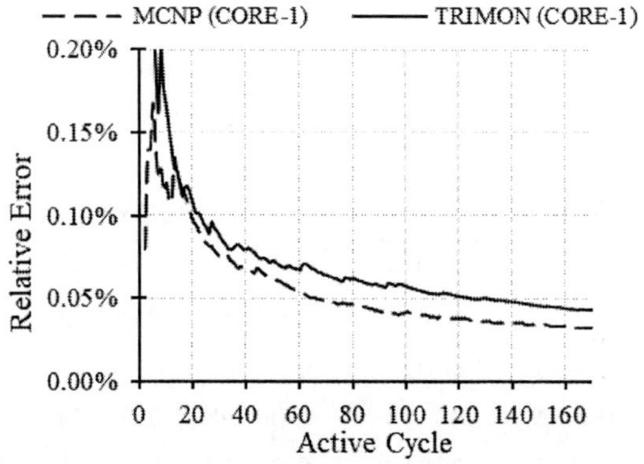

Figure 3. Relative error evolution for Core-1.

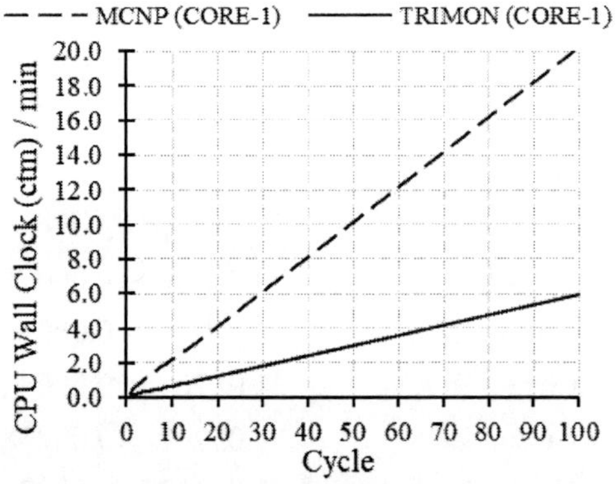

Figure 4. CPU wall clock of the first 100 cycles using TRIMON and MCNP.

4. Conclusions

In summary, the development of a Monte Carlo core calculation code for TRIGA reactors, TRIMON, is presented. TRIMON was first tested and applied to RTP. Essentially, the new code integrates local fuel burnup effect in Monte Carlo simulation, supplanting the existing state-of-the-art Monte Carlo codes which are still less optimal in terms of direct fuel burnup calculation and simulation performance.

The simulation results obtained from the first test run of the code indicates a good agreement with the experimental results obtained from the actual reactor operational data. Perhaps, the integration of diffusion-type homogenized neutron cross section with Monte Carlo method proposed in this paper can serve as a guideline for future researchers to explore the feasibility of using the similar technique in other nuclear reactor types.

5. References

[1] Peršič A, Žagar T, Ravnik M, Slavič S, Žefran B, Ćalić D, Trkov A, Žerovnik G, Jazbec A and Snoj L 2017 "TRIGLAV: A program package for TRIGA reactor calculations," *Nucl. Eng. Des.*, **318** p 24–34.

[2] Goorley T, James M, Booth T, Brown F, Bull J, Cox L J, Durkee J, Elson J, Fensin M, Forster R A, Hendricks J, Hughes H G, Johns R, Kiedrowski B, Martz R, Mashnik S, Mckinney, Pelowitz D, Prael R, Sweezy J, Waters L, Wilcox T and Zukaitis T 2016 Features G of MCNP6 *Ann. Nucl. Energy*, vol. 87, p 772–783.

[3] Rabir M H, Md Zin M R and Usang M D 2016 Neutron flux and power in RTP core-15 *AIP Conf. Proc.* **1704(1)** p 50018.

[4] Wang K, Li Z, She D, Liang J, Xu Q, Qiu Y, Yu J, Sun J, Fan X and Yu G 2015 RMC - A Monte Carlo code for reactor core analysis *Ann. Nucl. Energy* **82** p 121–129.

[5] Kuijper J C, Van Der Marck S C and Hogenbirk A 2007 Using Homogenized macroscopic group cross sections in continuous-energy Monte Carlo neutron transport calculations with MCNP *Jt. Int. Top. Meet. Math. Comput. Supercomput. Nucl. Appl.*

[6] Persic A, Slavic S, Ravnik M and Zagar T 1998 *TRIGLAV - a program package for research reactor calculations*.

[7] Romano P K and Forget B 2013 The OpenMC Monte Carlo particle transport code *Ann. Nucl. Energy* **51** p 274–281.

[8] Lieberoth J 1968 Monte Carlo Technique to solve the static eigenvalue problem of the boltzmann transport equation *Nukl. 11* **213(19)**.

[9] Cho N Z and Chang J 2009 Some outstanding problems in neutron transport computation *Nucl. Eng. Technol.* **41(4)** p 381–390.

IOP Conf. Series: Materials Science and Engineering **555** (2019) 012008 doi:10.1088/1757-899X/555/1/012008

A review on multilayer radiation shielding

Muhammad Arif Sazali[1], Nahrul Khair Alang Md Rashid[1] and Khaidzir Hamzah[1]

[1]Faculty of Chemical and Energy Engineering, Universiti Teknologi Malaysia, 81310 Johor Bahru, Johor

marif69@live.utm.my

Abstract. Radiation shielding is a body of material that is placed between a radiation source and an object to be protected with the aim of reducing the intensity of radiation at the object's location. It can be made from various materials. These materials can be stacked into a multilayer shield or they can be mixed into a composite shield. The main objective of the present study is to review the list of multilayer shield combinations that have been studied and to highlight the findings on material arrangement and consequent buildup factor. The scope of the study is limited to the results of the performed studies. It was observed that there was no clear method on arranging the layer. Buildup factor was also found to be complicated in multilayer shields. Future studies may focus on new multilayer shielding design with unlisted materials, complementary buildup calculations, and applications of metaheuristics in shielding optimization.

1. Introduction

As nuclear technology advances and is widely used in today's industries, more people are likely to be exposed to ionizing radiation. A shielding is needed to absorb and reduce the intensity of the radiation. It is one of the protection methods recommended by the International Commission of Radiation Protection (ICRP), other than minimizing operation time and maximizing distance. It requires a more careful design as it is usually permanent and not easily replaceable. Besides, it also needs to be able to fulfill other design requirements depending on where the shield is applied such as heat resistance and structural integrity.

Shielding can be divided into two configurations which are composite and multilayer. A composite shield is made up of a base material mixed with additives. This can improve the shielding capability of the material. For example, by varying the amount of additives in a concrete, the density of the material can be increased, resulting in a better performance [1]. However, it can be difficult to obtain a homogenous mixture of the component materials that can result in inconsistent shielding performances [2][3]. Uneven mixing of the composite can also cause pinholes which are pure regions that radiation can penetrate through [4]. This problem can be prevented by using multilayer shielding approach [5].

A multilayer shield consists of two or more layers of different materials. In this arrangement, the incoming radiation will have more chances to be scattered and absorbed by the shield. One of the first instances of using multilayer shield was in 1943 in which concrete and paraffinized wood were used for the graphite-moderated reactor in Chicago. This shield configuration is useful against mixed radiation. For example, a ^{252}Cf source emitting both fission neutrons and prompt gamma rays can be shield with layers of polyboron and concrete [6]. The composite, containing hydrogen and boron atoms, slows down the fast neutrons and absorbing them, while the concrete scatters the resultant gamma-rays. Often, a shield is made from the combinations of composite and multilayer configurations which can be observed later in this review. This is because each of the materials has different shielding properties that they can be mixed and matched to solve a particular problem depending on the application. Shultis and Faw [7] had provided an excellent review on the history of radiation shielding technology.

Currently, there are many types of materials and combinations that have been tested. Ongoing improvements and developments are still being carried out as nuclear technology progresses. The aim of this review is to list some of them and to highlight the trends discovered by the authors. It will focus on the results on the effect of material types and arrangements on the shielding performance and buildup factor. Suggestions for future works will also be provided.

2. Radiation Interactions through Multilayer Shield

As radiation passes through a shielding material, it has certain probability to interact with the atoms of the material. When this happens, it is either absorbed or scattered. In both scenarios, secondary radiation such as beta or gamma is produced. Besides, in the scattering process, the direction of the radiation particles may be changed. Eventually, some of scattered particles are transmitted through the shielding. This phenomenon is known as buildup and it is represented by the buildup factor, *B*. Neglecting this factor may cause huge errors in shielding calculations [8].

For a narrow beam geometry, the intensity of radiation *I* emitted through a single body of material is given by the Beer-Lambert Law. I_0 is the initial intensity, $\mu(E)$ is the attenuation factor of the material at the particular energy *E*, and *x* is the shield thickness. The $\mu(E)$ can be expressed as mass attenuation factor $\mu\ \rho(E)$ for gamma rays and as total macroscopic cross section $\Sigma_t(E)$ for neutrons. When a broad beam geometry is applied, a buildup factor *B* is added to Equation 1 as a correction factor to account for scattered beam.

$$I = I_0 e^{-\mu(E)x} \tag{1}$$

$$I = BI_0 e^{-\mu(E)x} \tag{2}$$

In the case of a multilayer shield as shown in Figure 1, the intensities of radiation emerging from the subsequent layers are dependent on the preceding layers. The resultant intensity I_3 can be derived into Equation 3. Although, based on the commutative property of multiplication in this equation, it would seem that there would be no effect if the order of materials is changed. However, this is observed to be the opposite in various studies as explained later.

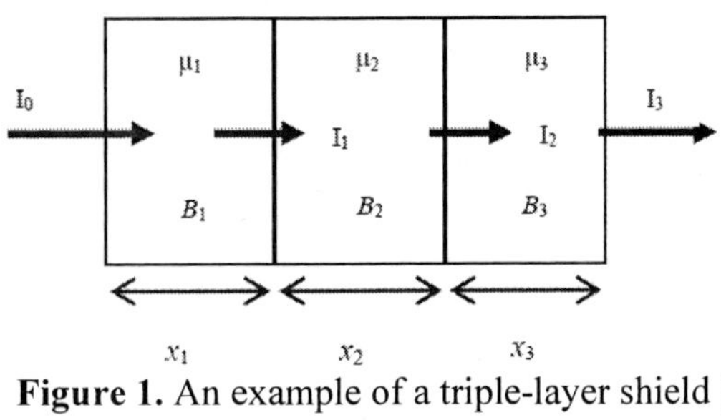

Figure 1. An example of a triple-layer shield [9].

$$I_3 = I_0\ B_1 B_2 B_3\ e^{-\ \mu_1 x_1 + \mu_2 x_2 + \mu_3 x_3} \tag{3}$$

On the other hand, Lamarsh and Baratta [10] had outlined four methods to approximate gamma buildup factors of double-layers when exact methods are unavailable. The first, if the materials differ in atomic number Z by only 5-10, is to use the buildup factor of the material with higher *B* value. It was based on observation that buildup do not vary rapidly with Z, except at low energy.

$$B(\mu x) = B_{max}(\mu_1 x_1 + \mu_2 x_2), \qquad \text{if } Z_1 - Z_2 = 5{,}10 \tag{4}$$

For any Z difference higher than in Method 1, the order of arrangement is considered. If the lower Z material is first, the buildup of the higher material is used. It is assumed that the lower energy photons transmitted from the first layer is absorbed in second layer.

$$B(\mu x) = B_2(\mu_2 x_2), \qquad \text{if } Z_1 - Z_2 > 10, \text{ low Z first} \tag{5}$$

Then, when the higher Z material is first, the buildup is determined by the energy of the gamma-rays E_γ. If E_γ is lower than 3 MeV, the buildup is given by Equation 6. As the flux emitting from the high Z material has a different energy profile from the original source, they are treated as a source in the second layer. When the E_γ is higher than 3 MeV, the buildup is similar to the previous case except that the buildup in the second layer is set at the minimum 3 MeV instead of higher energy. This is based on the assumption that the gamma-rays emerging from Layer 1 have energies about the minimum.

$$B(\mu x) = B_1(\mu_1 x_1) \times B_2(\mu_2 x_2), \qquad \text{if } Z_1 - Z_2 > 10, \text{ high Z first, } E_\gamma < 3 \text{ MeV} \tag{6}$$

$$B(\mu x) = B_1(\mu_1 x_1) \times B_2(\mu_2 x_2)_{min}, \qquad \text{if } Z_1 - Z_2 > 10, \text{ high Z first, } E_\gamma > 3 \text{ MeV} \tag{7}$$

Besides, there is also another formula aside from the ones suggested above. As briefly reviewed by Mann *et al.* [11], there is an empirical formula proposed by Kalos [12][13] for calculating buildup factor in a double-layer shield based on its component materials. The Kalos' formula is shown in Equation 8. It was also validated by Shin and Hirayama using Monte Carlo method [14].

$$B\ \mu x\ = B_2\ \mu_2 x_2\ +\ B_2\ \mu_1 x_1 + \mu_2 x_2\ - B_2(\mu_2 x_2)\ \times\ K\ \mu_1 x_1\ .C(\mu_2 x_2) \tag{8}$$

$$K\ \mu_1 x_1\ = \frac{B_1\ \mu_1 x_1\ - 1}{B_2\ \mu_1 x_1\ - 1} \tag{9}$$

$$C\ \mu_2 x_2\ = \begin{matrix} 1 & \text{'or high-Z/low-Z order} \\ exp\ -1.7\mu_2 x_2\ + (\alpha/K)[1 - exp\ -\mu_2 x_2 & \text{'or low-Z/high-Z order} \end{matrix} \tag{10}$$

$$\alpha = \frac{\mu_c\ \mu_{t\ 1}}{\mu_c\ \mu_{t\ 2}} \tag{11}$$

Using transport calculations, some modifications to the correction factor were made by Burke and Beck [15] as shown in Equation 12. In this case, γ is the ratio of Compton mass attenuation coefficient of the first layer to the second layer.

$$C\ \mu_2 x_2\ = \begin{matrix} exp\ -\mu_2 x_2/\gamma\ + 1.5[1 - exp\ -\mu_2 x_2\ ] & \text{'or high-Z/low-Z order} \\ exp\ -\mu_2 x_2 \times \gamma\ + (\alpha/K)[1 - exp\ -\mu_2 x_2\ ] & \text{'or low-Z/high-Z order} \end{matrix} \tag{12}$$

$$\gamma = \frac{\mu_c\ \rho\ _1}{\mu_c\ \rho\ _2} \tag{13}$$

The formula was further adjusted by Lin and Jiang [16] whom had approved the use of the empirical method for point isotropic source. This is shown in Equation 14-16. β is the ratio of total mass attenuation coefficient of the second material to the first material.

$$C\ \mu_2 x_2 = \begin{cases} exp\ -1.08\beta\mu_2 x_2\ + 1.13\beta\ell(\mu_2 x_2) & \text{for } high\text{-}Z/low\text{-}Z\ order \\ 0.8\ell(\mu_2 x_2) + (\gamma/K)exp(-\mu_2 x_2) & \text{for } low\text{-}Z/high\text{-}Z\ order \end{cases} \quad (14)$$

$$\beta = \frac{\mu_t\ \rho\ _2}{\mu_t\ \rho\ _1} \quad (15)$$

$$\ell\ \mu_2 x_2 = \frac{B_2\ \mu_2 x_2\ + 1}{B_1\ \mu_2 x_2\ + 1} \times [1 - exp\ -\mu_2 x_2\ ] \quad (16)$$

2.1. *Observations on Buildup Factor*

Unlike in the case of a single layer shielding, the buildup in a multilayer shielding is more complicated [17][18]. This is because a monoenergetic radiation flux that passes through the first layer may result in a flux with a spectrum of energy due to absorption and scattering. This makes it difficult to determine the attenuation factor (cross sections, linear attenuation, mass attenuation coefficient) for the shielding calculation. Therefore, many studies had been done to calculate the buildup factor for various multilayer shielding designs using simulations and experiments.

Buildup factor is dependent on the type of radiation, its energy, the type of shielding material, and the geometry involved [19]. It is also significantly affected by the material arrangement and the thickness of each layer [17]. While the buildup was observed to increase with thickness, the effects of changing the material order were varied. Abbas [20] deduced that a lead-water gamma shielding had a greater buildup than the opposite order at 1 and 2 MeV, but the case was reversed at 6 MeV. Al-Arif and Kakil [9] saw no considerable difference in measured buildup factor when the order of material was modified. They also noticed that the buildup factor had a strange dependency on the atomic number and the photon energy. At low energy, the buildup increased as atomic number decreased. At high energy, it increased as atomic number increased. Meanwhile, Mann *et al.* [11] stated that at fixed energy, a double layered shield in a low-Z/high-Z orientation always results in a lower buildup as compared to a single material with the same optical thickness. This difference would keep increasing as the optical thickness increases.

3. List of Material Combinations

Table 1 (a) and (b) below show the multilayer shielding designs found in literature. Most of them employed combinations of light and heavy materials to shield against neutrons and photons of gamma-rays or X-rays. Some of them focused on determining the buildup factor which is important in shielding calculations. However, not all of them involved simulations that were coupled with experiments.

IOP Conf. Series: Materials Science and Engineering **555** (2019) 012008 doi:10.1088/1757-899X/555/1/012008

Table 1(a). Summary of multilayer shielding combinations.

Author	Shielding Materials	Remarks/Application
Fuse *et al.* (1970) [21]	Iron | Water	• Slabs of iron movable in water for 2.5 MW reactor • MAC-RAD code was used as comparison to experiments
Kuspa (1972) [19]	Aluminum | Lead	• Double-layer slabs against 1, 4, 6, and 8 MeV gamma rays • Monte Carlo (MC) method was employed to calculate buildup
Mandour & Hassan (1987) [17]	Iron | Carbon	• Buildup calculation using MC method for 14.1 MeV neutrons
Shin & Hariyama (1998) [18]	Concrete | Iron | Water	• Buildup calculation using EGS4 MC code for 1 and 10 MeV gamma-rays
Hu *et al.* (2008) [22]	Iron | Polyamide composite | Lead	• Optimization using genetic algorithm metaheuristic coupled with MCNP • Combination of composite and multilayer shielding for ^{252}Cf neutron source and ^{60}Co gamma source
McCaffrey *et al.* (2009) [23]	Metal elastomers bilayers	• Air kerma attenuation measured using commercial metal/elastomer test layers • EGSnrc MC code for 30−150 keV X-rays
Hossain *et al.* (2010) [6]	Poly boron | Borax mixed concrete	• Shielding against ^{252}Cf neutron source with detection using BF_3 long counter detector • MCNP calculations for 14 MeV neutrons
Kim and Moon (2010) [24]	(2-layer) LiH | W (8-layer) LiH | W | depleted U | Fe | TiH_2 | ZrH_2 | Pb | B_4C	• Optimization using genetic algorithm for shielding in spaceship conditions • Dose calculation using ANISN code
Abbas (2012) [20]	Water | Lead	• Buildup calculation using EGS4 MC code for 1, 2, and 6 MeV gamma rays
Gaber *et al.* (2013) [25]	Boron oxide glass | Epoxy ilmenite	• Double layer shields for neutrons and gamma rays resulting from ^{252}Cf fission source • Simulation using MCNP-4C2 MC code
Kim *et al.* (2015) [4]	Sendust alloy polymer | Tungsten	• Lamination of tungsten and composite films to shield against X-rays of 150 keV
Al-Arif & Kakil (2015) [9]	Aluminium | Iron | Lead	• Various combinations of shielding layers for gamma energies 0.662 and 1.25 MeV • Measurements of attenuation coefficient experimentally using ^{60}Co, ^{137}Cs sources and NaI (TI) scintillation detector
Mann *et al.* (2016) [11]	Aluminium | Limestone	• Buildup calculation using Geometric Progression method for gamma energies 0.5, 1.0, 2.0, and 3.0 MeV

IOP Conf. Series: Materials Science and Engineering **555** (2019) 012008 doi:10.1088/1757-899X/555/1/012008

Table 1(b). Summary of multilayer shielding combinations (cont.).

Author	Shielding Materials	Remarks/Application
Whetstone and Kearfott (2016) [26]	Steel | Polyethylene	• Alternating layers of shielding materials for an active neutron interrogation system • MCNP5 simulations for 1.0, 2.5, 4.0, 6.0, 8.0, 10.0, 12.0, and 14.1 MeV neutrons
Hadad *et al.* (2016) [27]	HDPE | Lead	• Optimization of shielding for prompt gamma neutron activation analysis with ^{252}Cf neutron source by Monte Carlo analysis
Sariyer & Küçer (2018) [28]	Concrete | Iron-contained materials (FeB, Fe_2B, stainless steel)	• Double layer of 1 m of concrete and varying thickness of iron layer against neutrons generated from proton interactions • FLUKA code was used to calculate dose distribution.
Park *et al.* (2018) [5]	Polymer composites containing tungsten or bismuth-tin alloy | BiSn layered sheet	• Lamination of W composite film or BiSn composite film onto a BiSn-coated layered sheet for 150 keV X-ray shielding
Cai *et al.* (2018) [29]	Six combinations of polyethylene, iron, boron carbide, and lead	• Optimization using genetic algorithm to shield against fission spectrum of ^{235}U

Some of these multilayer designs were found to yield better results than that of when using them separately. This might be because the subsequent layer is able to absorb the scattered radiation of lower energy emanating from the preceding layer. Fuse *et al.* [21] observed that a heterogenous arrangement of thick iron-water-thin iron was superior than a homogenous mixture of iron slabs and water. Shin and Hirayama [18], when calculating buildup factors using EGS4 code, discovered that some multilayer configurations produced lower buildup factors than their single component materials. Moreover, using two complementary metal-elastomer bilayers could decrease the overall shielding weight while providing attenuation equivalent to that of pure lead [23]. This was found to be useful in shielding garments. A polymer-tungsten bilayer had been determined to achieve better attenuation than a single layer of 0.2 mm tungsten [4]. Mann *et al.* [11] concluded that in gamma-ray protection, dual layered shields were more effective than single layer shields. A more recent study showed that bilayers of iron-containing materials were better than single layer of concrete in shielding neutrons [28].

Another interesting note to be highlighted here is the work by Hu *et al.* [22], Kim and Moon [24], and Cai *et al.* [29]. Their studies implemented an optimization technique called the genetic algorithm (GA) metaheuristic to design the most optimal multilayer and composite shield from a given set of materials. They were able to determine the best thickness ratios of individual layers and come up with new composite materials that showed better attenuation than lead oxide. Promoted by John Holland in 1970s, the GA method was inspired by natural evolution. This algorithm randomly generates a solution from a set of components, selects the most fit solutions, and then, uses them to build new generations which improve upon the previous generation. This technique may be helpful for optimizing multilayer shields that require various goals and constraints.

3.1. *Effects of Layer Arrangement*

Currently, there is no general agreement on the effects of changing the layer arrangement on the shielding performance. Some authors found that there was a significant difference when the layers were interchanged. Fuse *et al.* [21] demonstrated that the arrangements of the iron slabs and water could affect the resultant neutron flux and gamma dose rates. It was proven that the difference between the most optimum and the poorest performing arrangement was noticeable. The thickness of the layers and the order that they were assembled could also modify the value of neutron buildup factors which had to be considered in neutron calculations [17]. For neutron shielding, placing a hydrogenous material first was found to be better than the opposite order [6]. Meanwhile, Whetstone and Kearfott [26] claimed that there was no significant difference in shielding performance when the arrangement of materials was changed. This might due to the thin alternating layers of steel and polyethylene in all designs which resulted in a similar transmitted neutron spectrum.

In the case of X-ray shielding, McCaffrey *et al.* [23] observed that low-Z upstream/high-Z downstream could yield up to five times more attenuation than that of the reverse order at 50 kVp. However, this difference would decline and vanished at 150 kVp. Although, for the 150 kVp X-rays, Kim et al. [4] had results showing that high-Z/low-Z order was better than the opposite arrangement. They argued that as the X-rays passed through the high-Z layer, they would lose their energy where they could easily be shielded by the low-Z layer. Nonetheless, the authors stated that there was no optimal combination that could provide effective shielding for a wide energy range, thus the choice of material would have to be tailored for the particular X-ray energy in any application.

4. Conclusions

The main purpose of this study is to review the multilayer shielding used in literature. The designs that have been evaluated were listed. Focus was given on the effect of changing the material arrangement and also observations on buildup factor. The buildup factor for a multilayer shielding was found to be complicated and there was no clear trend on how to effectively arrange the materials. Research opportunities may lie in testing new multilayer designs. They also need to be accompanied by complementary buildup factor calculations. Future studies may also investigate on the application of metaheuristics in multilayer shielding optimization which may be useful in improving radiation protection. Examples of other metaheuristics are simulated annealing, tabu search, and ant colony optimization.

5. References

[1] Waly E S A and Bourham M A 2015 Comparative study of different concrete composition as gamma-ray shielding materials *Ann. of Nucl. Energy* **85** p 306-310.
[2] Osman A, El-Sarraf M, Abdel-Monem A and Abdo A E S 2015 Studying the shielding properties of lead glass composites using neutrons and gamma rays *Ann. of Nucl. Energy,* **78** p 146-151.
[3] Wang P, Tang X, Chai H, Chen D and Qiu Y 2015 Design, fabrication, and properties of a continuous carbon-fiber reinforced Sm2O3/polyimide gamma ray/neutron shielding material *Fusion Eng. and Design* **101** p218-225.
[4] Kim Y, Park S and Seo Y 2015 Enhanced X-ray Shielding Ability of Polymer–Nonleaded Metal Composites by Multilayer Structuring *Industrial and Eng. Chem. Research* **54(22)** p 5968-5973.
[5] Park S, Kim H, Kim Y, Kim E and Seo Y 2018 Multilayer-Structured Non-leaded Metal/Polymer Composites for Enhanced X-ray Shielding *MRS Advances* p 1-9.
[6] Hossain M, Islam S, Quasem M, and Zaman, M. 2010 Study of shielding behaviour of multilayer shields containing PB and BX *Indian J. of Pure and App. Phys.* **48(12)** p 860-868.
[7] Shultis J K and Faw R E 2005 Radiation shielding technology *Health Phys.* **88(6)** p 587- 612.
[8] Shultis J K and Faw R E 2002 *Fund. of Nucl. Sci. and Eng.*

[9] Al-Arif, M S and Kakil D O 2015 Calculated-Experimental Model for Multilayer Shield *ARO-The Scientific J. of Koya University* **3(1)** 23-27.

[10] Lamarsh J R and Baratta A J 2001 *Introduction to nuclear engineering* **3** (Upper Saddle River, NJ: Prentice Hall).

[11] Mann K S, Heer M S and Rani A 2016 Gamma-ray double-layered transmission exposure buildup factors of some engineering materials, a comparative study *Rad. Phys. and Chem.* **125** p 27-40.

[12] Kalos M H 1956 Nuclear Development Associates (NDA) p 56–57.

[13] Kalos M H 1957 A Monte Carlo calculation of the transport of gamma rays (NDA) p 56-10.

[14] Shin K, Hirayama H 1995 Approximation model for multilayer gamma-ray buildup factors by transmission matrix methods: application to point isotropic source geometry *Nucl. Sci. Eng.* **120** p 211.

[15] Burke G de P, Beck H L 1974 Calculated and measured dose buildup factors for gamma rays penetrating multilayered slabs *Nucl. Sci. Eng.* **53** p 109.

[16] Lin U T , Jiang S H 1996 A dedicated empirical formula for gamma ray buildup factors for a point isotropic source in stratified shields *Radiat. Phys. Chem.* **48(4)** p 389–401.

[17] Mandour M A and Hassan M 1987 Neutron buildup factors for multilayered-media *Isotopenpraxis Isotopes in Env. and Health Studies* **23(12)** DOI: 10.1080/10256018708623872 p 437-442.

[18] Shin K and Hirayama H 1998 Description of multilayered gamma-ray exposure buildup factors up to 40 mfp by the approximating model *J. of Nucl. Sci. and Tech.* **35(12)** p 865-873.

[19] Kuspa, J. P. (1972). Calculation of buildup factors for multilayer slab shields using the Monte Carlo method. Masters Theses. 5114. Retrieved April 10, 2018, http://scholarsmine.mst.edu/masters_theses/5114.

[20] Abbas J 2012 Calculation of buildup factor for gamma-ray exposure in two layered shields made of water and lead *J. of Kufa-Physics* **4(1)**.

[21] Fuse T, Yamaji A and Miura T 1970 The optimum arrangement of laminated iron-water shields *Nucl. Eng. and Design* **13(3)** p 390-394.

[22] Hu H *et al.* 2008 Study on Composite Material for Shielding Mixed Neutron and γ-Rays *IEEE Trans. on Nucl. Sci.* **55(4)** p 2376-2384.

[23] McCaffrey J, Mainegra-Hing E and Shen H 2009 Optimizing non-Pb radiation shielding materials using bilayers *Med. Phys.* **36(12)** p 5586-5594.

[24] Kim B S and Moon J H 2010 Use of a genetic algorithm in the search for a near-optimal shielding design *Ann. of Nucl. Energy* **37(2)** p 120-129.

[25] Gaber F, El-Sarraf M and Kansouh W 2013 Utilization of boron oxide glass and epoxy/ilmenite assembly as two layer shield *Ann. of Nucl. Energy* **57** p 106-110.

[26] Whetstone Z D and Kearfott K J 2016 Layered shielding design for an active neutron interrogation system *Rad. Phys. and Chem.* **125** p 69-74.

[27] Hadad K, Nematollahi M, Sadeghpour H and Faghihi R 2016 Moderation and shielding optimization for a ^{252}Cf based prompt gamma neutron activation analyzer system *Int. J. of Hydrogen Energy* **41(17)** p 7221-7226.

[28] Sariyer D and Küçer R 2018 Double-layer neutron shield design as neutron shielding application *AIP Conf. Proc.* **1935(1)** (AIP Publishing) p. 180004.

[29] Cai Y, Hu H, Pan Z, Hu G and Zhang T 2018 A method to optimize the shield compact and lightweight combining the structure with components together by genetic algorithm and MCNP code App. Rad. and Isotopes **139** p 169-174.

Acknowledgement

We would like to thank Universiti Teknologi Malaysia for providing the facilities required to perform this review.

Identifying the possible source regions of elevated concentrations of Xenon-133 detected at the Takasaki Radionuclide Monitoring Station (JPX38) on 2 October 2017: Malaysian CTBT National Data Centre (MY-NDC) findings

F I A Rashid[1, a)], M Z Zolkaffly[1, b)]

[1]MY-NDC, Planning and International Relations Division, Malaysian Nuclear Agency, Bangi, 43000 Kajang, Selangor, Malaysia

[a]faisal_izwan@nuclearmalaysia.gov.my
[b]zulfakar@nuclearmalaysia.gov.my

Abstract. A couple of elevated concentrations of noble gas Xenon-133 were detected at the Takasaki Radionuclide Monitoring Station (JPX38) on 2 October 2017. Xenon-133 is among four relevant radioxenon isotopes that are used for monitoring and verification of nuclear explosions including Xenon-13, Xenon-133m and Xenon-131m. These radioxenon gases could be emitted either from nuclear explosion or civil nuclear facilities including medical isotopes production facilities, nuclear power reactors and nuclear research reactors. This paper presents the Malaysian CTBT National Data Centre (MY-NDC) findings in identifying the possible source regions for elevated concentrations of Xenon-133 detected at JPX38 on 2 October 2017. The findings will be useful in understanding the possible source of release for such Xenon-133 detection, whether coming from a late release of the 3 September 2017 North Korea nuclear test site or from civil nuclear facilities.

1. Introduction

The Comprehensive Nuclear-Test-Ban-Treaty (CTBT) is a multilateral treaty that banning all nuclear explosions – everywhere and by anyone. At the time of this study, the CTBT has not entered into force due to pending ratification from eight Annex 2 states of the CTBT namely, China, Egypt, India, Iran, Israel, North Korea, Pakistan, and United States [1]. A verification regime of the CTBT, which among others includes the International Monitoring System (IMS), was established to detect any nuclear explosion conducted on the planet. The IMS comprising a global network of 321 seismic, hydroacoustic, infrasound and radionuclide stations and 16 radionuclide laboratories is used to detect signals that could indicate a possible nuclear explosion [2].

Malaysia is one of the CTBT contracting states by signing it on 23 July 1998 and ratifying it on 17 January 2008 [3]. A Malaysian CTBT National Data Centre (MY-NDC) was established in December 2005 at the Malaysian Nuclear Agency (Nuklear Malaysia) and is tasked to provide technical information of CTBT related events to Nuklear Malaysia as the CTBT National Authority [4].

The IMS network of the CTBT inter-alia incorporates 40 noble gas monitoring stations that are capable to detect radioxenon gases in the atmosphere. In the case of underground nuclear explosion, noble gases like radioxenons are created in significant quantities and are the most likely radioactive signatures detected at IMS stations [5]. Due to their half-lives and fission yields, four radioxenon isotopes namely Xenon-135, Xenon-133, Xenon-133m and Xenon-131m are used for detecting nuclear explosion [6]. Table 1 shows the half-lives and the most intense γ-ray and X-ray of four radioxenon isotopes These isotopes are artificial isotopes which are chemically inert, unaffected to deposition, and very difficult to contain, consequently they have increased probability to escape from a nuclear test site [7].

Table 1. Half-life and the most intense γ-ray and X-ray of four radioxenon isotopes.

Radioxenon isotopes	Half-life	Energy X-ray (keV)	Intensity (%)	Energy γ-ray (keV)	Intensity (%)
Xenon-135	9.14 hours	30.80	2.1	249.77	90.0
Xenon-133	5.243 days	30.80	40.9	80.997	38.0
Xenon-133m	2.19 days	29.62	46.1	233.22	8.2
Xenon-131m	11.84 days	29.62	44.4	163.93	1.91

Apart from nuclear explosion, these radioxenon isotopes can also be released from civil nuclear facilities, such as medical isotopes production facilities, nuclear power reactors and nuclear research reactors. However, the magnitude of release of radioxenons is different for each nuclear facilities and nuclear explosion. Within the region surrounding the North Korean nuclear test site, there are numbers of operational civil nuclear facilities situated in China, Japan, North Korea, South Korea and Taiwan as shown in Figure 1. It is also worth to note that after Fukushima nuclear accident in March 2011, most of the nuclear power plants in Japan were not operating during September until October 2017.

On 3 September 2017, the North Korea conducted the sixth underground nuclear test which caused significant detectable earthquake in both North Korea and the neighboring countries [8]. The earthquake signal was picked up by over 100 IMS stations and the characterisation of the event was consistent with a man-made explosion [9]. From verification point of view, the explosion can only be classified as underground nuclear explosion only if the corresponding airborne radioactivity is detected, primarily in the form of noble gas radioxenon and radioargon [10].

Later, on 2 October 2017, two noble gas samples with elevated concentration of Xenon-133 were detected at one of the IMS radionuclide stations, namely the Takasaki Radionuclide Monitoring Station (JPX38), in Takasaki, Japan. JPX38 is equipped with automatic radioxenon collection and analysis system, known as the Swedish Unattended Noble Gas Analyzer (SAUNA) which employs scintillator beta-gamma coincidence detectors to detect airborne noble gas radioxenons [11]. Besides JPX38, there are other regional IMS radionuclide stations in the area surrounding the nuclear test site, namely CNX20 (Beijing, China), CNX22 (Guangzhou, China), RUX58 (Ussuriysk, Russia) and RUX60 (Petropavlovsk-Kamchatskiy, Russia) as shown in Figure 1.

Taking into account various possible sources of radioxenons, a robust analysis is critical to distinguish and identify the possible source of release of radioxenons detected at IMS stations. Hence, this paper aims at presenting MY-NDC findings in identifying the possible source regions of elevated concentrations of Xenon-133 detected at JPX38 on 2 October 2017. This study exhibits MY-NDC effort in understanding the possible source of release of such Xenon-133 detection at JPX38, whether they were coming from a late release of the 3 September 2017 North Korea nuclear test site or from operational civil nuclear facilities within the region.

IOP Conf. Series: Materials Science and Engineering **555** (2019) 012009 doi:10.1088/1757-899X/555/1/012009

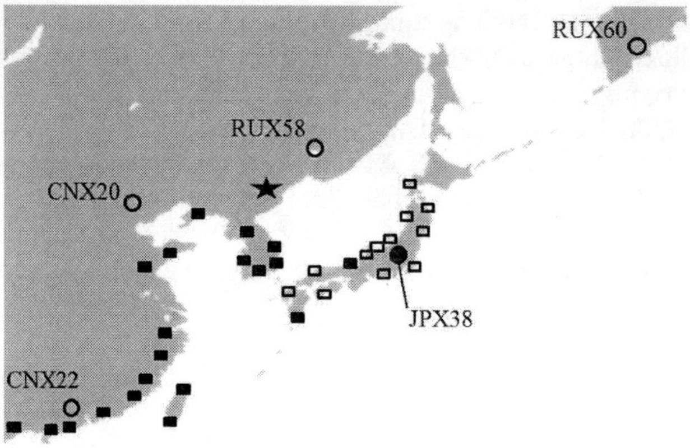

Figure 1. IMS radionuclide stations and civil nuclear facilities in the surrounding area of the North Korean nuclear test site. Filled circle marks certified radionuclide stations with radioxenon systems operational on 2 October 2017. Unfilled circles indicate uncertified radionuclide stations operational on 2 October 2017. The filled rectangles indicate operational civil nuclear facilities. The unfilled rectangles indicate inoperational civil nuclear facilities. The star marks the position of the nuclear test site in North Korea.

2. Methods

The methodology that we used in this study is summarised in Figure 2. Firstly, we measured the radioxenon isotopes of Xenon-135, Xenon-133, Xenon-133m and Xenon-131m from two noble gas samples of JPX38 with collection stop on 2 October 2017 0650 UTC and 1850 UTC using beta-gamma coincidence technique. In this relation, we used Norfy review tool, a specialised noble gas analysis software developed by the Provisional Technical Secretariat (PTS) of the Preparatory Commission for the Comprehensive Nuclear-Test-Ban Treaty Organization (CTBTO). As a result, we acquired the type of radioxenon isotopes present in the samples, along with its activity (in mBq) and concentration (in mBq/m^3).

Next, we performed atmospheric transport modelling (ATM) in backward mode using the result from the preceding beta-gamma coincidence analysis. We use the Web Connected Graphics Engine (Web-Grape) software developed by the PTS of CTBTO for backward mode ATM. The backward mode ATM is operated with meteorological data from the United States National Centers for Environmental Prediction (NCEP). The backward mode ATM provides an insight on field of regard (FOR) which denotes the possible source region for radioxenons detected within one single sample. We also run possible source region (PSR) model to identify the most consistent point source location that would explain the detection scenario of radioxenons at JPX38.

Lastly, we run forward mode ATM to predict the potentially affected radionuclide monitoring stations. For this purpose, we used the Lagrangian Particle Dispersion Model HYSPLIT, developed by the National Oceanic and Atmospheric Administration – Air Resources Laboratory [12]. The meteorological data from the Global Forecast System (GFS) was used for this purpose. Several ATM forward mode simulations were generated with hypothetical releases at different release points, including from the nuclear test site and a few civil nuclear facilities. The result from the forward mode ATM has helped us to shortlist the PSR identified from backward mode ATM and predict the other potentially affected radionuclide monitoring stations.

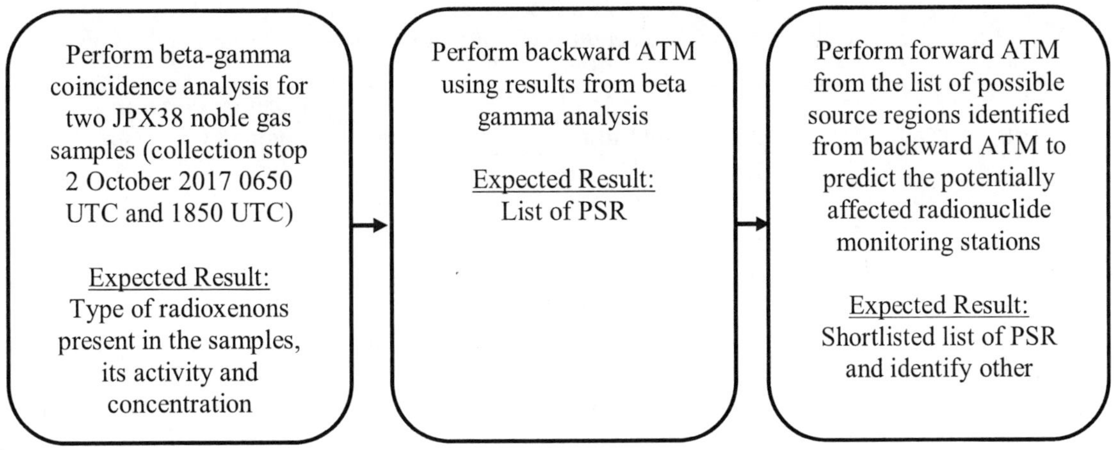

Figure 2. Methodology used by MY-NDC for the study.

3. Results and Discussion

3.1. Beta-gamma coincidence analysis

Both noble gas samples of JPX38 with collection stop on 2 October 2017 0650 UTC and 1850 UTC were analysed in this study using beta-gamma coincidence technique. Figure 3 illustrates beta-gamma coincidence histogram of both samples. We found that both samples detected Xenon-133. No traces of Xenon-135, Xenon-133m and Xenon-131m were detected from both samples. Table 2 shows the measurement category and activity concentration of Xenon-133 of both samples. Category C denotes anomalous Xenon detection at the station. We also acknowledged that no immediate releases of radioxenon gases were detected by the radionuclide monitoring stations within the region. This suggests that there are no direct pathways were created from the cavity to the surface to allow the propagation of particulates and noble gases during the test [13]. The absence of other radioxenon isotopes in the samples such as Xenon-131m has made the dating of nuclear fission not possible. If there are other radioxenon isotopes present in the samples, which in the correct isotropic ratio, it would have increased the evidence of a nuclear explosion source.

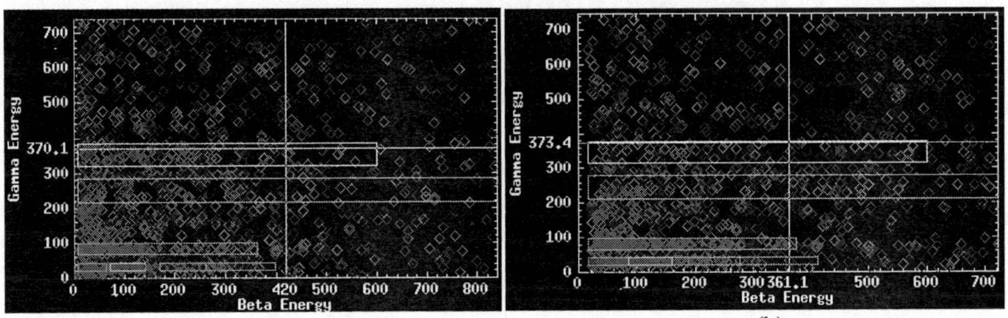

(a) (b)

Figure 3. Beta-gamma coincidence histogram of (a) Sample 1 and, (b) Sample 2.

Table 2. Measurement category and activity concentration of Xenon-133 for both samples.

Sample	Station	Collection Stop	Category	Concentration (mBq/m^3) Xenon-133	(+/-)
1	JPX38	2 Oct 2017, 0650 UTC	C	0.71	0.0619
2	JPX38	2 Oct 2017, 1850 UTC	C	0.95	0.06631

We also plotted activity concentration of Xenon-133 detected from both samples with historical detection of Xenon-133 at JPX38 from 1 January 2016 until 31 December 2017 as shown in Figure 4. We found that similar Xenon-133 activity concentration have been detected at JPX38 within that period. Within that period, most of Xenon-133 concentrations detected at JPX38 were between 0.1 to 0.4 mBq/m^3. Thus, in this case, concentration of Xenon-133 of sample 1 and 2 were far above the usual range and considered as significant. This argument is further justified based on the assignment of category C as anomalous Xenon detection to both samples.

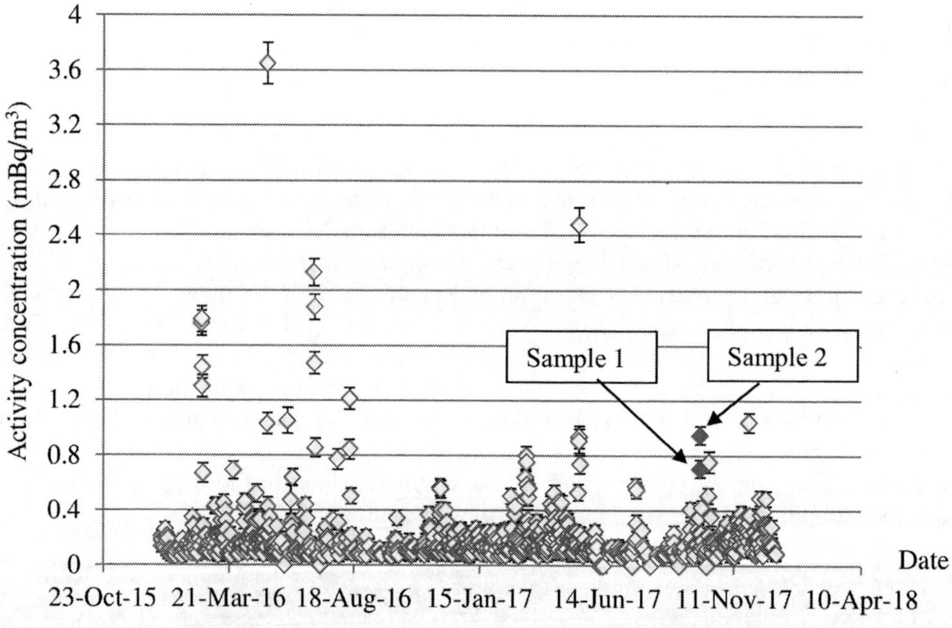

Figure 4. Historical detection of Xenon-133 at JPX38 from 1 January 2016 – 31 December 2017.

3.2. Backward mode ATM

We run backward mode ATM to identify the possible source regions for Xenon-133 detected at JPX38. Figure 5 shows the result of backward mode ATM from JPX38, starting from the initial detection of Xenon-133 on 2 October 2017 and backtracking until 26 September 2017. The simulations showed at early stage, the Xenon-133 plume circulated in the middle of Japan and later on moving towards nuclear test site in North Korea. At longer backward transport time, the plume started to spread out and cover the territory of China and Russia. The relatively high values of PSR were recorded for nuclear test site in North Korea which indicates the nuclear test site as a likely source for the Xenon-133 detections at JPX38. Based on the backward mode ATM performed in the context of this study, we also found that most of the regional nuclear reactors could be ruled out as PSR. Nevertheless, other regional source cannot be completely excluded.

IOP Conf. Series: Materials Science and Engineering **555** (2019) 012009 doi:10.1088/1757-899X/555/1/012009

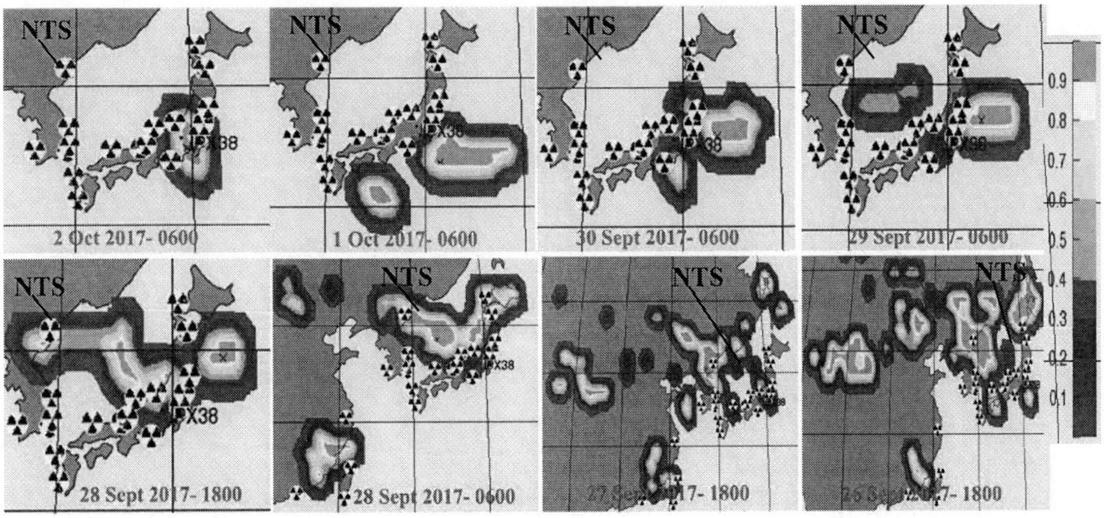

Figure 5. Backward mode ATM from JPX38. NTS denotes the nuclear test site in North Korea. The values of color coded scale on the right represent the value of PSR (correlation factor).

3.3. Forward mode ATM

The forward mode ATM was performed for PSRs, previously identified from backward mode ATM. Figure 6 shows the dispersion of Xenon-133 plume over seven days, assuming 24 hour hypothetical release on 28 September 2017 from nuclear test site, using 0.5 degree GFS meteorological data and 200,000 particle released.

The forward mode ATM assuming release on 28 September 2017 shows the prevailing wind directions to the north east. The forward simulations also indicated that RUX58 station would have detected the Xenon-133 plume as well. Our further investigation revealed RUX58 detected several peaks of Xenon-133 as well as Xenon -131m and Xenon-135 during October 2017. Despite RUX58 detected such peaks, the station was a non-certified IMS station. Hence, calibration parameters of this non-certified noble gas system might be an issue. Further, RUX58 also was not in operation for quite a long time and just went back online in between, at the days indicated by the forward simulations. Consequently, special precaution must be made as Xenon-131m signal was on the edge of detection limit. From forward simulation, a late release from nuclear test site remains one of the PSR. Nevertheless, magnitude and phase of detections predicted from the forward simulation did not match with the actual detections at JPX38 as shown in Figure 7. Based on the forward simulation, we were also of the view that other regional source might also be associated with the Xenon-133 detection at JPX38.

We also noted that most of nuclear power plants in Japan were not in operation and therefore can be excluded as possible source of release of radioxenons.

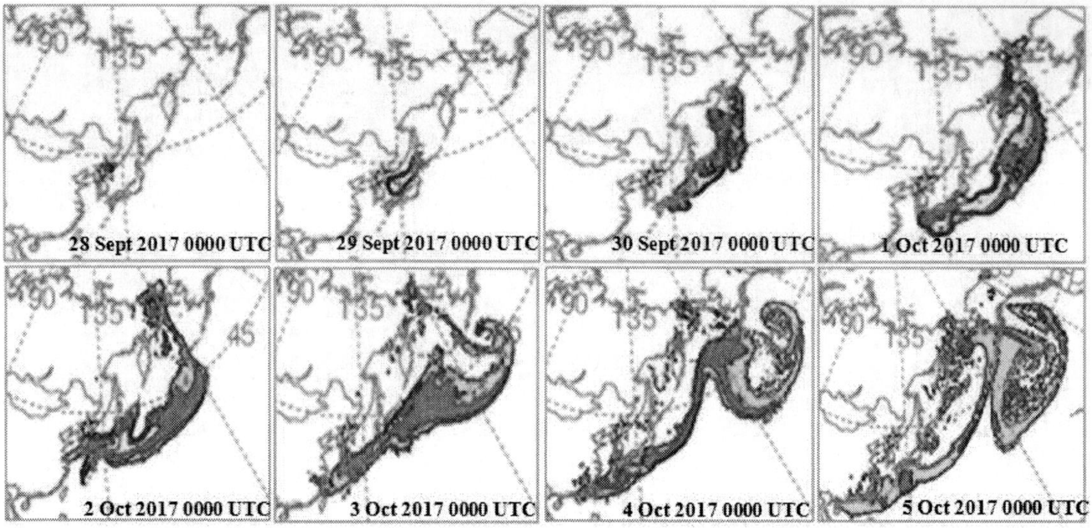

Figure 6. Forward mode ATM from nuclear test site.

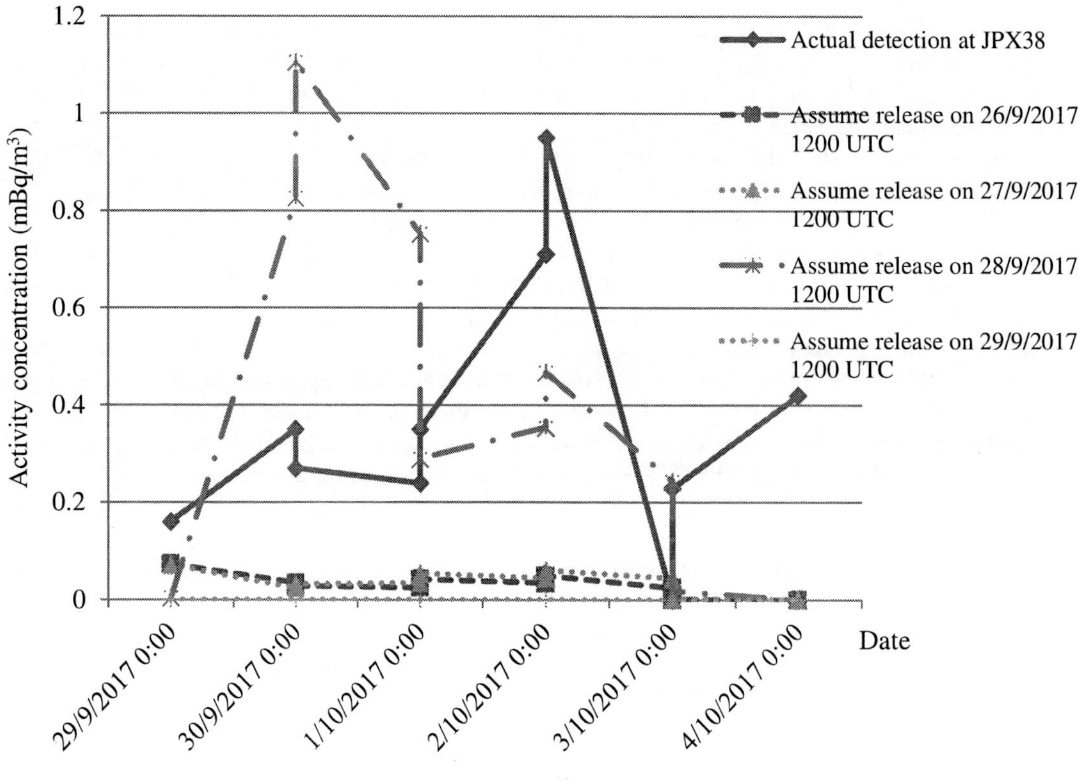

Figure 7. Magnitude and phase of detections predicted from the forward simulation with different release time assumptions.

4. Conclusion

Results from beta gamma coincidence analysis confirmed significant activity concentrations of Xenon-133 in both samples at JPX38 on 2 October 2018. No other radioxenon isotopes were detected from both samples. Backward ATM simulations suggest the nuclear test site as possible source location with high correlation PSR factor. However, other regional sources could not be completely ruled out as possible source location. Forward ATM simulation shows strong correlation with nuclear test site as well as others regional civil sources. In conclusion, we found that the results were not conclusive with regard to a possible association to the explosion on 3 September at the nuclear test site. Other regional civil sources could also explain such detection at JPX38.

5. References

[1] Nuclear Threat Initiative 2018 Comprehensive Nuclear-Test-Ban Treaty available url: https://www.nti.org/learn/treaties-and-regimes/comprehensive-nuclear-test-ban-treaty-ctbt/ (retrieved on 31 July 2018).

[2] Kalinowski M B, Becker A, Saey P R J, Tuma M P and Wotawa G 2008 The complexity of CTBT verification taking noble gas monitoring as an example *Complexity* **14** 89–99.

[3] Rashid F I A and Zolkaffly M Z 2018 Stakeholder engagement for promoting the Comprehensive Nuclear Test-Ban Treaty (CTBT): Malaysia's experience *IOP Conf. Ser. Mater. Sci. Eng.* **298** 012054.

[4] Rashid F I A and Zolkaffly M Z 2017 National data centre preparedness exercise 2015 (NPE2015): MY-NDC progress result and experience *AIP Conf. Proc.* **1799** doi: 10.1063/1.4972923 p 030013-1-9.

[5] Kalinowski M B, Axelsson A, Bean M, Blanchard X, Bowyer T W, Brachet G, Hebel S, Mcintyre J I, Peters J, Pistner C, Raith M, Ringbom A, Saey P, Schlosser C, Stock T, Taffery T and Ungar K R 2010 Discrimination of nuclear explosions against civilian sources based on atmospheric Xenon isotopic activity ratios *Pure Appl. Geophys* **167** p 17–39.

[6] Geer L E D 2001 Comprehensive nuclear-test-ban treaty: relevant radionuclides *Kerntechnik* **66(3)** p 113-20.

[7] Geer L E D 1996 Atmospheric radionuclide monitoring: A Swedish perspective *Monitoring a Comprehensive Test Ban Treaty* ed Huseby, E.S. and A.M. Dainty (Kluwer Academic Publishers) p 157-77.

[8] Zhang L, Hou X, Cheng P, Chen N, Fan Y and Liu Q 2018 Impact of North Korean nuclear weapons test on 3 September, 2017 on inland China traced by ^{14}C and ^{129}I *J. of Radioanalytical and Nuclear Chem.* **316(1)** (doi: 10.1007/s10967-018-5747-y) p 383-88.

[9] CTBTO 2017 CTBTO Executive Secretary Lassina Zerbo on the unusual seismic event detected in the Democratic People's Republic of Korea available url: https://www.ctbto.org/press-centre/press-releases/2017/ctbto-executive-secretary-lassina-zerbo-on-the-unusual-seismic-event-detected-in-the-democratic-peoples-republic-of-korea/ (retrieved 3 September 2017).

[10] Lowrey J D, Biegalski S R, Bowyer T W, Haas D A and Hayes J C 2015 Consideration of impact of atmospheric intrusion in subsurface sampling for investigation of suspected underground nuclear explosions *J. of Radioanalytical and Nuclear Chem.* **307(3)** (doi 10.1007/s10967-015-4462-1) p 2439-44.

[11] CTBTO RN38 Takasaki Japan available url: https://www.ctbto.org/verification-regime/featured-stations/types/radionuclide/rn38-takasaki-japan/

[12] Stein A F, Draxler R R, Rolph G D, Stunder B J B, Cohen M D and Ngan F 2015 NOAA's HYSPLIT Atmospheric Transport and Dispersion Modeling System *Bulletin of the American Meteorological Soc.* **96** (https://doi.org/10.1175/BAMS-D-14-00110.1) p 2059–77.

[13] Gaebler P, Ceranna L, Nooshiri N, Barth A, Cesca S, Frei M, Grünberg I, Hartmann G, Koch K, Pilger C, Ross J O and Dahm T 2018 A multi-technology analysis of the 2017 North Korean nuclear test *Solid Earth Discuss* (https://doi.org/10.5194/se-2018-102).

IOP Conf. Series: Materials Science and Engineering **555** (2019) 012010 doi:10.1088/1757-899X/555/1/012010

The Comprehensive Nuclear-Test-Ban Treaty (CTBT): seismic monitoring

M Z Zolkaffly [1, a)], **F I A Rashid** [1, b)]

[1]Malaysian CTBT National Data Centre (MY-NDC), Planning and International Relations Division, Malaysian Nuclear Agency, Bangi, 43000 Kajang, Selangor, Malaysia

[a]zulfakar@nuclearmalaysia.gov.my
[b]faisal_izwan@nuclearmalaysia.gov.my

Abstract. The Comprehensive Nuclear-Test-Ban Treaty (CTBT) is briefly characterized as a global arms control and disarmament initiative, which complements the goal of the Non-Proliferation Treaty (NPT). The verification regime of the CTBT is designed to monitor countries' compliance with the CTBT by detecting any nuclear explosion conducted on Earth - underground, underwater or in the atmosphere. This paper briefly goes through the International Monitoring System (IMS) of CTBT's verification regime that uses four different technologies - seismic, hydroacoustic, infrasound and radionuclide - to monitor the planet for nuclear explosions. It pays particular attention to the seismology monitoring techniques of CTBT IMS, including the practical steps in discriminating underground explosions and earthquakes using data mainly from the IMS seismic network. It also highlights the seismic analyses performed by the Malaysian CTBT National Data Centre (MY-NDC) that in principle, applies the practical steps of CTBT nuclear explosion seismology monitoring. In summary underground explosions produced seismic waves with unique characteristics which allowed the discrimination between explosions and earthquakes.

1. Introduction

After a half century nuclear weapons were developed, tested, and used, the Comprehensive Nuclear Test Ban Treaty (CTBT) banned all nuclear weapon test explosion or any other nuclear explosion, has been negotiated and opened for signature in September 1996. The underlying goal of the CTBT is to inhibit the further development of more sophisticated nuclear weapons, thus complementing the goal of the Non-Proliferation Treaty (NPT) that inhibits the spread of nuclear weapons (horizontal proliferation) [1]. As of November 2018, the CTBT has been signed by 183 nations and ratified by 166. The Treaty, however, cannot enter into force until it is fully ratified by 44 specific nations listed in the Annex II of the Treaty. Eight of which have yet to do so – China, Egypt, India, Iran, Israel North Korea, Pakistan and the United States.

Article IV of the Treaty elaborates on the global verification regime to monitor compliance with Treaty provisions. The Treaty establishes an extensive verification regime that is designed to monitor phenomena worldwide so as to detect the occurrence of nuclear explosions anywhere – be it underground, underwater or in the atmosphere, and to facilitate the resolution of ambiguous events. The breadth of its coverage is intended to create a significant deterrent against possible efforts to evade the ban on testing [2]. This verification regime consists of the following six elements:

- International Monitoring System (IMS)
- International Data Centre (IDC), iii)
- Global Communications Infrastructure (GCI)
- Consultation and clarification
- On-Site Inspection (OSI), and
- Confidence-building measures.

Article IV also provides that the verification regime shall be capable of meeting the requirements of the Treaty when it enters into force. This requires a substantial program of preparation in advance of the Treaty's entry into force. To make such necessary preparations, the Preparatory Commission for the CTBT Organization (CTBTO) was established in 1997, made up of CTBT States Signatories and supported by a Provisional Technical Secretariat (PTS). The key tasks of the CTBTO include the establishment of IMS, IDC in Vienna, GCI, as well as complete procedures for the operation of those facilities and for the conduct of OSI where concerns are raised about a possible nuclear explosion [3].

2. International Monitoring System (IMS)

The IMS uses four complementary verification methods, utilizing the modern technology available – seismic, hydroacoustics, infrasound and radionuclide technologies. Seismic, hydroacoustic and infrasound stations monitor the underground, the large oceans and the atmosphere respectively. Radionuclide stations detect radioactive debris from atmospheric explosions or vented by underground or underwater nuclear explosions. In addition, radionuclide laboratories assist radionuclide stations in identifying these radioactive substances [4].

The IMS is designed to be non-discriminatory and does not single out any country or region for enhanced monitoring [5]. There more than 300 IMS stations built worldwide to monitor the planet for any sign of a nuclear explosion. Data from these stations are transmitted to the IDC in Vienna through the GCI, a global private data network, which is largely based on satellite links. As stated in the Treaty, States Parties will have equal and direct access to all IMS data, raw or processed, for verification as well as civilian uses. However, also as stated in the Treaty, the responsibility for determining Treaty compliance and judgements about the nature of events detected by the IMS network rest with the States Parties, not with the CTBTO. Thus, States Parties can utilise IMS data, along with any additional sources of objective information available to them, to monitor the Treaty compliance [6].

The IMS network provides generally uniform coverage across the globe, and when completed, it will consist of:

- 50 primary and 120 auxiliary seismic monitoring stations to monitor shockwaves in the earth that could be caused by a nuclear test
- 11 hydro-acoustic stations listening for sound waves traveling through the oceans that could be caused by a nuclear test explosion
- 60 infra-sound stations on the earth's surface will be able to detect ultra-low frequency sound waves caused by large explosions – these sound waves are inaudible to the human ear.
- 80 radionuclide stations using air samplers to detect radioactive particles released from atmospheric explosions and/or vented from underground or under-water explosions
- 16 radionuclide laboratories for analysis of samples from the radionuclide stations

3. CTBT seismic monitoring

A seismic event, either from a natural earthquake or manmade explosion, generates body waves or compressional waves (P-waves) and surface waves or shear waves (S-waves) that are potentially detectable by wave sensors of infrasound, hydroacoustics, and seismic. Both P-waves and S-waves signals are crucial for the analysis of a suspicious Treaty-violation event. They provide essential information on the location, strength and nature of an event. P-waves are primary waves that alternately compress and expand the ground in the direction of the wave's propagation. They are the fastest travelling wave and can move through any material. S-waves are secondary waves in the ground that move perpendicular to the direction of the wave's propagation. They travel along the surface of the Earth are slower than P-waves waves and can only move through solids as this kind of movement is impossible in liquid or gaseous materials [7].

The direction from which the waves were emitted can be identified through the measurement of the azimuths of P-waves and S-waves. Since both waves travel at different speeds, it is also possible to determine the distance to the source by measuring different arrival times of the waves. During analysis, surface wave measurements help to identify the depth and magnitude of an event.

The Treaty calls for two global seismic monitoring networks – a primary seismic network with 50 stations and an auxiliary one with 120 stations. The stations of the primary seismic network transmit continuous authenticated data within 5 minutes of being recorded to the IDC in Vienna which is used in automatic processing. The auxiliary seismic network takes advantage of existing seismic stations which are being upgraded to meet the IMS technical standards, and do not send data in real time but upon request only.

4. The practical steps of CTBT nuclear explosion seismology monitoring

Practically, the CTBT nuclear explosion seismology monitoring can be organized in six practical steps as described in Table 1.

Detection of seismic resulting from either natural or man-made event is typically done with arrays of sensors (or seismometers) deployed as a group and spread out over an area, which can be on the order of tens or hundreds of square kilometres. Data recorded by the arrays can be interpreted to estimate the direction from which signals arrive.

Association is the process to identify sets of signals recorded from different stations, which originate from the same event. In such cases, data collected by arrays will provide directional information that is helpful in resolving which signals correspond to which event.

Table 1. The six practical steps of CTBT nuclear explosion seismology monitoring.

	Steps	Objective
1	Detection	Detects signals recorded by each sensor of a particular network
2	Association	Associate into a single group the various signal recorded by different sensors that originate from a common source/event
3	Location estimation	Estimate the location and time of that event and the uncertainty of the location estimate
4	Identification	Identify the nature of the event—whether suspicious or not in the context of CTBT monitoring
5	Yield estimation	Estimate how big was the event
6	Attribution	Determine what country carried it out if it was a nuclear test

Location estimation is needed for all the events for which a set of detections can be associated, since an interpretation of the location (including the event depth) is commonly used to reach preliminary conclusions on whether an event could possibly be an explosion [8]. Location estimation of the event is performed by measuring the arrival times of various seismic signals from the recorded waveforms. The identified arrival times are then used to estimate the latitude, longitude, depth, and origin time of each detected event. For any type of seismic wave that the station might observe, it is important to know the travel time from any assumed source-location to any particular seismographic station. The source-station distance and azimuth can be roughly estimated from data at a single station. The accuracy of a location estimate is best characterized with a confidence interval. Typically, detections at three or more stations are needed for confidence intervals to be calculated.

As mentioned earlier, the primary seismic stations send continuous near-real-time data to the IDC in Vienna. The IDC will search for signals of potential Treaty violation from these continuous waveform data of primary seismic, where lists of detections are formed. These detections are then associated with events, and the event location, depth, and origin time, including other associated parameters, such as magnitude, are estimated. The IDC will distribute those seismic data and derived products to States that are party to the Treaty [9]. The IDC is also required by the Treaty to routinely screen-out events considered to be consistent with natural phenomena or non-nuclear, man-made phenomena.

Identification is performed once the detections from a seismic event have been associated and an accurate location estimate has been obtained. The identification of the nature of a seismic event on the basis of its seismic signals is a process of making a determination from seismograms as to whether it could be a nuclear explosion, or a natural earthquake, or a manmade event. This process involves a large subject in view of the many possibilities as seismic event generate many different types of seismic waves, in various different frequency bands. Discriminating between earthquakes and explosions are performed based on interpretation of the event location, including its depth, on the relative excitation of a variety of P-waves and S-waves; and on properties of the signal spectrum associated with each of these two different types of source. With regard to these three broad categories, many different methods have been tried, with various degrees of success. No single method of event identification based on seismological data is fool proof, but in combination these methods have proven highly reliable [1].

In addition to the use of teleseismic signals (i.e. those propagating to distances of 1500 km and more, via paths that can reach down substantially more than 100 km into the Earth's interior) to monitor effectively for large explosions, there has been growing recognition of the merits of regional seismic waves (i.e. those propagating at shallower levels) for monitoring down to lower magnitudes explosions, often well below magnitude three (3). Regional methods are typically based upon the general observation that explosion signals, when compared to earthquakes, have much stronger P-waves at high frequency, whereas those from earthquakes have stronger S-waves. The method has been applied by many authors to seismic signals recorded from the three nuclear explosions conducted in the 21st century by North Korea. Figure 1 shows an example of comparing regional signals of a very small earthquake and a small explosion [10].

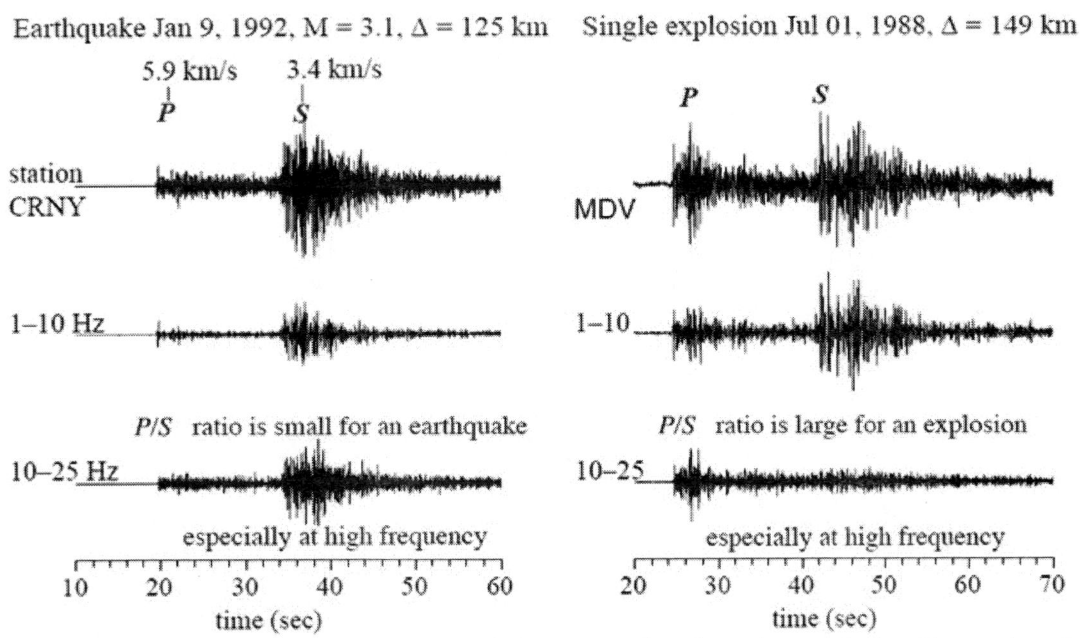

Figure 1. Typical vertical-component records from an earthquake and an explosion. Traces plotted are: (top) unfiltered; (middle) low-frequency bandpass filtered; and (bottom) high-frequency bandpass filtered. Arrivals of P and S waves are shown on the unfiltered traces [10].

IOP Conf. Series: Materials Science and Engineering **555** (2019) 012010 doi:10.1088/1757-899X/555/1/012010

Since the CTBT bans nuclear explosive testing at all levels of yield, thus, all tests are violations. The capability to estimate yield, however, does not directly arise in the limited context of deciding whether or not a detected test would be a treaty violation. The interest in yield estimation is generally because of the need to interpret a characterization of seismic monitoring capability expressed in terms of seismic magnitude, to a monitoring capability expressed in terms of explosive yield. Estimating the size of a detected event in terms of nuclear yield could be derived from seismic magnitude [11]. A single relationship between magnitude and yield does not exist as explosions of a given yield generate different amplitudes and magnitudes of seismic waves depending upon:

- the efficiency of seismic wave propagation from source to recording station,
- the rock type at the source,
- depth of the explosion, and
- whether the explosion is well coupled or decoupled.

Formulas relating the body-wave magnitude, *mb*, to the yield, *Y*, based on data from past underground nuclear explosions are of the form:

$$mb = A + B\,log(Y) \tag{1}$$

where *Y* is in kilotons, and *A* and *B* are constants that depend on features 1) to 4).

Most past tests of yield greater than about 1 kiloton were detonated at greater depths as yield was increased so as to ensure containment. Their data are well fit by B = 0.75 [12].

Attribution denotes the ability of States Parties to identify the nation or organization responsible for a nuclear explosion, if it has been identified. In the CTBT context, attribution can occur in three generic ways either alone or in combination:

- National Technical Means (NTM) or intelligence assets identify the nation or organization associated with the explosion;
- the identified explosion occurs at a location that can be demonstrated to be under the control of a particular nation or organization; and/or
- analysis of the debris from an identified event reveals characteristics that can be associated with a specific nation or organization.

In principle, each of these steps or any combination of them for the attribution process is possible. It requires the development of analytical processes and data bases.

The responsibility of identification, yield estimation and attribution, within the context of the CTBT, are left to each State Party. This is because of the serious consequences of concluding that a nuclear test has taken place; and that it occurred on the territory of a particular nation [13]. There are likely to be strong technical components associated with making such calls. In this regard, the IDC assists States Parties by carrying out standard analytical procedures (screening) to screen out those events that could not be nuclear explosions. The CTBT Protocol indicates that the IDC may apply "standard event screening criteria" based on "standard event characterization parameters". If the IDC can screen out most events, the State Parties then can give more attention on the remainder. For example, an event may have its depth estimated with high confidence as 50 km. Such an event would be screened out by a criterion that eliminates events confidently estimated as deeper than 10 km.

For verification purpose, using seismic data alone, however, cannot be the only basis for verifying that a nuclear explosion has taken place. Only the measurement of relevant radionuclides from radionuclide monitoring technology can provide the 'smoking gun' that provides positive confirmation of nuclear explosion [14]. For example, in the case of the North Korea tests in 2006, 2009, and 2013, all of which were announced by the DPRK as nuclear, objective evidence for the nuclear nature of the

2006 and 2013 explosions came from detections of radionuclides that are diagnostic of a nuclear explosion.

5. Application of practical steps for CTBT nuclear explosion seismology monitoring by MY-NDC

MY-NDC, which was established in December 2005, is tasked to perform the CTBT data management and provide relevant information on CTBT related events to the Malaysian Nuclear Agency (Nuklear Malaysia), the CTBT National Authority [15]. In carrying out CTBT seismic monitoring and analysis, MY-NDC in principle applying most of the practical steps discussed above.

Taking MY-NDC participation in the National Data Centre Preparedness Exercise 2015 (NPE2015) as an example, MY-NDC has applied those practical steps to monitor seismic events occurring in the region of fictitious State of ENPEDOR from 1st October until 30th November 2015. The seismicity in the fictitious State of ENPEDOR was investigated in detail to identify the nature of the events, either as natural or induced within the specified time interval.

As for location estimation, Figure 2 illustrates the location of seismic event in ENPEDOR on 15th October 2015 at 08:05:44.3 UTC using data from three IMS seismic stations, namely FINES, AKASG and GERES. Subsequently, as for the identification step, Figure 3 shows the comparison of spectral characteristics of recorded seismic signal on 15th October 2015 (08:05:44.3 UTC) with the past earthquake recorded in the same location on 16th August 2015 and the seismic signal generated from underground explosion detected at seismic station of GERES. It shows low signal of P-waves that often recorded by an earthquake event, instead of high signal of P-waves of underground explosion event. However, since the event is also recorded a low magnitude, verifying the nature of the event whether it is an earthquake or explosion is even more complicated. On that basis, the association of seismic event with underground explosion could not be completely ruled out [16].

To estimate the yield of explosion, MY-NDC use the following formula proposed by L.R. Syres et.al which draw the relationship between seismic magnitude and yield of nuclear explosion [17].

$$mb = 4.262 + 0.973\, log(Y) \tag{2}$$

where, *mb*: Richter seismic scale, and *Y*: yield of nuclear explosion (kt).

The International Seismic Centre has recorded a reading of 3.1 on the Richter scale for that seismic event. A reading of 3.1 would correspond to an explosive yield of between 0.06 and 0.07kt of Trinitrotoluene (TNT) - equivalent, according to the above equation. TNT is a chemical compound, also known as trinitrotoluene, and is commonly used as an explosive for military and industrial applications. The explosive yield of TNT is considered the standard measure of strength of bombs and other explosives.

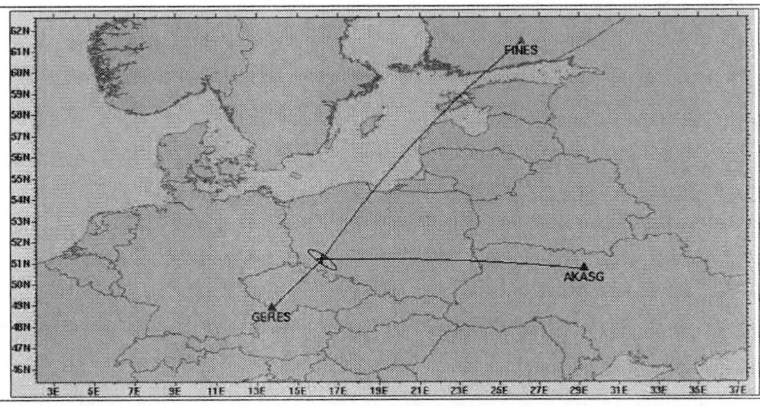

Figure 2. Location of seismic event in ENPEDOR on 15th October 2015 at 08:05:44.3 UTC using data from three IMS seismic stations.

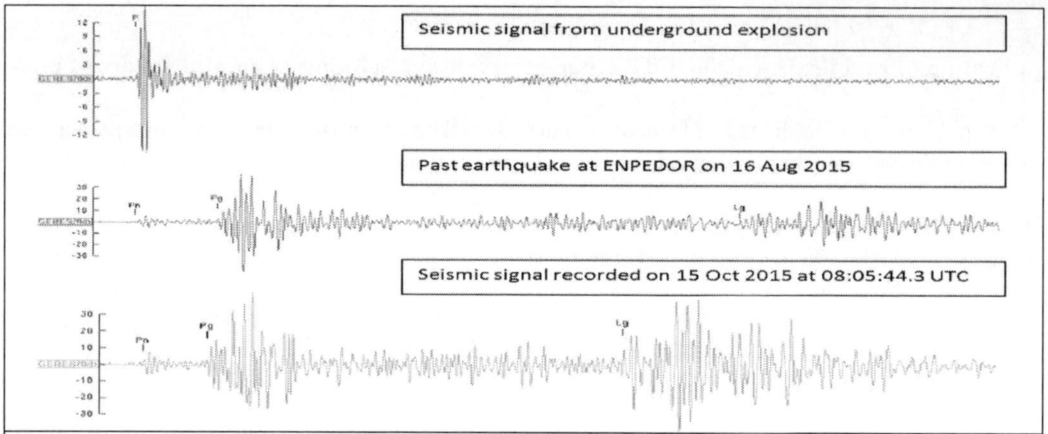

Figure 3. Comparison of seismic signals from explosion event (top), past earthquake at ENPEDOR (middle), and event recorded on 15th October at 08:05:44.3 UTC (bottom) detected at seismic station GERES.

Site forensic was carried out the get an insight of the seismic location. From site forensic, MY-NDC found that the seismic location is within fault zone and has a history of numerous seismic activities. In addition, there are also huge active mining activities within that area, which could involve man-made explosion activity. Several geological researches are being conducted at that area to study mining-induced seismic event.

MY-NDC preliminary concluded that the results of NPE2015 revealed that the seismic event recorded on 15th October 2015 at 08:05:44.3 UTC is most likely the origin of reported radionuclide release. Due to low magnitude event, MY-NDC could not rule out the association of seismic event with underground explosion. The detection of numerous relevant radionuclides and radioxenon isotopes suggest that the event is indeed a nuclear explosion, and not related to release from nuclear reactor event [16].

6. Summary

In summary, underground explosions produce seismic waves with unique characteristics which allow the discrimination between explosions and earthquakes. The CTBT nuclear explosion seismology monitoring can be organized into six practical steps. The monitoring and discrimination capability continues to improve, due to the growth in numbers of stations acquiring relevant data streams and due to the development of better analysis of available data.

7. References

[1] Richards 2014 PG Comprehensive Nuclear-Test-Ban Treaty seismic monitoring: 2012 USNAS report and recent explosions, earthquakes, and other seismic sources *AIP Conf. Proc.* **1596(1)** p 79-95.

[2] Article-by-article analysis of the Comprehensive Nuclear Test-Ban Treaty [available url: https://fas.org/nuke/control/ctbt/text/artbyart/index.html].

[3] Australia Safeguards and Non-Proliferation Office 2014 *ASNO Annual Report 2013–2014* (RG Casey Building Australia: John McEwen Crescent BARTON ACT 0221) [available url: http://www.dfat.gov.au/asno/annual_reports.html] p 39.

[4] Overview of the verification regime [available url: https://www.ctbto.org/verification-regime/background/overview-of-the-verification-regime/].

[5] CTBT: International monitoring system [available url: https://2009-2017.state.gov/t/avc/rls/212176.htm].

[6] National Research Council 1997 Research required to support comprehensive nuclear test ban treaty monitoring (National Academies Press) p 8.
[7] An Educational Site for Budding Seismologist [available url: http://www.geo.mtu.edu/UPSeis/waves.html].
[8] Garwin R L 2003 Technical Issues Related to the Comprehensive Nuclear Test Ban Treaty *APS April Meeting Abstracts* p 43
[9] Bratt SR 2001 The International Data Centre of the Comprehensive Nuclear-Test-Ban Treaty: vision and progress (Kerntechnik) 66 p 134–42
[10] Kim W Y, Simpson D W and Richards P G 1993 Discrimination of earthquakes and explosions in the eastern United States using regional high-frequency data *Geophysical Research Letters* **20** p 1507–1510.
[11] US National Academy of Sciences 2012 The Comprehensive Nuclear Test Ban Treaty: Technical Issues for the United States [available via http://www.nap.edu/catalog.php?record_id=12849] (The National Academies Press).
[12] Ringdal F, Marshall P D and Alewine R W 1992 Seismic yield determination of Soviet underground nuclear explosions at the Shagan River test site *Geophysic. Journal International*, v **109** p 65–77.
[13] M E Pasyanos and W R Walter 2009 Improvements to Regional Explosion Identification Using Attenuation Models of the Lithosphere *Geophysic. Research Lett.* **36** (doi:10.1029/2009GL038505) p 5
[14] Hein H and Läslo E 2018 Infrasound as a tool for CTBT verification.
[15] Rashid F I A and Zolkaffly M Z 2018 Stakeholder engagement for promoting the Comprehensive Nuclear Test-Ban Treaty (CTBT): Malaysia's experience *IOP Conf. Ser.: Mater. Sci. Eng.* **298** p 012054
[16] Rashid F I A and Zolkaffly M Z 2017 National data centre preparedness exercise 2015 (NPE2015): MY-NDC progress result and experience *AIP Conf. Proc.* **1799** doi: (10.1063/1.4972923) p 030013-1-9
[17] Syres L R and Ekstrom G 1989 Comparison of seismic and hydrodynamic yield determination for the Soviet joint verification experiment of 1988 *Proc. Natl. Acad. Sci. USA* **86** p. 3456-60.

IOP Conf. Series: Materials Science and Engineering **555** (2019) 012011 doi:10.1088/1757-899X/555/1/012011

Biochar-coated aluminium electrodes for Cr, Mn and Fe removal in electrochemical treatment for polluted river water remediation

Nabiha Mohd Yussuf[1,2,a)], Zaidi Embong[1,2,b)], Mohd Idrus Mohd Masirin[1,3], Adnan Zainorabidin[1,3], SaifulAzhar Ahmad Tajudin[1,3] ,Norshuhaila Mohamed Sunar[1,4], Salawati Ahmad[5], Omeje Maxwell[6] and Joel Emmanuel[6]

[1]Research Centre for Soft Soil (RECESS), Office for Research, Innovation, Commercialization and Consultancy Management (ORICC),UniversitiTun Hussein Onn Malaysia
[2]Faculty of Applied Science and Technology (FAST), Universiti Tun Hussein Onn (UTHM), Pagoh Campus, Km 1, Jalan Panchor, 84600, Muar, Johor
[3]Faculty of Civil and Environment Engineering, Universiti Tun Hussein Onn Malaysia, Parit Raja, Batu Pahat, Johor
[4]Faculty of Engineering Technology (FTK), UTHM Pagoh Campus, Km 1, Jalan Panchor, 84600, Muar, Johor
[5]Centre for Language Study, Universiti Tun Hussein Onn Malaysia
[6]College of Science and Technology, Department of Physics, Covenant University, Ota, Ogun State, Nigeria

[a]nabihayussuf@gmail.com
[b]zaidi@uthm.edu.my

Abstract. Biochar (a by-product of biomass pyrolysis) is developed as a renewable, low-cost, and promising carbon-based electrode for electrochemical remediation of Cr, Mn and Fe in polluted river water. The biochar - electrode were prepared with two types of tailored sieved size medium (2.0 mm) and fine (150 µm). The effect of the different sieved size on the electrochemical reactor was set at 60V of voltage supply, 6 cm inter-electrode distance within 105 minutes electrolysis operational time. The results indicated that the highest removal efficiency for Cr was achieved at 95% after 105 minutes followed by Mn at 99.9% in 65 minutes and Fe at 92.76% in 105 minutes. Interestingly, the electrode with larger sieved size showed the highest removal efficiency for all three elements. These results emphasize the importance of tailoring the porous structure of the biochar electrode material as a function of the specific size of adsorbate ions to improve the electrochemical water treatment performance for polluted river water.

1. Introduction

Water pollution caused by heavy metals is a global problem and have received worldwide attention. Since late 20th century, contaminated river water has become a very critical task that needs to be taken care of, due to the public expectation of producing water that are clean, free of color, turbidity, taste, odor and other harmful metal ions. Heavy metal containing water should be managed properly, or it could lead to serious damage to the environment which will result in long term effect on human health and other living creature [14]. There are many techniques used in water remediation of contamination, they are physical, biological, photolytic, chemical, and electrochemical.

Of these, electrochemical processes have gained lots of attentions and interests in recent years as a cost efficient and accomplished technology for the treatment of contaminated water [20]. Meanwhile, it offer several advantages over other techniques [19] due to its operation at ambient temperature and pressure as well as robust performance and capability to adjust to variations in the influent composition and flow rate. The method itself is considered to be capable of degrading a notable wide range of contaminants, including phosphate [22], PFAAs [23], phenolic compounds [8], arsenic [4], nitrate [14], methyl orange (MO) [18] and heavy metal [2][9][15][16][17][19][21][25]. The nature of the electrochemical process involves the application of electricity to pass a current through an aqueous metal solution, which also contains a cathode plate and an anode. The treatment is the precipitation of the heavy metals in a weakly acidic or neutralized electrolyte as hydroxides. In that manner, the choice of the electrode material does not only provide specific application options but also play a significant role in improving the method proficiency against various type of contaminated compounds [2].

The biochar, otherwise known as charcoal, are widely used as adsorbent material for heavy metal removal in water [3] due to its considerable heavy metal removal capacity. It has feature such as cost effective, large surface area, stable chemical properties, high conductivity, availability, structure adjustable and is emerging as an economical substitute to the activated carbon to remove diverse organic and inorganic contaminants. Biochar normally used in adsorption for water treatment but the disadvantages of this technique alone are its low efficiency [9]. Accordingly, it should be very interesting to use the biochar fabricated with aluminium electrode in an electrochemical cell to investigate its removal performance. The main objectives of the research were to:

- study the efficiency in removing Cr, Mn and Fe using integrated EC-adsorption system and,
- investigate the effects of different biochar sieved size on the removal of metals.

2. Experimental

2.1. Electrode preparation

A mixture of biochar (150 μm and 2.0 mm sieved size) with 9:1 mass ratio was prepared using methanol (99.99% assay; Fisher Scientific) as solvent. The mixture was sonicated for 1 hour and then binded with cyanoacrylates using a doctor knife on a 7 × 7 cm aluminium plate then placed in an oven for 4 hours. Two sets of electrodes were made from each tailored biochar sample to verify removal efficiency for Cr, Mn and Fe metal from polluted river water.

2.2. Analytical technique

The polluted river water used in this research was obtained from Senggarang river, Johor, Malaysia. The pH of the river water was determined to be 3.51 by using pen type pH meter (PH-8061, Huixia). Quantitative analysis of Cr, Mn and Fe in the river water before and after the electrochemical process were determined by using an atomic absorption spectrometer (AAS, Perkin Elmer A Analyst 800 (Flame and Furnace AAS); USA). The linearity and sensitivity check was done for Cr (5 ppm; 4 ppm), Mn (2 ppm; 2.5 ppm) and Fe (5 ppm; 5 ppm); respectively. The river water was diluted 10 times prior to analysis with three replicate measurements to quantify metal content in the samples. Both anode and cathode were dissembled and dried overnight after electrochemical operation, and then surface structure was analysed by scanning electron microscopy (SEM, HITACHI SU1510). The elemental composition change on the electrode surface was analysed by energy-dispersive spectroscopy (EDS).

2.3. Determination of Cr, Mn and Fe removal efficiency

The removal efficiency of Cr, Mn and Fe at time t, were determined using following equation:

$$\text{Removal efficiency} = \frac{C_O - C_t}{C_O} \times 100\% \quad (1)$$

where C_o and C_t are the metal concentration in water (mg/L) before and after (at time t) the electrochemical experiment (mg/L), respectively.

2.4. Electrochemical reactor setup

An electrochemical cell was designed with biochar coated aluminium electrodes for both anode and cathode to treat the capacity of a 500 mL river water containing 0.49 mg/L of Cr, 0.78 mg/L of Mn and 10.7 mg/L of Fe. The experimental setup is shown schematically in Figure 1. All experiments were carried out at room temperature, normal pressure; an original pH value of the water (pH value of 3.51), supply energy of 60 V current and no additives were added. During the process, hydrogen and oxygen were generated at cathode and anode due to the half-cell reaction. Those generated gasses pushed most of the created sludge to the surface of the aqueous phase while the other remained in the body water (Tran et. al., 2017). The different sizes of sieved biochar affect the value of the process and result in different removal efficiency. During the 2 h of treatment, the samples were drawn every 5 minutes for monitoring the metal ion concentration of the water together with measuring collected heavy metals of the cathode.

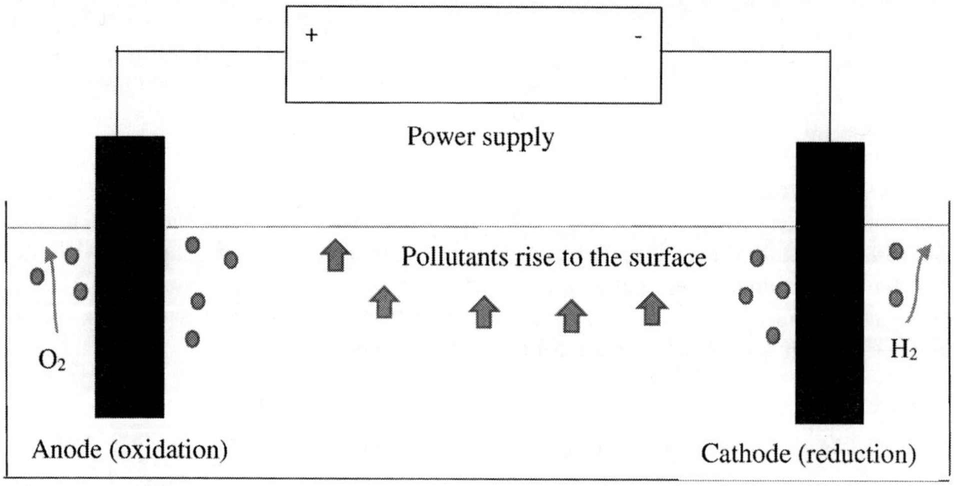

Figure 1. Schematic diagram of the electrochemical cell.

3. Results and discussion

3.1. Characterisation of biochar electrode

Figure 2 shows the surface morphology of the 150 μm and 2.0 mm sieved size biochar electrodes after heavy metal removal studies. While the 150 μm biochar electrode surface cover with darker surface area, the 2.0 mm surface can be seen accumulated with more shining area. EDS analysis (Figure 3) shows that significant amounts of Cr, Mn and Fe element was detected on the surface of each used electrodes, confirming the deposition of the metals while Table 1 represents the atomic percentages of concerned elements from EDS analysis.

IOP Conf. Series: Materials Science and Engineering **555** (2019) 012011 doi:10.1088/1757-899X/555/1/012011

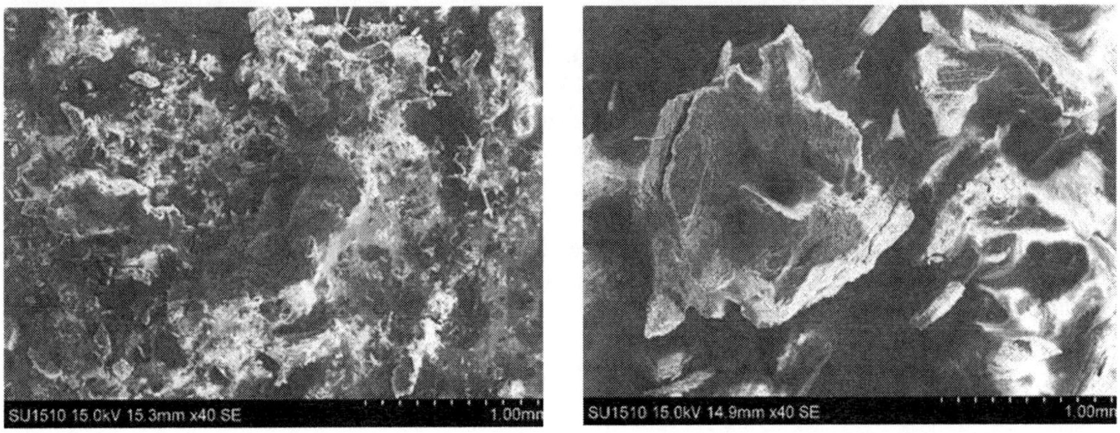

(a) (b)

Figure 2. SEM image for (a) coated biochar with 150 µm sieved size and, (b) coated biochar with 2.0 mm sieved size.

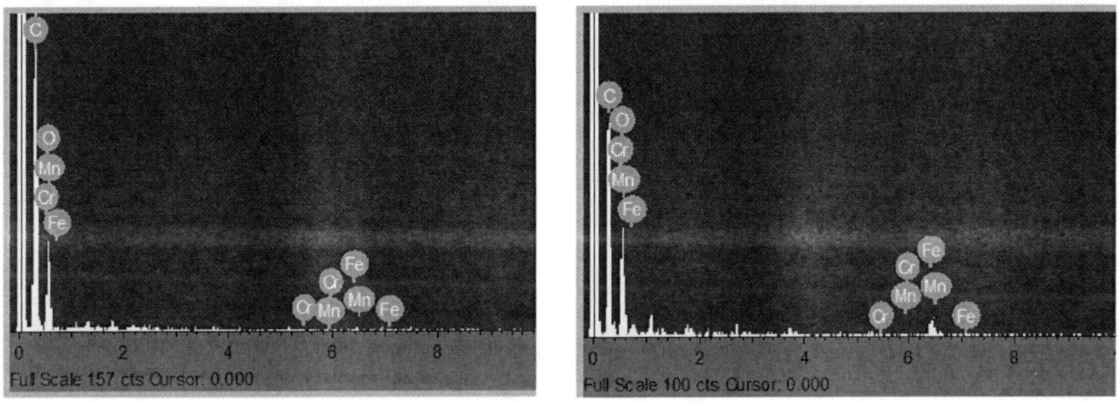

(a) (b)

Figure 3. EDS spectrum for (a) 150 µm sieved size biochar electrode and, (b) 2.0 mm sieved size biochar electrode after the electrochemical treatment.

Table 1. Listed the atomic percentages of each heavy metal based on EDS spectrum analysis.

Element	150 µm	2.0 mm
C	68.10	62.17
O	31.30	35.30
Cr	-0.04	-0.04
Mn	0.08	0.36
Fe	0.56	2.22

3.2. Elemental analysis in water

As could be shown in Figure 4, with increasing the size of biochar, the efficiency of the metal removal with aluminium electrodes increases too. Explicitly, with 150 µm biochar electrode, Cr, Mn and Fe were removed 82.54%, 90.55% and 84.03%, respectively. All the metal removal could be enhanced by increasing the sieved size of biochar. It shows that higher Cr, Mn and Fe removal efficiency of 95%, 99.9% and 92.76% respectively, was achieved with the 2.0 mm biochar coated electrode. Furthermore, with increasing time, the removal efficiency increases too. The highest removal is attributed to the Mn, whereas the lowest is for Cr and Fe. This can be explained by the element with higher reduction potential has a tendency to gain an electron and be reduced by the oxidation process. Due to the more positive standard reduction potential value, manganese would be treated and removed out of the solution followed by chromium and iron [21].

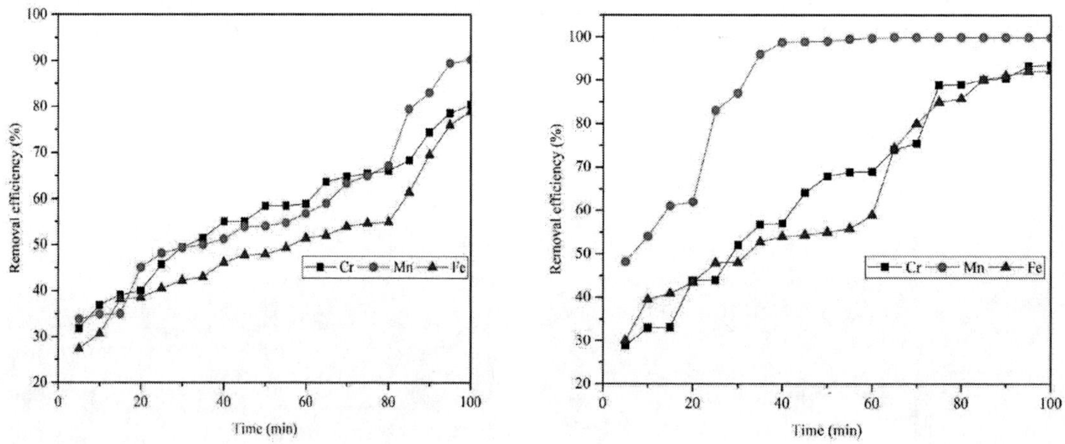

(a) (b)

Figure 4. Variation of Cr, Mn and Fe removal efficiency during electrochemical every 5 minutes sample collection for 105 minutes (a) biochar coated 150 µm (b) biochar coated 2.0 mm.

4. Conclusion

The biochar electrode with 2.0 mm sieved size removed more metals compared to the 150 µm sieved size after 105 minutes operating time. Of the three elements (Cr, Mn and Fe) Mn was found to be the highest removal for both biochar electrodes. This happened based on the value in standard reduction potential table. Mn has the power of reducibility 1.51 V which is higher than Cr (1.33 V) and Fe (0.77 V) that makes Mn get deposited first and removed faster. This study was conducted with a single cell (one cathode and one anode) to demonstrate the suitability of the integrated system for heavy metal removal. It is recommended that future studies use multiple cells connected in series to continuously and more efficiently remove heavy metals.

5. References

[1]. Attour A, Touati M, Tlili M, Ben Amor M, Lapicque F and Leclerc J P 2014 Influence of operating parameters on phosphate removal from water by electrocoagulation using aluminium electrodes. *Sep. and Purif. Tech.* **123** p 124 – 129.

[2]. Al-Shannag M, Al-Qodah Z, Bani-Melhem K, Qtaishat M R and Alkasrawi M 2015 Heavy metal ions removal from metal plating wastewater using electrocoagulation: Kinetic study and process performance. *Chem. Eng. J.* **260** p749 – 756.

[3]. Altun T and Kar Y 2016 Removal of Cr (VI) from aqueous solution by pyrolytic charcoals *New Carbon Mat.* **31** p 501 – 509.

[4]. Balasubramanian N, Kojima T, Ahmed Basha C and Srinivasakannan C 2009 Removal of arsenic from aqueous solution using electrocoagulation *J. of Hazard. Mat.* **167** p 966 – 969.

[5]. Emamjomeh M M and Sivakumar M 2009 Review of pollutants removed by electrocoagulation and electrocoagulation/flotation processes *J. of Env.. Manage.* **90** p **1663 – 1679**.

[6]. Hakizimana J N, Gourich B, Chafi M, Stiriba Y, Vial C, Drogui P and Naja J 2017 Electrocoagulation process in water treatment: A review of electrocoagulation modelling approaches *Desalination* **404** p 1 -21.

[7]. Holt P, Barton G and Mitchell C 1999 Electrocoagulation as a wastewater treatment. The 3rd *Annual Australian Environ. Eng. Research Event.* (Castlemaine, Victoria: 23-26 November 1999).

[8]. Hernández-Franciso E, Peral J and Blanco-Jerez L M 2017 Removal of phenolic compunds from oil refinery wastewater by electrocoagulation and Fenton/photo-Fenton processes *J.. of Water Process Eng.* **19** p 96 – 100.

[9]. Hussin F, Abnisa F, Issabayeva G and Aroua M K 2017 Removal of lead by solar-photovoltaic electrocoagulation using novel perforated zinc electrode *J. of Cleaner Prod.* **147** p 206 – 216.

[10]. Jones D A 1996 Principles and prevention of corrosion (Upper Saddle River: Prentice Hall Inc)

[11]. Jaworek A, Krupa A and Czech T 2007 Modern electrostatic devices and methods for exhaust gas cleaning: a brief review *J. of Electrostatics* **65** 133–155.

[12]. Jiao R, Fabris R, Chow C W K., Drikas M, Leeuwen J V, Wang D and Xu Z 2017 Influence of coagulation mechanisms and floc formation on filterability *J. of Env. Sci.* **57** p 338 – 345.

[13]. Kim K.S, Kim S and Jun T H 2015 Activated carbon-coated electrode and insulating partition for improved dust removal performance in electrostatic precipitators *Water Air Soil Pollut.* **226** p 367.

[14]. Kumar N S and Goel S 2010 Factors influencing arsenic and nitrate removal from drinking water in a continuous flow electrocoagulation (EC) process. J. of Hazard. Mat. **173** p 528 – 533.

[15] Lu J, Li Y, Yin M, Ma X and Lin S 2015 Removing heavy metal ions with continuous aluminium electrocoagulation: A study on back mixing and utilization rate of electro-generated Al ions *Chem. Eng. J.* **267** p 86 – 92.

[16]. Lu J, Wang Z R, Liu Y L and Tang Q 2016. Removal of Cr ions from aqueous solution using batch electrocoagulation: Cr removal mechanism and utilization rate of in situ generated metal ions *Process Safety and Env. Protect.* **104** p 436 – 443.

[17]. Mouedhen G, Feki M, De Petris Wery M and Ayedi H F 2008 Behavior of aluminium electrodes in electrocoagulation process J. of Hazard. Mat. **150** p 124 – 135.

[18]. Pi K W, Xiao Q, Zhang H Q, Xia M and Gerson A R 2014 Decolorization of synthetic Methyl Orange wastewater by electrocoagulation with periodic reversal of electrodes and optimization by RSM *Proc. Safety and Env. Protect.* **92** p 796 – 806.

[19]. Rincón G J and Motta E J L 2014 Simultaneous removal of oil and grease, and heavy metals from artificial bilge water using electro-coagulation/ flotation *J. of Env. Manage.* **144** 42 – 50.

[20]. Tran T K., Chiu K F, Lin C Y and Leu H J 2017 Electrochemical treatment of wastewater: Selectivity of the heavy metals removal process Int. J.. of Hydrogen Energy **42** p 27741 – 27748.

[21]. Un U T and Ocal S E 2015 Removal of heavy metals (Cd, Cu, Ni) by electrocoagulation *Int. J. of Env. Sci. and Dev.* **6(6).**

[22]. Vasudevan S, Lakshmi J, Jayaraj J and Sozhan G 2009 Remediation of phosphate-contaminated water by electrocoagulation with aluminium, aluminium alloy and mild steel anodes. *J. of Hazard. Mat.* **164** 1480 – 1486.

[23]. Yang B, Han Y, Yu G, Zhuo Q, Deng S, Wu J and Zhang P 2016 Efficient removal of perfluoroalkyl acids (PFAAs) from aqueous solution by electrocoagulation using iron electrode *Chem. Eng. J.* **303** p 384 – 390.

IOP Conf. Series: Materials Science and Engineering **555** (2019) 012011 doi:10.1088/1757-899X/555/1/012011

[24]. Yussuf N M, Embong Z, Abdullah S, Masirin M I M, Tajudin S A .A , Ahmad S, Sahari S K, Anuar A A and Maxwell O 2017 The enhancement of heavy metal removal from polluted river water treatment by integrated carbon-aluminium electrodes using electrochemical method. *Mat. Sc. and Eng.* **298**.

[25]. Zewail T M and Yousef N S 2014 Chromium ions (Cr^{6+} & Cr^{3+}) removal from synthetic wastewater by electrocoagulation using vertical expanded Fe anode. *J. of Electro. Chem.* **735** p123 – 128.

IOP Conf. Series: Materials Science and Engineering **555** (2019) 012012 doi:10.1088/1757-899X/555/1/012012

A study on detection water leakage of underground metal and PVC pipes using ground penetrating radar

T S T Amran[1,a)], M S M Amin[1], M R Ahmad[1], S Sani[1], N A Masenwat[1], C E A Bundak[2] and C J Singoi[2]

[1]Non-Destructive Testing - Material Structure Integrity group (NDT-MSI), Industrial Technology Division, Malaysian Nuclear Agency, 43000 Kajang, Selangor, Malaysia
[2]The Faculty of Science, Universiti Putra Malaysia (UPM), 43400 Serdang, Selangor, Malaysia

[a]sarah@nm.gov.my

Abstract. Ground Penetrating Radar (GPR) is a high resolution electromagnetic techniques that is designed primarily to investigate the shallow subsurface of the earth, building material, roads and bridges. GPR was also able to detect water leaks in the underground distribution system. A series of laboratory experiments were conducted to determine the validity and effectives of GPR technology in detecting water leakage using two different types of pipes which are metal and polyvinyl chloride (PVC) pipes. Experiment was conducted using GPR 800 MHz antenna using 4 inch pipe with diameter of hole is ¼ inch to stimulate the leakage. GPR identify leaks in buried water pipes either by detecting underground voids created by the leaking water as it erodes the material around the pipe or by detecting anomalous change in the properties of the materials around pipes due to the water saturation.

1. Introduction

Water loss in town and suburban is currently a significant issue which reflect the performance of water supply management in Malaysia. Poor infrastructure leading to water loss also contributes to water shortages. Even though the rate of non-revenue water (NRW) (Table 1) increased only 0.1 percent from 35.2 percent in 2016 to 35.3 percent in 2017, it is still considered a major inefficiency in the industry that must be addressed [1]. Malaysia hopes to reduce non-revenue water rate to 25 percent by 2020. In order to find the specific location needing repair, a leak detection survey must be firstly performed as proposed by Hunaidi *et al.* [2]. Surface geophysical methods are noninvasive, trenchless tools used to characterize the physical properties of the subsurface material. Many geophysical techniques have been suggested as candidates for detecting water leakage, including ground penetrating radar (GPR), acoustic devices, gas sampling devices and pressure wave detectors. Various techniques for detecting a mains water leaks are available but mostly are time-consuming, disruptive and expensive. GPR can be used as tool for water leakage detection because low cost, high resolution and real time targeting [3]. In principle GPR can identify leaks in buried water pipes either by detecting underground voids created by the leaking water as it erodes the material around the pipe, or by detecting anomalous change in the properties of the material around pipes due to water saturation. GPR could have a higher potential of avoiding difficulties encountered with commonly used acoustic leak detection methods as it applies to plastic pipes [4]. GPR could also be used as a supplement to these methods to increase accuracy in high risk areas such as high traffic streets and large structures.

This paper focuses on studying the detection water leakage of underground pipes using two different material of pipes which are metal (iron) pipe and PVC pipe using GPR with frequency 800 MHz antenna.

Table 1. Non-Revenue Water (NRW) 2016-2017 (SPAN, 2017)

State	2016				2017			
	System Input Volume	Billed Authorised Consumption	NRW	NRW (%)	System Input Volume	Billed Authorised Consumption	NRW	NRW (%)
	(MLD)				(MLD)			
Johor	1,737	1,286	450	**25.9**	1,753	1,320	433	**24.7**
Kedah	1,362	725	637	**46.7**	1,370	719	651	**47.5**
Kelantan	471	238	232	**49.4**	475	240	234	**49.3**
F.T. Labuan	72	50	22	**30.5**	71	48	23	**32.0**
Melaka	500	405	95	**19.0**	513	413	101	**19.6**
N. Sembilan	773	520	253	**32.7**	770	519	245	**32.6**
Pulau Pinang	1,054	827	227	**21.5**	1,058	826	231	**21.9**
Pahang	1,111	579	532	**47.9**	1,109	582	528	**47.5**
Perak	1,318	916	402	**30.5**	1,314	907	406	**30.9**
Perlis	243	96	148	**60.7**	242	89	152	**63.1**
Sabah	1,221	586	634	**52.0**	1,261	582	679	**53.8**
Sarawak	1,328	850	479	**36.0**	1,399	870	529	**37.8**
Selangor	4,807	3,260	1,547	**32.2**	4,842	3,316	1,526	**31.5**
Terengganu	628	440	189	**30.0**	613	427	186	**30.4**
MALAYSIA	16,625	10,77	5,846	35.2	16,789	10,860	5,929	35.3

2. Experimental works

The water leakage was simulated inside a test bed with dimension of 70 cm (L) x 61 cm (H) x 45 cm (W). A 4 inch L-shape PVC pipe with a pre-drilled hole (1/4 inch) at the centre was buried in the middle of the test bed with a depth of 60 cm. The pipe was covered by dry sand and the hole faced downwards to stimulate leakage. The pipe was placed in perpendicular direction with the GPR scanning direction. The GPR used was MALA RAMAC/GPR 800MHz shielded antenna. The distance between elbow and the first hole (1/4 inch) was marked, then route for the GPR antenna was created with long white string as shown in Figure 1. The frequency used for the GPR scanning was 800 MHz (Figure 2) and the setting was shown in Table 2. The GPR data for PVC pipe was measured before the water injection. Then, the 800 MHz antenna was pulled simultaneously with injection of 15 litres of water. The GPR scanning during the water leakage was taken about 5 minutes to get proper GPR data. Subsequently, the GPR scanning data was collected after 30 minutes, 1 hour and 24 hour after water injection. As soon as the data of PVC pipe was completely collected using GPR and it was replaced with metal (iron) pipe for the same purpose to observe the underground water leaking.

Table 2. Antenna setting.

Parameter	Value
Antenna	800 MHz shielded
Trigger	Distance
Trig int	0.005 m
Measuring wheel	Meas.wheel 300-800 MHz
Number of samples	502
Sampling frequency	8800 MHz
Time window	57 ns
Antenna separation	0.14 m

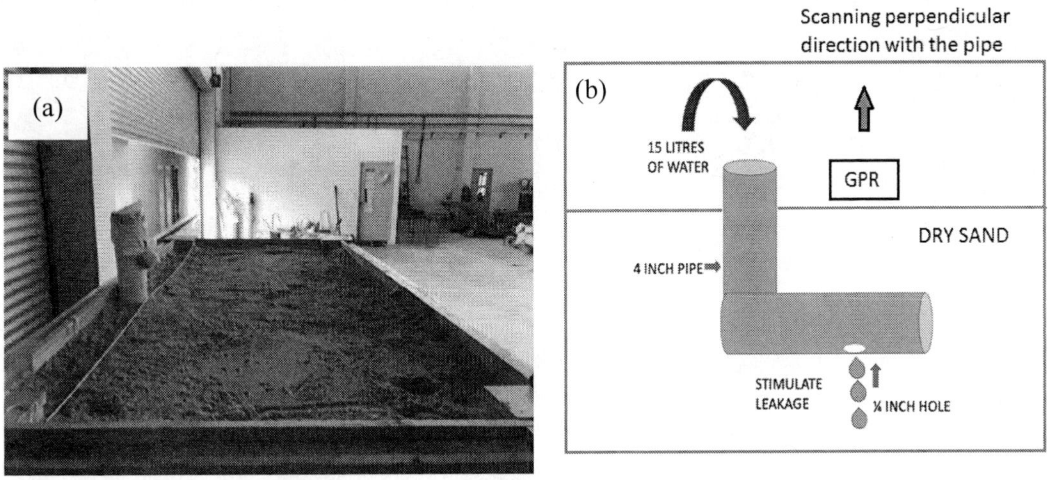

Figure 1. (a) Experimental set up of test bed with a long white string for antenna road during GPR scanning for ¼ inch hole diameter (b) Schematic diagram of the experimental set up.

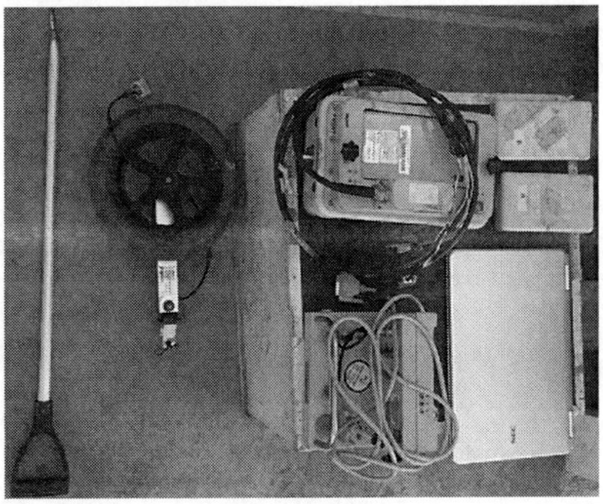

Figure 2. The GPR equipment set for frequency 800 MHz.

3. Result and analysis

In this experiment, the PVC and metal (iron) pipe were buried in the test box to stimulate underground leakage. One small hole (1/4 inch) was drilled in the middle of the pipe and the open end was sealed. Each pipe was injected with 15 litres of water and the GPR data were collected before, during, 30 minute, one hour and 24 hour after water injection. The GPR scanning was done perpendicular with the direction of the pipe and the hole was facing downward.

IOP Conf. Series: Materials Science and Engineering **555** (2019) 012012 doi:10.1088/1757-899X/555/1/012012

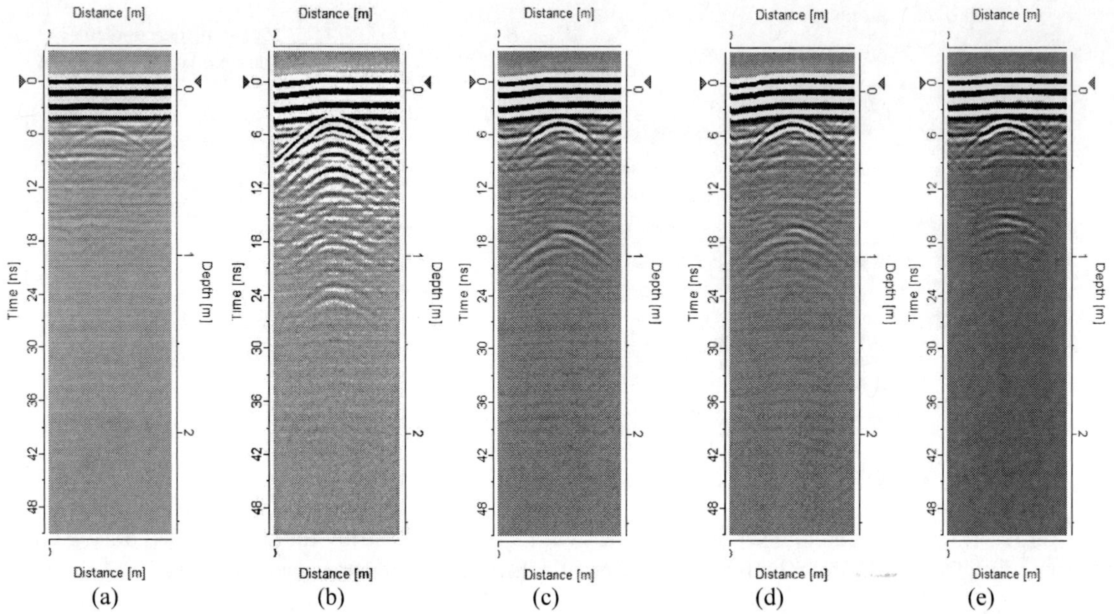

Figure 3. GPR radar profile for PVC pipe for time interval (a) before water leaking, (b) during water leaking, (c) 30 minutes after water leaking, (d) 1 hour after water leaking, (e) 24 hours after water leaking.

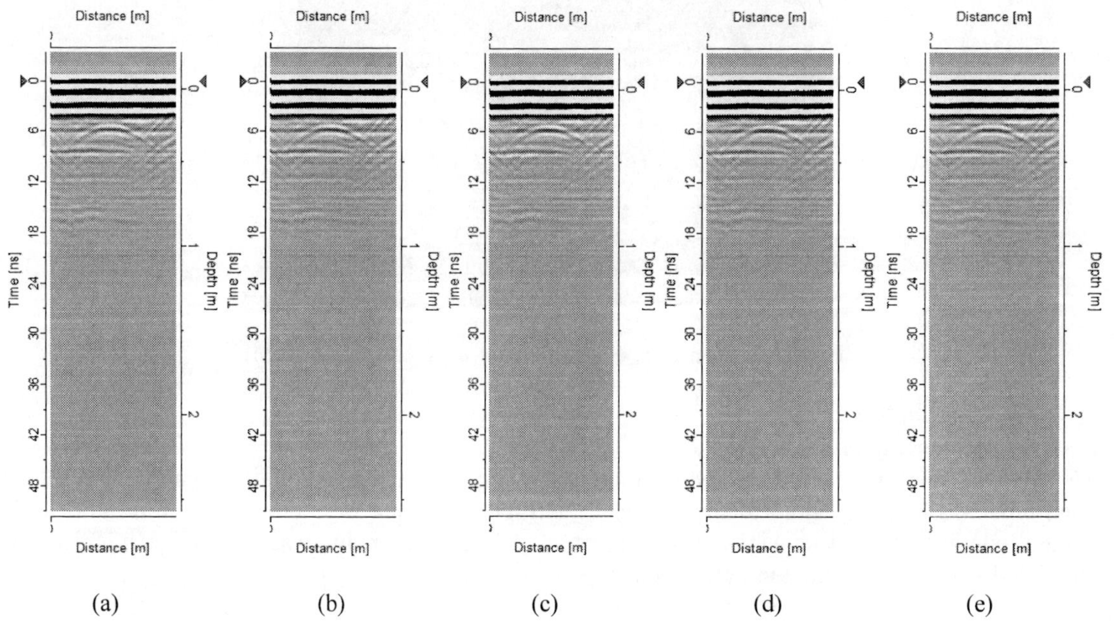

Figure 4. GPR radar profile for metal (iron) pipe for time interval (a) before water leaking, (b) during water leaking, (c) 30 minutes after water leaking, (d) 1 hour after water leaking, (e) 24 hours after water leaking.

Figure 3 shows the GPR profiles collected using 800 MHz antenna for PVC pipe. It can be seen from Figure 3 (b) during water leaking, the reflection pattern was totally clear and seem massive below the PVC reflection line. The change in hyperbola pattern was influenced significantly by changes in dielectric medium, since water leakage occurred the 'ringing signal' effect appears. However, as the time passes until 1 hours, the hyperbola pattern can be seen at the bottom of the sand. This is due to the concentration of water at bottom of the sand. The soil-water mixtures generate hyperbolic anomalies similar to finite-size reflectors because of the change in dielectric properties [5]. The reflection of water leakage has been blurry after 24 hours. The result indicates that the saturation water was already dissipated into the dry sand.

Figure 4 illustrates the radargram images for water leakage simulation of metal (iron) pipe. The metal (iron) pipe can be clearly seen in the radargram of Figure 4 (a). During the water leaking the reflection pattern was not clearly see like PVC pipe profile (Figure 3 (b)). The iron pipe was detected but the leak not clearly seen due to the metal that considered to be complete reflector. Besides the inability to locate the leak, some limitations of the GPR technique have been identified in the sand. This is due to the vast majority more of radar-wave energy reflected from the top of the piping and only less energy penetrated the top of the iron pipe. The radar-wave speed decreased after passing through the iron pipe due to the transmitted energy was weak [6]. There was no obvious difference anomalous pattern throughout the experiment until 24 hours.

4. Conclusion

In this research the GPR with frequency 800 MHz has been applied as a tool to detect leakage of water in 4 inch PVC and metal (iron) pipes respectively that being used to transport fresh water. This experiments has been built to measure and monitor the leakage of water (before, during and after the injection of water). GPR is effectively to detect the water leakage in PVC pipes but poorly in metal (iron) pipes. The metal (iron) pipe considered to be complete reflector and not allow much amount of signal to pass through. Metal contribute to high electrical conductivity rapidly attenuate the radar energy, restrict penetration depths, and severely limit the effectiveness of GPR. The effectiveness and limitation of GPR used in detection of water leakage in pipes is clearly illustrated in this experiment.

5. References

[1] Non-Revenue Water (NRW) 2017 *National Water Services Commission (SPAN)* retrieved from https://www.span.gov.my.

[2] Hunaidi O, Chu W, Wang A, and Guan W 2000 detecting leaks in plastic water distribution pipes *J. AWWA* **92** 2 p 88-94.

[3] Daniels D J 2004 Ground penetrating radar second edition (London United Kingdom: The Institution of Electrical Engineers*)*.

[4] Hunaidi O and Giamou P 1998 Ground penetrating radar for detection of leaks in buried plastic water distribution pipes *GPR98 Conf. Kansas USA* p 783-786.

[5] Nakhkash M and Mahmood-Zadeh M R 2004 water leak detection using ground penetrating radar *Tenth Int. Conf. on Ground Penetrating Radar* p 525-528.

[6] Lee K F, Wang T K, Kang Y M, Wang C S and Lin K A 2013 Identification of pipelines from the secondary reflect wave travel time of ground-penetrating radar waves *J. of Marine Sci&Tech* **21(4)** p 417-422.

Acknowledgments

The authors would like to extent their appreciations and sincere thanks to Ministry of Energy, Sciences, Technology, Environment and Climate Change (MESTECC) and Agensi Nuklear Malaysia as well as the Material and Structure Integrity Group (NDT-MSI) for the support and co-operation.

IOP Conf. Series: Materials Science and Engineering **555** (2019) 012013 doi:10.1088/1757-899X/555/1/012013

Investigation on different compositions of flyash concrete on gamma and neutron radiation

R Ramli[1,a)], Faridah Mohamad Idris[2], M F K M Yapandi[3], Juniza Md Saad[1], S Akmal Biyamin[3]

[1]Mechanical Engineering Department, Universiti Tenaga Nasional, Kajang, Selangor, Malaysia.
[2]Malaysia Nuclear Agency. Kajang, Selangor, Malaysia.
[3]TNB Research, Kajang, Selangor, Malaysia.

[a]rogemah@uniten.edu.my

Abstract. : Fly ash is a solid waste from a coal combustion product and about 43% is recycled. Fly ash is often used as a replacement of partial replacement for Portland cement in concrete production. Concrete has been widely used as radiation materials due to its lower cost. Five samples of fly ash concrete mixture were prepared and investigated by using small angle scattering (SAN) for neutron irradiation and gamma radiation test for gamma irradiation. It was found that 20% of fly ash concrete mixture absorbed more neutron radiation whereas 40% of flyash concrete mixture absorbed more gamma radiation.

1. Introduction

Concrete has been widely used as radiation shielding for neutron and gamma radiation because of its low price and good shielding performance [1]. Cement is the main component used in production of concrete. The manufacturing of cement creates huge amount of CO_2 as a by-product that possess a serious threat to the environment and human health [1][2]. Fly ash, is a non-combusted by product of coal-fired power plants, can be substituted for large amount of Portland cement [1]. This can improved concrete environmental characteristic and also reduced the cost production of concrete.

The substitution ratio for fly ash to cement is typically 1:1 to 1.5:1 which shows that a greater or equivalent weight of cement is substituted by fly ash [3]. Fly ash also requires lesser water than cement which reduces permeability, bleeding and crack problems [4]. The microstructure variation is important to value the mechanical properties and deformation factor [5]. In order to increase the absorption of neutron and gamma radiation, barite $BaSO_4$ is good as gamma absorber and colemanite $CaB_3O_4(OH)_3.H_2O$ is good as neutron absorber. Both materials were used to replaced fine and course aggregate in manufacturer of concrete [6].

Even though research had been made on fly ash concrete since 1981, it is still considered to be at the early stage and the application of fly ash concrete is still expected to be studied and implemented. In term of nuclear industry, fly ash concrete is said to perform as a nuclear shielding between the nuclear source and the surroundings [1].This paper investigated the effect of different composition of fly ash concrete on gamma and neutron radiation, also to determine the mechanical properties of fly ash concrete mixture.

2. Methodology

2.1. Materials

Ordinary Portland cement (OPC) Type I from a local manufacturer was used. Fly ash was obtained from TNB Research came from Sultan Azlan Syah Power Plant, TNB Janamanjung Sdn. Bhd. The chemical properties of fly ash are shown in Table 1. Others materials used are fine aggregate (sand and colemanite), course aggregate (barite), water and plastisizer. Sand, barite and colemanite were provided by Malaysian Nuclear Agency (MNA).

Table 1. Properties of fly ash by weight percentage.

Chemical	Contents
SiO_2 (%)	46.46
Al_2O_3 (%)	18.07
Fe_2O_3 (%)	11.65
CaO (%)	6.35
MgO (%)	4.28
N_2O (%)	0.80
K_2O (%)	1.10
TiO_2 (%)	1.36
MnO_4 (%)	0.10
SO_3 (%)	8.88
P_2O_3 (%)	0.42
Specific gravity (kg/m^3)	2.32

The concrete mixture batches were prepared based on the six different ratios of cement and fly ash as shown in Table 2. The additional compositions of each mixture are consisted of 2.42 kg of sand, 0.121 kg of colemanite, 4.598 kg of barite, 0.242 kg of plasticizer and 0.484 kg of water. The water to cementitious ratio was kept at 0.266. The calculation of each mixture was based on concrete mix design provided by Malaysia Nuclear Agency (MNA). The cube mould is made of cast iron and the size is 15 cm^3. The concrete mixture is dried for 48 hours and afterward is submerged in water at room temperature for curing purpose for at least 28 days.

The standard barite-colemanite concrete sample, F produced at Malaysia Nuclear Agency (MNA) using the formula given by Farhi *et al.* as follows: Baryte-colemanite heavy concrete: 4% wt. water, 79% wt. baryte $BaSO_4$, 10% wt. colemanite $CaB_3O_4(OH)_3 \cdot H_2O$ and 7% wt. Portland cement [6]. The above composition is used to mix concrete cement that is used as the controlled sample. In concrete samples A to E, a portion of 7% wt. Portland cement portion is reduced to include fly ash as given in Table 2.

Table 2. Concrete mixture composition.

Sample	Cement	Flyash		Water	w/(c+fa)	Sand	Colemanite	Barrite	Plasticizer
		%	Weight						
A	1.45	20	0.36	0.484	0.266	2.42	0.121	4.598	0.242
B	1.09	40	0.73	0.484	0.266	2.42	0.121	4.598	0.242
C	0.73	60	1.09	0.484	0.266	2.42	0.121	4.598	0.242
D	0.36	80	1.45	0.484	0.266	2.42	0.121	4.598	0.242
E	0	100	1.82	0.484	0.266	2.42	0.121	4.598	0.242
F	1.82	0	0	0.484	0.266	2.42	0.121	4.598	0.242

2.2. Characterization

2.2.1. Neutron scattering test

Neutron scattering test was done at Small Angle Neutron Scattering machine at Malaysia Nuclear Agency. The samples were put at the front of the neutron beam from the machine and covered with other shielding material such as concrete to reduce the exposure of radiation during collecting the data. The sample was aligned at 180 degrees with the collimator. A neutron survey meter is placed at the front of the sample in order to record the radiation reading after it passes through the each of the samples.

2.2.2. Gamma absorption test

The gamma absorption test was performed using Cobalt-60 radioactive substance. The count rate of detected gamma rays as a function for the mass thickness of various concrete samples that were located between the detector and the source, was recorded by using Geiger-Mueller ratemeter. The attenuation coefficient of gamma ray was calculated by using Lambert-beer law. Based on Lambert-Beer law:

Intensity of beam,

$$I = I_0 e^{\mu_\rho X} \tag{1}$$

Mass attenuation coefficient,

$$\mu_\rho = \frac{\ln(I_0/I)}{X} \tag{2}$$

Density

$$\rho = \frac{\text{mass}}{\text{volume}} \tag{3}$$

Linear attenuation coefficient,

$$\mu = \mu_\rho \times \rho \tag{4}$$

2.2.3. Scanning electron microscope (SEM)
The samples were coated first with gold coating by using BIO-Rad SEM Coating System before proceeding with the microstructure image. The samples undergone the SEM testing under 15.0 kV accelerating voltage. Energy Dispersive X-ray (EDX) was also used to test the samples to identify the composition of fly ash in the concrete mixture.

2.2.4. Compression test
Compression test was performed to investigate the strength and hardening of the concrete samples. The samples were done with the self-compacting concrete method under standard laboratory conditions.

3. Results and discussion

3.1. *Effect of flyash addition on radiation.*
Based on Figure 1, the average equivalent neutron radiation scattered for sample A (20% fly ash) is lower when compared to the control sample. For sample B, C, D and E, the radiation shows higher values than the control sample with sample C peaking up to 127.78 μSv. Figure 2 shows the percentage of neutron radiation absorbed is the highest from sample A which surpasses the control barite colemanite concrete. Sample B, C, D and E show a much lower percentage of neutron radiation absorbed compared to the control sample. The lowest percentage is discovered to be from sample C (60% fly ash) which is 96.9%.

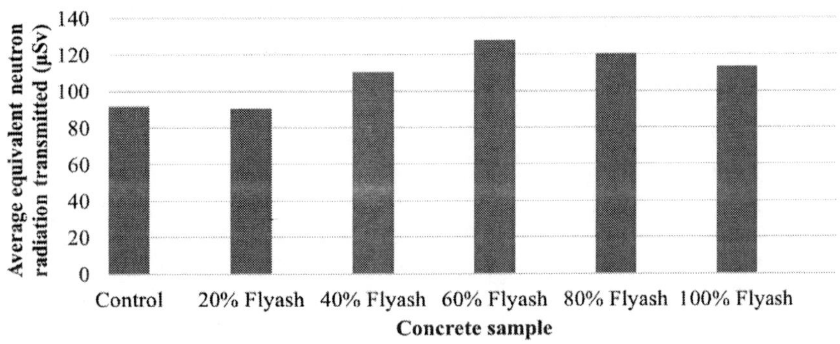

Figure 1. Average equivalent neutron radiation detected vs. concrete samples.

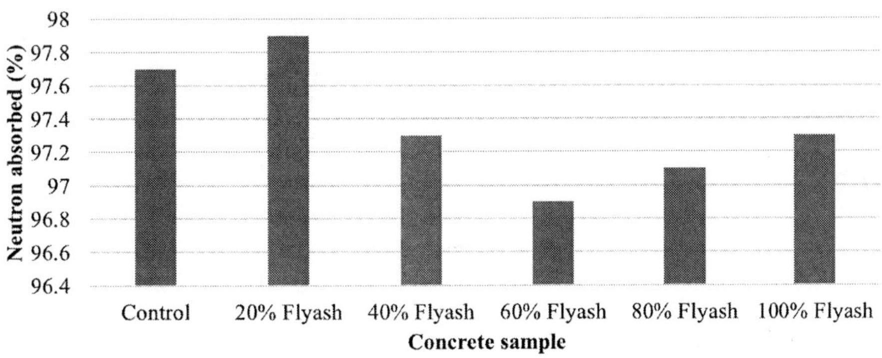

Figure 2. Percentage of neutron radiation absorbed over different samples ratio.

For neutron scattering test, the result shows that the control barite colemanite concrete sample, which contained a significant amount of neutron absorption particles, could attenuate neutrons by 97.7% which is more than ideal for radiation shielding purposes. However, based on the experiment, it is shown that sample A (20% fly ash) composite concrete held up to be better than the standard sample for obtaining 97.9% of neutron particles scattered. The finer particles of fly ash filled up the pores between the concrete to reduce the void ratio and increase the strength of the sample.

Based on Figure 3, the average gamma radiation emitted for sample B, C and D (20%, 40% and 60% fly ash) are lower when compared to the control sample with sample B having the lowest reading at 33844 counts per minute. For sample A and E, the radiation shows higher values than the control sample with sample A having the peak reading of 38936 counts per minute. Figure 3 and Figure 4 shows the relationship between average gamma waves radiation count rate and the amount of gamma radiation absorbed over the different ratios of concrete samples ranging from 0-100% fly ash composites. Five different readings were tabulated per each sample and the average values for the sample is calculated.

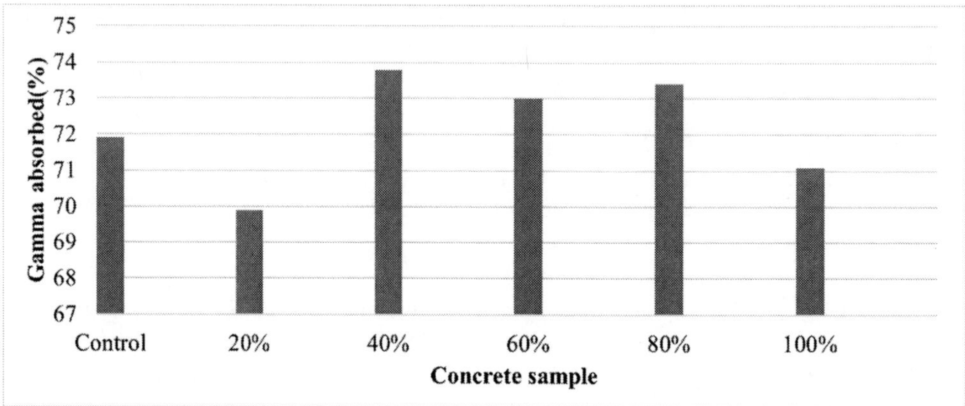

Figure 3. Amount of gamma absorbed over different samples ratio.

Figure 3 shows the percentage amount of gamma radiation absorbed is the highest from sample B (73.8%) followed by sample D and sample C which surpasses the control barite colemanite concrete. Sample A and E show a much lower percentage of gamma radiation absorbed in comparison to the control sample. The lowest percentage is discovered to be from sample A (20% fly ash) which is 69.9%.

Approximately, Figure 3 and Figure 4 show the exact opposite results as to the reason that the percentage amount of gamma ray is absorbed by the concrete sample, obtaining a different result to the amount of gamma radiation emitted onto the Geiger Mueller rate meter. The higher the amount of gamma ray absorbed, the lower the average count rates of gamma waves detected by the rate meter.

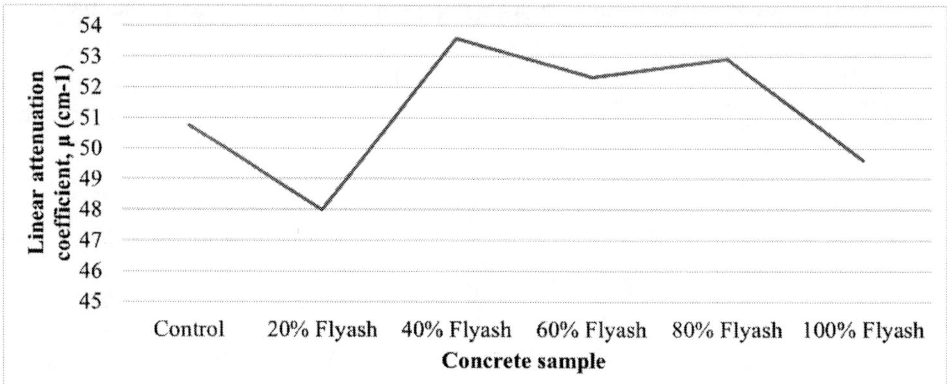

Figure 4. Linear attenuation coefficient for each concrete samples.

3.2. *Effect of flyash addition on surface morphology*

Figure 5 and Figure 6 show SEM images with two different magnifications, 750 and 5000. The SEM images obtained from sample A, B and C show that the fly ash components were bonded to the concrete mixture. It is also suggested that sample B and C may have a smaller fly ash particles which have a higher reaction when comparing with the larger ones. This can be associated with difference in their relative surface area. This is because the ratio for Ca/Si mol adjacent to the center smaller and larger particles are packed closely together [7]. Therefore, the reduction to the chock points can cause the gamma absorption to be improved.

The image obtained from all the concrete that contains fly ash shows that most of fly ash particles have a quite smooth surfaces on them and only partially contains foreign components sticking. The flyash in those composition ratios act as fillers (Figure 7) which will occupy the voids in between the concrete mixture [7]. This will cause the density of the concrete mixture to increase, hence leads to the increment of the attenuation coefficient of neutrons.

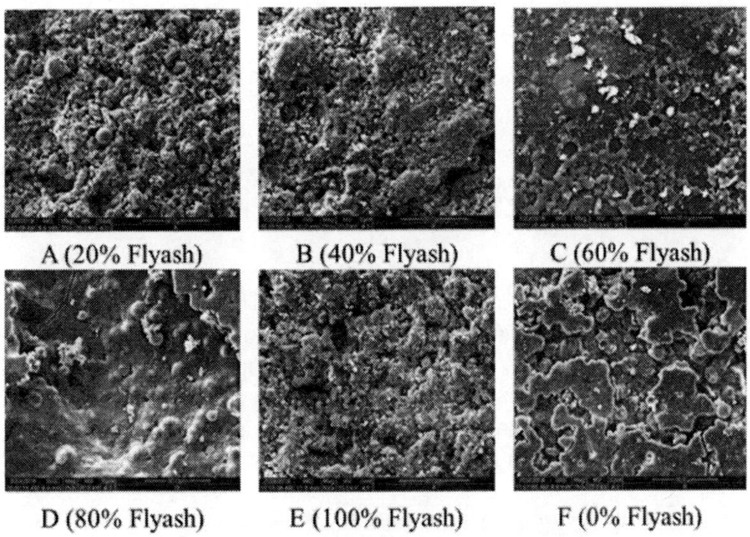

Figure 5. SEM results (x750 magnification).

Figure 6. SEM results (x5000 magnification).

IOP Conf. Series: Materials Science and Engineering **555** (2019) 012013 doi:10.1088/1757-899X/555/1/012013

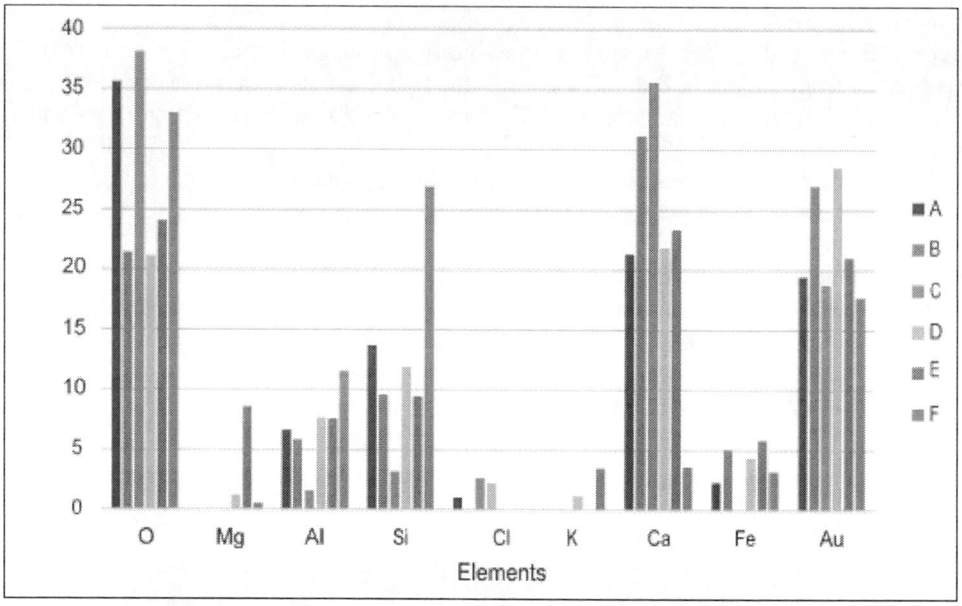

Figure 7. Weight percentage of every elements in each concrete sample from EDX.

3.3. *Effect of flyash addition on the hardening properties*

The compressive strength results are as shown in Table 3. The compressive strength obtained from sample A is much lower when compared to the controlled sample where the maximum compressive force that can be withstood is 340 kN. For sample B, the compressive strength is about proportionate with the controlled sample. In case of sample C, D and E, no compressive strength was recorded, i.e. these samples were failed during test. In conclusion, there is a decrease in compressive strength as more cement is replaced by fly ash exceeding 40%.

Table 3. Compressive strength results for different samples.

Sample	Maximum compressive force (kN)	Compressive strength (N/mm^2)
A	340	15
B	540	24
C	-	-
D	-	-
E	-	-
F (control)	560	25

4. Conclusion

This study has demonstrated that varying the ratios of cement to fly ash mixture in concrete sample has a significant influence on the gamma and neutron radiation. The fly ash in those composition ratios act as a binder and as a filler which occupied the voids in between the concrete mixture. This will cause the density of the concrete mixture to increase and lead to the increase in the attenuation of coefficient. It is found that 20% of flyash is suitable for neutron scattering and 40 – 80% of flyash is suitable for gamma absorption. It is also suggested that the decrement of the cement amount in the aggregate resulted on lower binding factor. In addition, the linear attenuation coefficient changes with fly ash replacement as it is dependent on the density of mixtures.

5. References

[1] Singh S, Ghumman S S, Singh C, Thind K S, Mudalar G S 2010 Buildup of gamma ray photons in flyash concrete: a study Ann. of Nucl. Energy **37** p 681-684.

[2] Singh K, Singh S, Dhaliwal A and Singh G 2015 Gamma radiation shielding analysis of lead-flyash concretes *App. Rad. and Isotopes* **95** p 174-179.

[3] U.S. Department of Transportation 2017 Federal highway administration https://www.fhwa.dot.gov. [Accessed August 2018].

[4] Rodriguez J. 2018 Uses, benefits and drawbacks of fly ash in construction *The Balance Small Business* https://www.thebalancesmb.com. [Accessed August 2018].

[5] Mehta P K, Monteiro P J M 2006 *Concrete: microstruture, properties and materials* (The mcGraw Hill Companies, Inc.) **3**.

[6] Farhi E *et al.* 2015 Neutron scattering instrumentation at low power reactors for science, engineering and education J. of Neutron Research.

[7] Wang S, Baxter L and Fonseca F 2008 Biomass fly ash in concrete: SEM, EDX and ESEM analysis: fuel **87** p 372-379.

Acknowledgement

The work reported in this paper is partially taken from an Undergraduate Final Year Project report submitted for Bachelor of Mechanical Engineering at Universiti Tenaga Nasional. The authors wish to thank to all who provided assistance in this research.

IOP Conf. Series: Materials Science and Engineering **555** (2019) 012014 doi:10.1088/1757-899X/555/1/012014

Investigation on the gamma sources integrity inside a dry pit storage

Rosli Darmawan[1,a)]

[1]Malaysian Nuclear Agency, Bangi, 43000 Kajang, Selangor, Malaysia.

[a]roslid@nm.gov.my

Abstract. One of the primary concerns in the operation of nuclear related facilities is the safety of the system. Worldwide publicity on a few nuclear accidents as well as the notorious Hiroshima and Nagasaki bombing have always brought about public fear on anything related to nuclear. Thus, investigation and analysis on the safety system shall be continuously conducted to determine the safety condition of the facilities. In the early days, analytical and experimental methods were employed. However, this approach is limited to low risk experimentation. With the advancement of computer technology, specific computer codes were used to predict and analyse high risk safety issues such as the fall-out from a nuclear explosion or analysis of melt down occurrence in a reactor core. This paper discusses a prediction model on a gamma irradiation facility source pit which having failure on its air cooling system. The CFD model predicts the temperature of the gamma sources under such situation to determine its integrity.

1. Introduction

While the stricken Fukushima Daiichi nuclear power plant is bracing for the worst of the reactor unit 1, 2 and 3's melting core, another frightening development unfolded at reactor unit 4. The spent fuel pool containing 1535 fuel assemblies may have been exposed and started to heat up [1]. The worst that can happen is that the exposed spent fuel assemblies heat up and catch fire, which may end up melting the cladding and releasing large amount of radioactive materials into the environment. The above phenomenon was part of the frightening development during Fukushima Daiichi crisis, which also became a major concern that similar cases might happen to any nuclear and radioactive sources storage all over the world.

U.S. based Union of Concerned Scientist elaborated the danger that it might happen in any spent fuel pools of unit 4, 5 and 6 of Fukushima Daiichi nuclear power plant. There is also concerned on the difficulty to conduct repair works on broken cooling mechanism of the pools, since the workers may be exposed to lethal radiation dose for a very short period of time [2]. The spent fuel pool condition of reactor unit 4 was modeled by Wang *et al.* using mass and energy balance to estimate the rate of evaporation of water, water level and the spent fuel temperature. It revealed close agreement with data provided by TEPCO and predicted the time taken to total water loss and fuel degradation [3]. Study on other radioactive waste dry storage facility inside the proposed Yucca Mountain was conducted by Pepper *et al.* [4] through CFD modeling. The model showed that the decay heat from the waste's casks was removed by ventilating air through the drift and conduction through drift walls; whereas thermal heat from radiation found to have little effect on the overall cooling mechanism.

IOP Conf. Series: Materials Science and Engineering **555** (2019) 012014 doi:10.1088/1757-899X/555/1/012014

In this study, the condition of a dry storage pit of a gamma-irradiation facility's radioactive source was modeled to assess its integrity under abnormal situation. Under normal shut-down condition, the gamma source is stored inside the dry storage pit and the cooling air system inside the pit will be operated. The air cooling system is kept running to ensure the decay heat from the source is properly dissipated to keep the source temperature under controlled. Without the air cooling system, there is concern on the fate of the source, since the decay heat may increase the pit's temperature. If the temperature exceeded the melting point of the source capsule, the radioactive material may be exposed and released to the surrounding. Therefore, the temperature inside the bunker and the source pit need to be determined to assess the condition of the source. A CFD simulation was conducted to predict the temperature of the source under no-cooling condition as elaborated further in the next section.

2. Methodology

The simulation of the source, rack and air condition inside the pit and bunker was conducted using CFD technique. The physical condition of the system is as illustrated in Figure 1 and the schematic of the modeling system is shown in Figure 2. The system being modeled consists part of the bunker, storage pit, source rack, baffle plate and the sources. All of the walls are considered adiabatic as the bunker and pit's walls are made of high density concrete. The baffle plates are modeled as a single-square entity and the frames holding the sources are ignored for simplicity to reduce computational load. The temperature of the air inside the pit and bunker is set at 30°C according to the actual measurement [5]. The pressure inside the bunker is assumed at ambient. No external force or pressure is imposed on the air, thus it is assumed to be in stationary and in laminar flow condition. All of the walls are considered smooth and no frictional force is exerted on the fluid. The top part of the model is considered as convective wall exposed to free stream air at room temperature. The convective heat transfer coefficient in free stream air is set to $10W/m^2$ K as used by Rodriguez *et al.* [6] and Gastelurutia *et al.* [7]. Other fluid properties of the model are shown in Figure 2. The simulations assumed heat source from the gamma sources which is dissipated through conduction and convection via the source rack, baffle plates and air inside the pit to the bunker. Natural convection modeling was conducted using Boussinessique approximation at 30°C condition. The validation of the models was conducted on modeling program verification and numerical solution verification aspects similar to Rosli *et al.* [8].

An actual physical measurement of the temperature inside the storage rack on the actual facility was conducted by M Arif [5]. The measurement was taken with the cooling system of the facility turned off for more than 24 hours. The temperature inside the bunker and storage pit was measured using a thermocouple type K. The locations and the results of the measurement are shown in Figure 3 and Table 1. Physical measurement inside the pit proved to be difficult due to the hindrance by the source rack structure and the narrow gap between the pit's wall and the rack. The thermocouple could only get inside the pit up to the middle of the baffle plates (L4), thus leaving the rest of the pit's condition unknown. The modeling attempted to predict the sources' surface temperature based on the known measured values. The modeling was done only on axial distribution since the width of the storage pit is very narrow that the temperature variation is negligible.

IOP Conf. Series: Materials Science and Engineering **555** (2019) 012014 doi:10.1088/1757-899X/555/1/012014

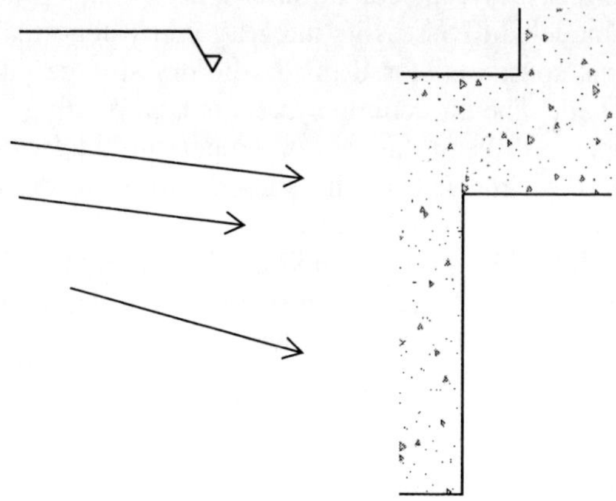

Figure 1. Sectional views of the storage pit showing the baffle plates and the source rack.

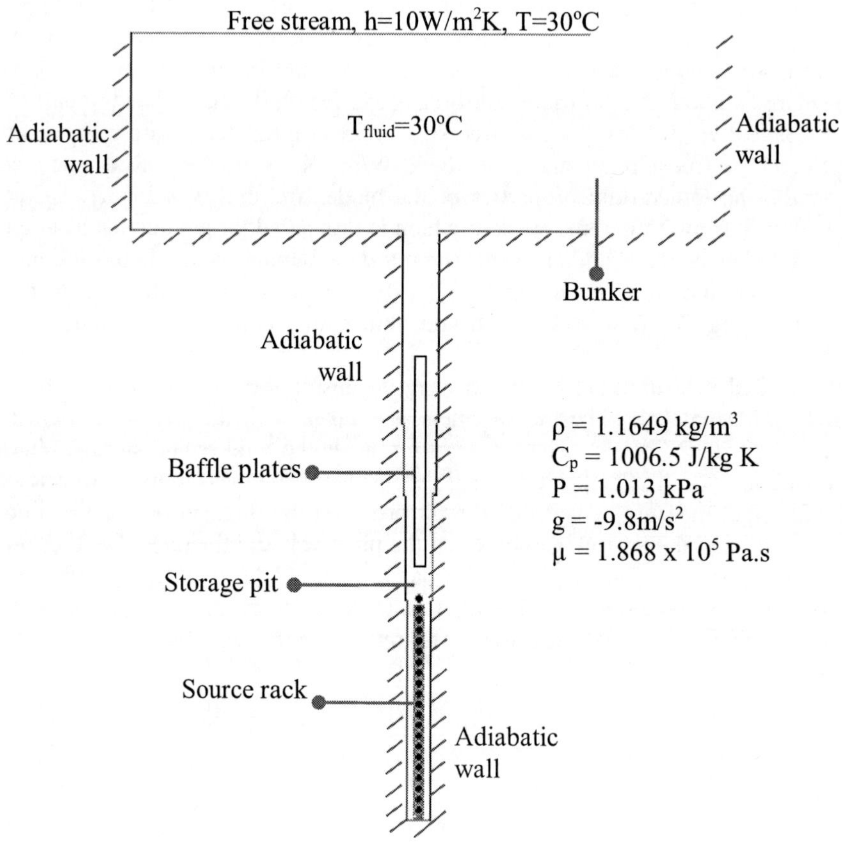

Figure 2. The schematics of the bunker, pit, baffle plates and source rack modelling including the walls conditions and air parameters.

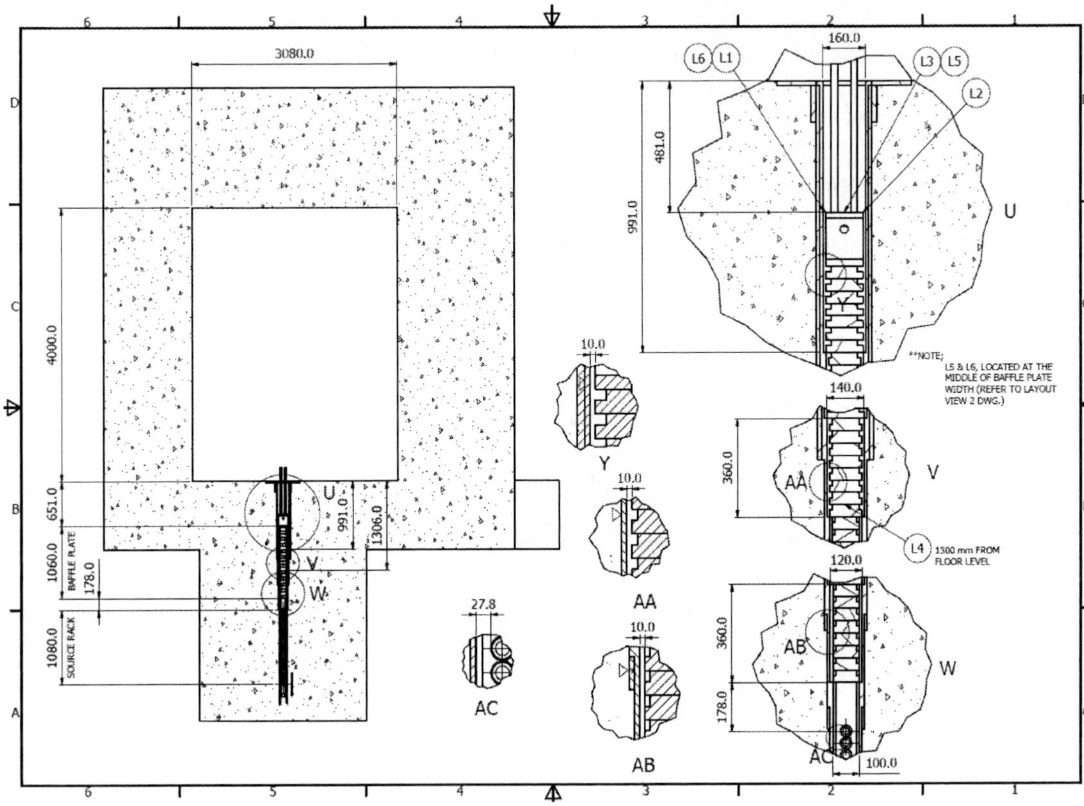

Figure 3. The cross section of the bunker, pit, baffle plates and source rack, showing the physical measurement position for L1 to L6 [5].

Table 1. Measured temperature inside the bunker and storage pit [5]

Location	PROBE 1 - T1 (°C)	PROBE 2 - T2 (°C)	Notes
Ambient	28.7	28.7	Outside bunker/room
Room	30.2	30.2	Inside bunker/room
L1	29.5	29.3	Edge of top plate
L2	29.3	29.5	Edge of top plate
L3	29.6	29.6	In between guide rail
L4	30.0	30.8	Deepest position achieved
L5	29.1	29.2	Middle of Top baffle plate
L6	30.6	30.7	Edge of Top baffle plate

3. Results and discussion

The simulation of the source, rack and air condition inside the pit and bunker was conducted on three different models as shown in Figure 4, 6 and 8. The first model was without the baffle plates to simulate a simpler version of the modeling. The second model was with the full baffle plates and the third model was with a simulated baffle plates. The full baffle plates model represented the whole length of the baffle plates above the source rack. However, due to the 2D nature of the modeling, it did not represent the true geometry of the baffle plates, where this model only represented by a single square configuration. As will be discussed further, this representation did not result in a trend similar to the measurement made in the actual facility. Therefore, the third model; a modified version of the baffle plates was modeled which showed close agreement with the actual measurement.

The result of the simulation without the baffle plates is shown in Figure 4, Figure 5 and Table 2. The temperature profile shows that the temperature inside the bunker is around 30^{0}C, almost similar to the measured value. The temperature remains the same at L3, which is at the top of the baffle plates. The temperature started to increase near L4 which is at the middle of the baffle plates, and continue to rise to about 32.8^{0}C at the sources' surface. Although, the results show close temperature values between the simulation and the measurement, however, the values show different trend as shown in Table 2. The actual measurement shows lower temperature at L3 compared to the temperature inside the bunker and at deepest position L4; whereas the simulation results shows that the temperatures inside the bunker and L3 are the same before increase at L4 and at the sources. Thus, this model's prediction does not represent the actual condition at the bunker and storage pit; and the predicted temperature value of the source at 32.85 ^{0}C might not be accurate.

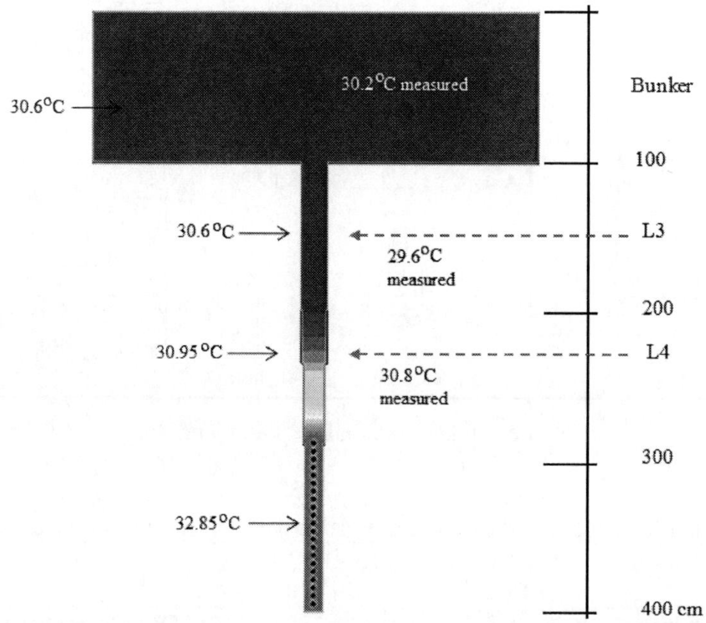

Figure 4. Temperature contour of the model without baffle plates showing measurement positions at bunker, L3 and L4 for comparison.

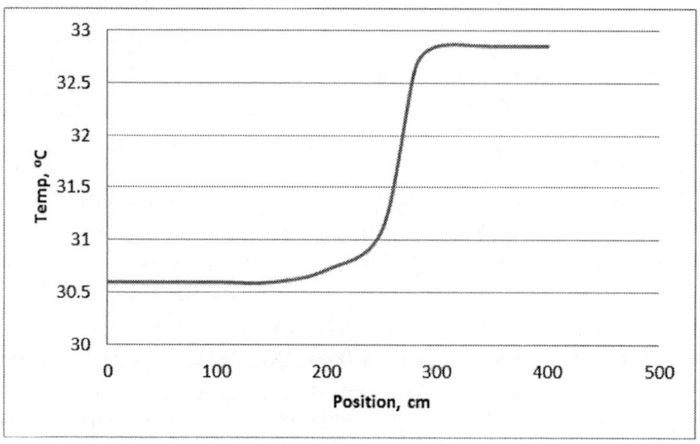

Figure 5. Temperature profile of the model without baffle plates.

Table 2. Comparison on temperatures between simulation and actual measurement for the model without baffle plates.

Position	Temperature, °C	
	Simulation	Measured
Bunker	30.60	30.2
L3	30.60	29.6
L4	30.95	30.8
Source	32.85	NA

Table 3. Comparison on temperatures between simulation and actual measurement for the model with full baffle plates.

Position	Temperature, °C	
	Simulation	Measured
Bunker	29.60	30.2
L3	25.49	29.6
L4	25.22	30.8
Source	33.85	NA

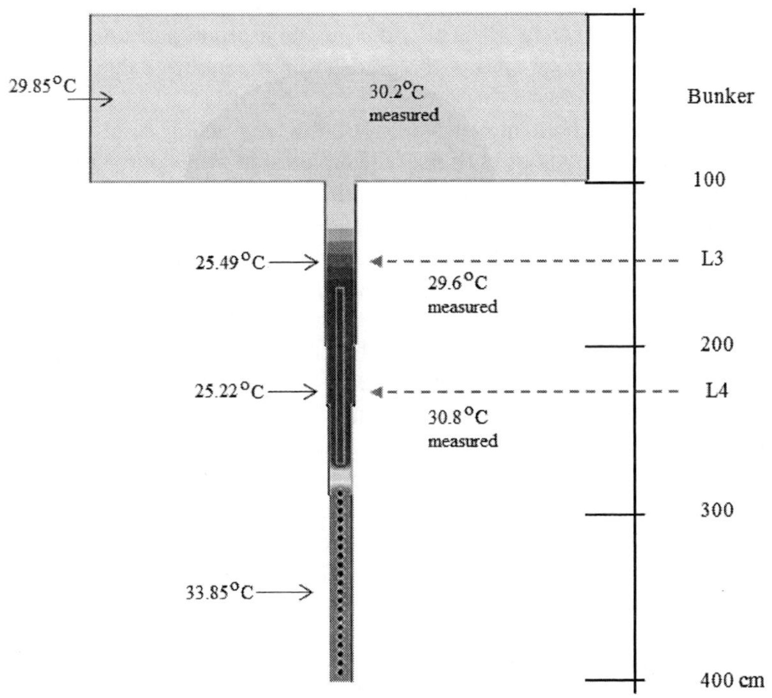

Figure 6. Temperature contour of the model with full baffle plates showing measurement positions at bunker, L3 and L4 for comparison.

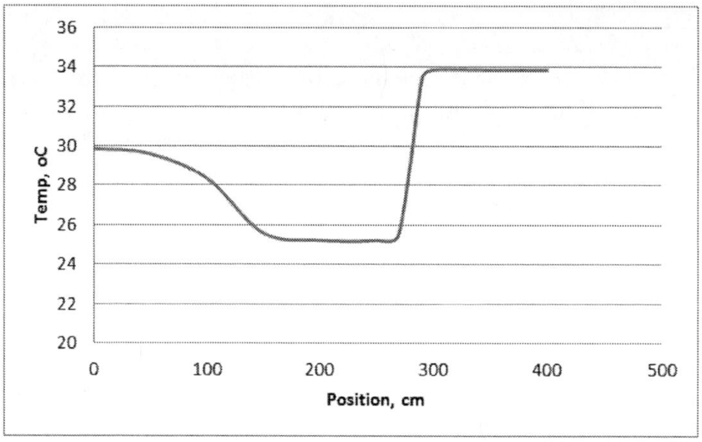

Figure 7. Temperature profile of the model with full baffle plates.

The result of the simulation with the full baffle plates is shown in Figure 6, 7 and Table 3. The temperature profile shows that the temperature inside the bunker is at 29.6⁰C, which is a little bit lower than the measured value. The temperature starts to decrease entering the pit, and predicted at 25.49°C at L3 which is at the top of the baffle plates. The temperature decreases further to 25.22°C at L4 which is at the middle of the baffle plates and then suddenly rises up to 33.85⁰C at the sources' surface. The results of the simulation deviate further from the actual measurement, which shows lower temperatures inside the bunker, at position L3 and L4. The trend shown in Table 3 also differed between the simulation results and the actual measurement. The measured values show that the temperature decrease at L3 and increase again at L4. The predicted values on the other hand shows decreasing trend from the bunker to L4. Thus, this model's prediction does not represent the actual condition at the bunker and storage pit and the predicted temperature value of the source at 33.85 ⁰C might not be accurate.

As discussed earlier, since both models did not show promising results, another modelling was attempted. A model which consists of both the baffle plates and the source rack was created. In order to simulate the temperature profile similar to the actual measurement, a simulated baffle was created. Several configurations of the baffle plates were modelled and the results were compared with the actual measurement. After several attempts, simulated model in Figure 8 found to have a similar trend with the measured values. This was accomplished by modifying the baffle plates' dimension and position inside the pit. The result of the simulation with the simulated baffle plates is shown in Figure 8, 9 and Table 4. Two simulation results with different sources' temperature are presented. Table 4 shows that both of the simulations have similar trend with the actual measurement. The temperatures at L3 are lower than the temperatures inside the bunker and L4. The temperature profile for simulation 1 and simulation 2 shows that the temperature inside the bunker is at 30.55⁰C and 30.1°C respectively. These predicted results of the bunker's temperature are very close to the measured value of 30.2°C. The temperature starts to decrease entering the pit and predicted at 29.25°C for simulation 1 and 28.94°C for simulation 2 at L3 which is at the top of the baffle plates. These predicted results at L3 are also very close to the measured value of 29.6°C. The temperature starts to increase near L4 and predicted at 30.58°C for simulation 1 and 30.52°C for simulation 2 at L4. These predicted results at L4 are also very close to the measured value of 30.8°C. Finally, with the promising result of the model inside the bunker, L3 and L4; which shows very close agreement with the measured values; the sources' surface temperature are predicted at about 33.85°C for simulation 1 and 34.85°C for simulation 2. These two predicted temperatures are much lower than the melting temperature for stainless steel cladding of the gamma sources. Therefore, the simulation shows that without the cooling mechanism inside the storage pit, the integrity of the gamma sources are still intact and sufficiently cooled by natural convection.

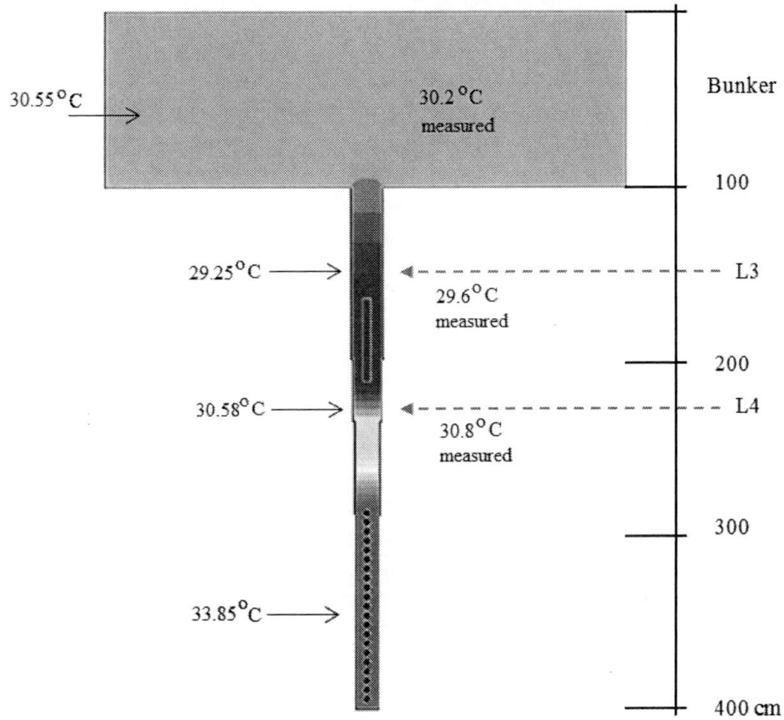

Figure 8. Temperature contour of the model with simulated baffle plates showing measurement positions at bunker, L3 and L4 for comparison.

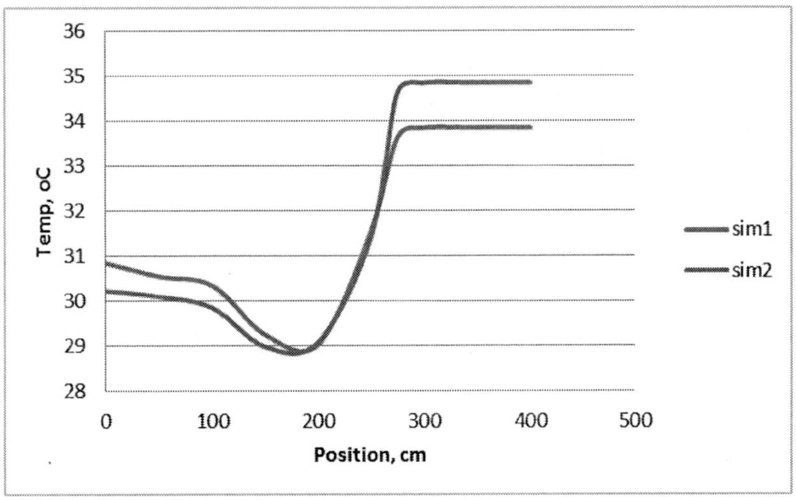

Figure 9. Temperature profile of the model with simulated baffle plates.

Table 4. Comparison on temperatures between two simulations and actual measurement for the model with simulated baffle plates.

	Temp, °C		
Position	Simulation 1	Simulation 2	Measured
Bunker	30.55	30.1	30.2
L3	29.25	28.94	29.6
L4	30.58	30.52	30.8
Source	33.85	34.85	NA

4. Conclusion

A CFD modelling of natural cooling through natural convection was simulated inside a dry storage pit of a rack of gamma sources. The model successfully predicted the surface temperature of the gamma sources under ambient condition with the cooling system switched off. The simulation shows a close agreement with the measured values inside an actual gamma irradiation facility. The model predicted that the ambient air natural convection may keep the sources' surface temperature at about 34°C, which is much lower than the melting temperature of the stainless steel cladding, thus ensure the integrity of the sources.

5. References

[1] Dvorak P 2012 Fukushima Daiichi's Unit 4-spent fuel pool: safe or not? *Japan Real Time, May 2, 2012.*

[2] Wright D 2011 Spent fuel pool at Fukushima *Union of Concerned Scientist, March 16, 2011.*

[3] Wang D, Gauld I C, Yoder G L, Ott L J, Flanagan G F, Francis M W, Popov E L, Carbajo J J, Jain P K, Wagner J C and Gehin J C 2012 Study of Fukushima Daiichi nuclear power station Unit 4 spent-fuel pool *Nuclear Tech.* **180(2)** p 205-215.

[4] Pepper D W and Chen Y 2005 Heat transfer analysis of waste casks stored in the Yucca Mountain repository *J. of Numerical Heat Transfer* **47(7)** p 671-690.

[5] Arif M H 2017 *Ujian bacaan suhu punca Co-60 – Loji Raymintex* (Agensi Nuklear Malaysia: *April 5*).

[6] Rodríguez I, Castro J, Pérez-Segarra, Oliva C D 2009 Unsteady numerical simulation of the cooling process of vertical storage tanks under laminar natural convection *Int. J. of Thermal Sci.* **48** p 708–721.

[7] Gastelurrutia J, Ramos J C, Larraona G S, Rivas A, Izagirre J, del Río L 2011 Numerical modeling of natural convection of oil inside distribution transformers *Appl. Thermal Eng.* **31** p 493-505.

[8] Rosli D, Nor Mariah A, Noraini A A, M Khairol Anuar M A 2017 Investigation on alternative method to enhance the cooling capacity of an open pool research reactor *AIP Conf. Proc.* **1799**(10:1) p 28-38.

IOP Conf. Series: Materials Science and Engineering **555** (2019) 012015 doi:10.1088/1757-899X/555/1/012015

Study on natural rubber absorption of selected actinides

Alkhaledi, Nahhar[1,a),]Khoo Kok Siong[2]

[1]Centre for Frontier Sciences, Faculty of Science and Technology, The National University of Malaysia
[2]Centre for Frontier Sciences, Faculty of Science and Technology, The National University of Malaysia

p81906@siswa.ukm.edu.my

Abstract. This paper contains the research done concerning natural rubber. This included its origin, economic importance, and how the functional groups in the natural rubber reacted with selected actinides. The selected actinides discussed were Thorium, Uranium and Plutonium. The information was retrieved from different sources including information from laboratory experiments. The results obtained were gathered when conducted under different conditions. The conditions varied based on the temperatures and the pH. The functional groups exhibited different outcomes when reacted with different actinides under different conditions. The functional groups namely amine carboxyl and hydroxyl groups have different efficiency and capacity of adsorption. Besides economic importance, the environmental significance of rubber also has been discussed.

1. Natural rubber

Natural rubber which is also known as caoutchouc is an addition of polymers of an organic compound called isoprene. It was discovered by Christopher Columbus in 1495-1496 after he saw American Indians playing with balls that were made from trees juice and were bouncing. It was named caoutchouc by Charles de la Condamine, a French scientist which simply means 'weeping tree' [4]. It was first grown in London by Henry Wickham. Another attempt was done in Sri Lanka but the results were no pleasant. In Singapore which was thought as the basis of the industry in Southeast Asia another plantation was set [3]. Table 1 below shows the leading countries in the production of Natural Rubber and the amount of production in tons per year (Association of Natural Rubber Producing Countries).

IOP Conf. Series: Materials Science and Engineering **555** (2019) 012015 doi:10.1088/1757-899X/555/1/012015

Table 1. Leading countries in the production of Natural Rubber and the amount of production in tons per year (Association of Natural Rubber Producing Countries).

Country produce (tons)	Country produce (tons)	Country produce (tons	Country produce (tons)
Thailand 4,305,069	Malaysia 826,421	Nigeria 149,052	Gabon 22,691
Indonesia 3,107,544	Philippines 444,818	Sri Lanka 130421	Ghana 21,440
Viet Nam 946,865	Guatemala 356,392	Liberia 74,750	Cambodia 20,099
India 900,000	Coted'ivoire289,563	Cameroon 55,605	Ecuador 18,249
China 864, 806	Brazil 185,725	Mexico 51,397	

2. Economic importance

Natural rubber is one of the commodities the nation's growing the product use for export. In Thailand, it holds the fifth position of the ten export product groups. It shares 5.7% of the income the nation of Thailand gets from the exports. Rubber is used to manufacture products like tire, balls, bands, shoes which are used as commodities of trade [11]. Rubber cultivation acts as a source of employment. The rubber tree is becoming a significant source of wood as a result of improved wood technology.Rubberwood can be used to substitute timber from natural forest since it is a by-product of latex and can be grown in renewable plantations [6].

3. Composition

Rubber consists of elastic polymers or elastomers which are arranged in chainlike molecules (Table 2). The molecules can stretch and regain their original shape again. The monomer isoprene (2-methyl-1, 3-butadiene) is a joined diene hydrocarbon. In the isoprene chain, NR has C=C, a bonding which makes it difficult to be oxidized [8]. The rubber from the first opening of a mature tree consists of the solvent fraction which is insoluble. The fraction is formed through carbon-carbon bonding also called hard-gel. The 80-90 percent of the solid rubber is formed by the hard-gel. The residual solvent fraction which is soluble has low molecular weight and is formed by oxidative degradation [8].

Table 2. Composition of fresh Natural Rubber Latex.

Constituent	%, Content
Minerals	0.7-0.9
Rubber	30-40
Resins	1.5-3.0
Proteins	1.0-1.5
Water	55-60
Carbohydrates	0.8-1.0

The weight of a molecule of Latex or cis-1, 4-polyisoprene ranges from 100,000 to 1000,000 Daltons. Natural rubber contains other materials. These are resins, proteins, fatty acids some inorganic materials which salts are. Synthetic natural rubber can be obtained from the synthesis of polyisoprene (Rolere et al. 69). Shown below is the chemical structure of cis-polyisoprene and which is the major component of the natural rubber.

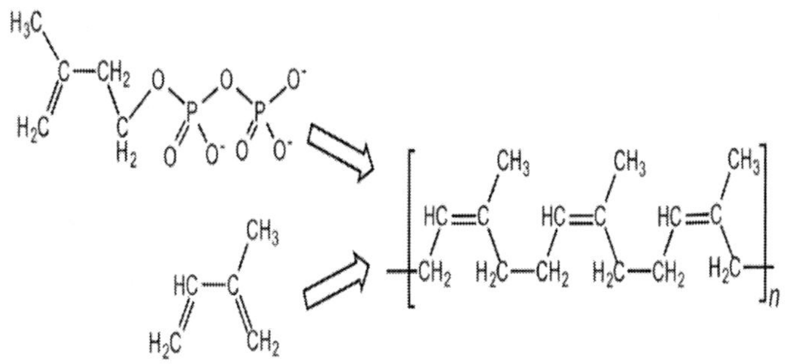

Figure 1. Natural rubber functional groups.

Natural rubber contains functional groups which include fatty acids, ester among others. Through oxirane group, an amine binds to the rubber molecule or the binding can be done by neutralizing the latex before the amine is added. The 4-amino-para-phenyl diamine (PAA) has a faster reaction than that of meta-Nitro-Aniline (mNA). The least reactive is the N-isopropyl-N'-phenyl-para-phenyl diamine (IPPD) [1]. In many polyisoprenes, the groups of the fatty acid ester are the α-terminal group structures. The long chain of fatty acid ester persists even after deproteinization and acetone extraction methods for its purification. Each one molecule of natural rubber contains two molecules of the ester with long-chain fatty acid.The glyceride and $-CH_2OP$ groups at the chain ends of the Natural rubber can be removed by saponification [1].

By adding 1wt% mixture of saturated fatty acid, which acts as a nucleating agent and the addition of a plasticizer, that is, unsaturated fatty acid can speed the crystallization of synthetic cis-1,4-polyisoprene. The addition of hydroxyl group by the process hydroboration esterifies the Natural Rubber at the 3,4-isoprene with stearic acid.The stearoyl group at the temperatures of -25°C, stimulates the crystallization of the natural rubber. When the linked fatty acids are used in the place of the stearoyl group, no stimulation occurs on the crystallization. On adding methyl linoleate to the stearoyl, the esterified rubber, the crystallization occurs at the highest rate at -25°C.

A drying agent such as sodium hydroxide or phosphorus pentoxide when used to remove water from the natural rubber, phospholipids and proteins at the end of the rubber chain make branching points. During the storage, the gel fraction is formed as the two functional groups of the natural rubber which are active react. The formation of a micelle structure through hydrogen bonding of polar groups found in the molecules of phospholipids brings together the phospholipids [9]. Most of the phospholipids in non-polar and polar solvents don't undergo solubilization and form the micelle structure. The hydrogen bonding does not permit the mobility of the phosphate groups which may be found in the Natural rubber.

Hydrogen peroxide and sodium nitrite are used as reagents can be used to obtain Liquid natural rubber from natural rubber latex through the degradation chemical reaction. The samples of Liquid natural rubber contain hydroxyl and carbonyl as functional groups at their ends when preparing in acidic and basic mediums. Chemically the reagents do not react in an alkaline medium and the chains of the natural rubber split through the process of oxidation. In acidic medium, the reagents react forming peroxynitrous acid which further decomposes to form the radicals of hydroxyl. The hydroxyl radical split the natural rubber chains to form hydroxyl groups at the ends [1]. The hydroxyl group (OH) at the end of the long chain natural rubber molecule can be esterification. The peptide bond formation is one of the most important reactions of the amine group. During these reactions, the water molecules α-NH_3^+ group found in amino acids and COO^- which is in the second amino acid.

3.1 The functional groups remove heavy metals from solutions

The functional groups found in natural rubber have the ability to absorb some heavy metals from the liquid solution. The metals that can be absorbed include, cadmium and lead. This is done onto the biomass of the blue-green alga Anabaena sphaerica from an aqueous solution which can be affected by the contact time, pH, the concentration of the metal ion and biosorbent dosage [1]. There are some peptides of amino acids which are as a result of –COOH and $-NH_2$ group which combine with heavy metals like Mn^{2+}, Cu^{2+}, Co^{2+} and Ca^{2+}. The Hg^{2+}, Cd^{2+} and Pb^{2+} ions are highly toxic and can denature the proteins found in the Natural rubber especially the enzymes. The absorption of mercury and lead ions (Figure 3) can be minimized through their combination with proteins. The ions of these metals with proteins form precipitates when linked with amino groups and the carboxyl groups. The Mg^{2+} form ionic bonding between the molecules of rubber.

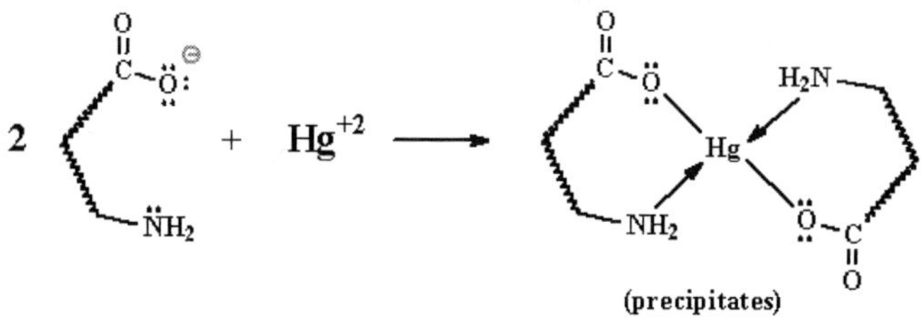

Figure 2. Absorption of mercury ions through their combination with proteins.

The chemical reaction of the amino group is determined by α-NH_2 and α-COOH (carboxyl group). The Carboxyl group participates in forming an ester. The processes like decarboxylation, reduction and oxidation can be used to detect the presence of an amine group [10].The adsorption capacity of Natural rubber increases from Pb^{2+} to Cd^{2+} and then Cu^{2+} [7]. The adsorption of the functional groups of rubber increases with the increase in the amount of PAMPS. There are optimum conditions which facilitate the adsorption process of Pb^{2+}, Cu^{2+} and Cd^{2+} by the hydroxyl, carbonyl, and amine groups. These are the temperature of 30-40°C and pH range of 4-5. The efficiency of desorption and the capacity of adsorption do not significantly decrease [10].

The active functional groups that interact with the ions in the aqueous is affected by the amount of pathogen-associated molecular patterns (PAMPS). PAMPS also affects the swelling of rubber and the process of cavities generation in the adsorbent [10]. There is an increase in metal ions adsorption with an initial metal ion concentration increase. In the initial concentration of Pb^{2+} and Cd^{2+} of approximately 500mg of metal ion per 25mL make the exchange of the ions constant. Above 150mg of metal ion/25mL concentration of Cu^{2+} the adsorption of the copper ions remains constant [5].

The functional groups Carboxyl (COOH) and amine (NH_2) react with plutonium, cesium, and strontium using photoelectron spectroscopy method. The sorbent has the highest capacity of adsorption in plutonium. The cesium, strontium react with the pairs of electron of nitrogen and oxygen absorbent of Amine and carboxyl groups. The degrees of oxidation of plutonium are equal for the bisorbent [10].

There are complexes of uranium and thorium. Thorium complex is gotten when Thorium reacts with the hydroxyl group. The complex form increases the solubility and adsorption of Thorium. When the thorium reacts with the hydroxyl group in the process of adsorption, the complex form obtained is $Th(OH)^0_4$ which is obtained above pH 5 [10].

Thorium reaction with a hydroxyl group to give Thorium hydroxide is as shown in Figure 3.

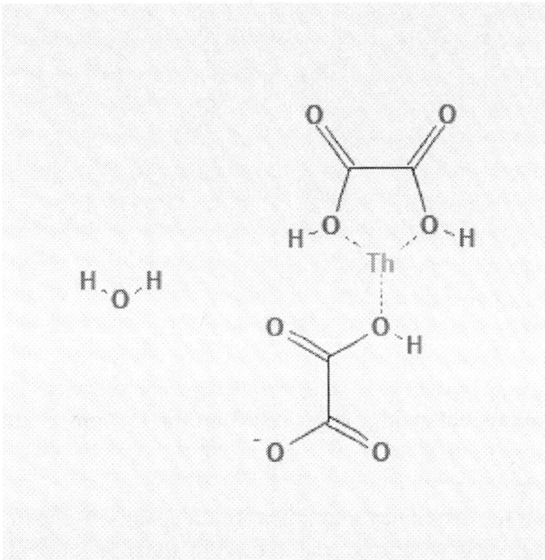

Figure 3. Thorium reaction with a hydroxyl group to give Thorium hydroxide.

As with the thorium in the reaction described above, it will occur with uranium and six of hydroxyl groups forming Uranium hydroxide in the same mechanism almost as shown in the following Figure 4.

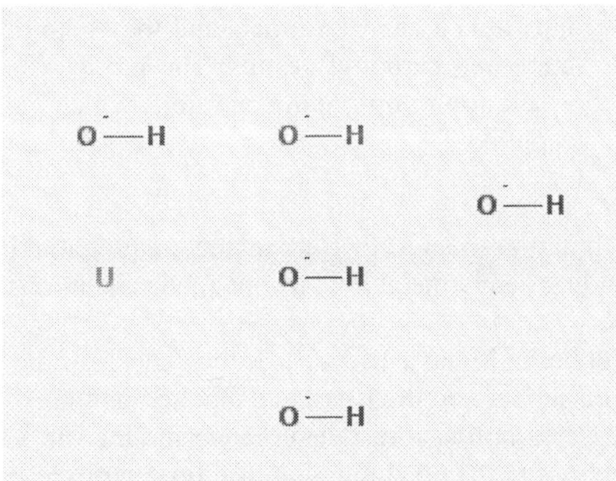

Figure 4. Hydroxyl groups forming Uranium hydroxide.

It would be very logical to assume that the plutonium would behave like the other actinides, such as uranium and thorium, and that is what happened. Plutonium reacts with the hydroxyl group following the same pattern and mechanism forming plutonium hydroxide as shown in Figure 5.

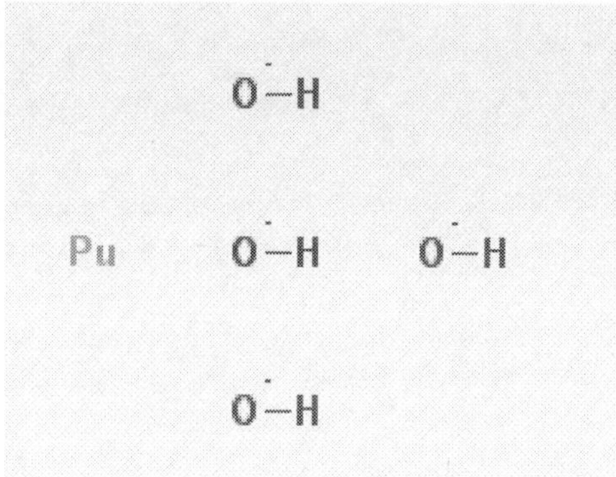

Figure 5. Plutonium reacts with the hydroxyl group, forming plutonium hydroxide.

4. Recommendations

Further research on the pH on the efficiency and adsorption capacity of the heavy metals by the functional groups associated with natural rubber. More research on how these functional groups react with other metal apart from cadmium, copper, and lead and how the reaction behaves under different conditions. Also, more research on the adsorption of other metals in the actinide group can be done.

5. Conclusion

The Natural which is a renewable product has a very great economic importance to the countries' economy and has an environmental advantage over the synthetic products. Natural rubber molecule has functional groups linked at its terminals and they take part in reaction with heavy metals, that is, hydroxyl, amine and carbonyl groups. The groups react with the metal ions in water solution or an aqueous solution in optimum conditions which gives nearly constant results if no external interference is applied.

6. References

[1]. Ahmed M J K and Ahmaruzzaman M 2016 A review on potential usage of industrial waste materials for binding heavy metal ions from aqueous solutions.*J.l of Water Process Eng.* **10** p 39-47.

[2]. Drabble J H 2015 Rubber in Malaya 1876-1922: the genesis of the industry.

[3]. Gaut B S 2017 Natural rubber and the Russian dandelion genome *National Sci. Rev.*

[4]. Grogan K *et al.* 2015Cross-border forest disturbance and the role of natural rubber in mainland Southeast Asia using annual Landsat time series. *Remote Sensing of Env.* **169** p 438-453.

[5]. Kitaura T *et al.* 2018 Characterization of Natural Rubber End Groups Using High-Sensitivity NMR *Macromolecular Chem. and Phys.* **219(3)** p 1700331.

[6]. Khin A *et al.* 2018 Forecasting technology using for dynamic natural rubber production models in selected Asian countries and world market *Adv. Sci. Lett.* **24(5)** p 3368-3373.

[7]. Park J Hwan *et al.* 2016 Competitive adsorption of heavy metals onto sesame straw biochar in aqueous solutions. *Chemosphere* **142** p 77-83.

[8]. Rolere S *et al.* 2015Investigating natural rubber composition with Fourier Transform Infrared (FT-IR) spectroscopy: A rapid and non-destructive method to determine both protein and lipid contents simultaneously *Polymer Testing* **43** p 83-93.

[9]. Tao J *et al.* 2018Natural rubber particle modified fabrics with catalytic activity and hydrophobicity *Composites Sci. and Tech.* **162** p 123-130.

IOP Conf. Series: Materials Science and Engineering **555** (2019) 012015 doi:10.1088/1757-899X/555/1/012015

[10].Yusof N H *et al.* 2015 Preparation and properties of natural rubber with filler nano-matrix structure *Colloid and Polymer Science* **293(8)** p 2249-2256.

[11].Yi Z F *et al.* 2014 Developing indicators of economic value and biodiversity loss for rubber plantations in Xishuangbanna, southwest China: A case study from Menglun township *Ecological Indicators* **36** p 788-797.

[12].Schwantes J M; Douglas M; Bonde S E; Briggs J D 2009 Nuclear archeology in a bottle: Evidence of pre-Trinity U.S. weapons activities from a waste burial site *Analyt. Chem.***81(4)** p 1297–1306.

IOP Conf. Series: Materials Science and Engineering **555** (2019) 012016 doi:10.1088/1757-899X/555/1/012016

Monte Carlo simulation for designing collimator of neutron diffractometer facility in Malaysia

Muhammad Aliff Ashraff Rosdi[1], Pei Sean Goh[1,a)], Safwan Shalbi[1], Faridah Idris[2], Rafhayudi Jamro[2], Norriza Mohd Isa[2]

[1]Nuclear Engineering Department, School of Chemical and Energy Engineering, Faculty of Engineering, University Teknologi Malaysia, 81310 Skudai, Johor Bahru, Malaysia.
[2]Malaysia Nuclear Agency, 43000 Kajang, Selangor, Malaysia

[a]peisean@petroleum.utm.my

Abstract. Neutron diffraction (ND) is an application of non-destructive test which involves the process of interference between neutron and atoms within materials. This technique has drawn many attentions especially in industrial, metallurgical and nuclear sectors. Neutron diffractometer system (NDS) is a system which includes all the instrumentation to perform ND technique. One of the most important instruments in NDS set-up is neutron collimator which is used to reduce the size and steer the direction of the beam. Generally, neutron collimator used in NDS is based on step convergent collimator type. Currently, the NDS facility is planned to be built at one of the radial beam port of TRIGA MARK II PUSPATI research reactor (RTP) in Malaysia. The aim of this research is to characterize the materials suitable for neutron collimator and to obtain the ideal geometry for neutron collimator design. In order to achieve these aims, several neutron collimators with different geometries have been designed using Monte Carlo Simulation Codes MCNPX. The results are then compared to obtain the best design with high thermal neutron flux and low gamma contamination at the object plane. This research may be useful in the selection of neutron collimator design for NDS facility at TRIGA research reactor.

1. Introduction

Reactor TRIGA PUSPATI (RTP) is a pool type light water research reactor using enriched uranium zirconium hydride ($UZrH_{1.6}$) fuel with enrichment of 19.9% U^{235} and is located in Malaysia. RTP has a nominal thermal power of 1MW with maximum brilliance of 1.897×10^{13} neutrons $cm^{-2}s^{-1}$ obtained at 750kW [1]. RTP is composed of a thermal column and four neutron beam ports including three radial and one tangential beam ports as illustrated in Figure 1. Since its first commissioning in June 1982, various research works have been conducted using the neutrons produced from the RTP. However, until now, only two of the existing beam ports were used which are beam port 3 for neutron radiography (NR) and beam port 4 for small angle neutron scattering (SANS). In order to maximize the utilization of beam port facility of RTP, previous researches claimed that NDS can be developed at radial beam port 1 [2].

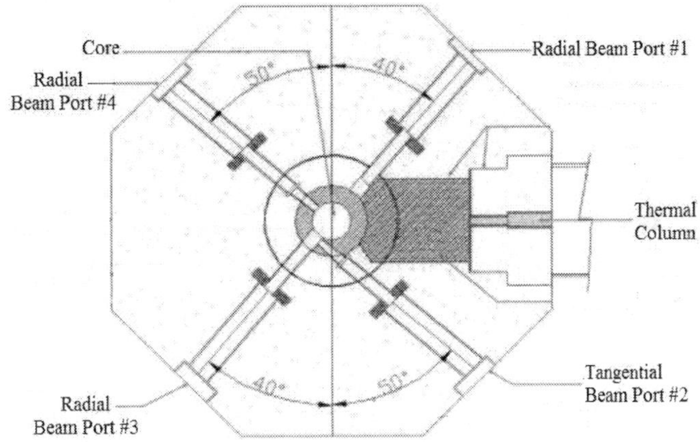

Figure 1. Cut-away view of RTP [2].

Neutron diffraction (ND) is one of the neutron scattering applications which has been widely used to determine atomic structure of material, measuring strain within material and study the composition of crystal and amorphous materials. In order to implement ND technique, thermal neutron with intensity at least 10^4 neutron $cm^{-2}s^{-1}$ is necessary [2]. Besides, the size and operational condition of the system need to be optimized and properly shielded by following diffraction rules and regulations. Generally, neutron diffractometer system (NDS) consists of several vital instrumentations including collimator. Neutron collimator is a beam forming assembly where function is to reduce the cross section of neutron beams and to align the direction of neutron beams in a specific direction. Basically, neutron collimator contains the assembly of gamma and fast neutron absorbing plates in order to filter the contaminations. Neutron collimator is composed of the combination of neutron and gamma absorber materials such as borated paraffin, cadmium, borated polyethylene, indium, gadolinium oxide, boron-aluminium plate, lead and bismuth to shield neutron and gamma particles respectively. Neutron collimator utilized in Romanian research reactor was constructed using boron-aluminium plate, indium and lead materials while collimator in HANARO research reactor in South Korea is composed of boron-based materials and bismuth [4][5]. Besides, GGR-1 research reactor located in Greece uses the neutron collimator fabricated from the assembly of borated polyethylene, lead, and Mylar films coated with gadolinium oxide [6]. Researchers from China proposed a collimator design consisted of bismuth, series of leads and boron based materials [7]. In addition, placement of sapphire crystal in the collimator is highly recommended as it has strong attenuation coefficient toward epithermal and fast neutrons while low attenuation coefficient toward thermal neutrons [8]. Selection of neutron collimator geometry for ND technique needs to be emphasized based on the uniformity and maximum flux produced at the end of the collimator outlet.

In this study, Monte Carlo MCNPX code is used to obtain the most optimal neutron collimator geometry for neutron diffractometer system with respect to neutron and gamma radiation. Several collimators with different combination of neutron and gamma absorbent materials were designed and tested in order to obtain the ideal thermal neutron flux at the end of the collimator tube. The findings of this study may be useful to the development of neutron collimator design combining neutron and gamma shielding materials.

2. Methodology

Calculation of radiation transport and dosimetry inside the neutron diffractometer collimator was performed using the Monte Carlo MCNPX code. The neutron collimator for NDS facility has an overall length of 166cm with two sections as shown in Figure 2. The collimator is divided into two parts, first part has a diameter of 18cm with length of 50cm, and second part has a diameter of 20cm with length of 115cm. There are an additional length of 1cm extra with diameter of 24cm on the second part of collimator which functions as step to ease the handling of neutron collimator. Basically, material used for the exterior part of neutron collimator is made from aluminium as it is a low activation element as compared to steel. Furthermore, the neutron collimator was coated with lead and boron flex in order to reduce scattering of neutron and gamma radiations.

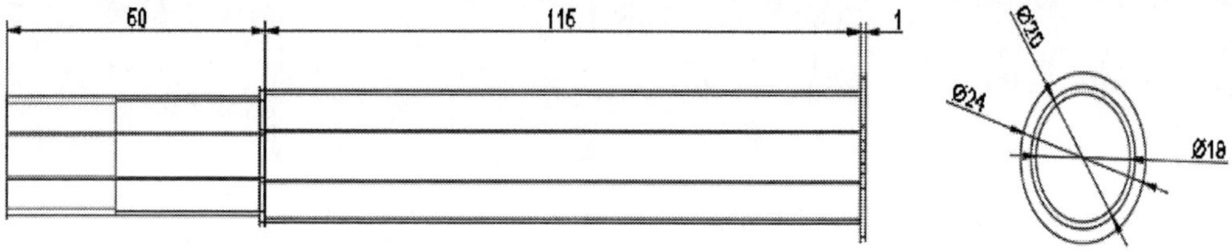

Figure 2: External design of neutron diffractometer collimator.

The main purpose of this research is to design and simulate several step convergent collimator in order to achieve high thermal neutron flux and low gamma contamination. Step convergent collimator type is chosen in order to produce small size of thermal neutron beam. This is because beam from neutron diffractometer collimator should be sharp to allow the investigation of specific location on object. The characteristic of an ideal neutron beam are parallel, higher flux intensity, mono-energetic, uniform cross-section, and lower contamination of unwanted radiations [9]. In this study, a total of 8 designs of neutron collimator with different materials selection and geometry were investigated. Arrangement and position of the materials were made inside of the collimator. The following materials in Table 1 were used: high density polyethylene (HDPE), bismuth, lead, borated paraffin, borated polyethylene and sapphire crystal. Geometry of the materials is shown in Figure 3 and Table 2.

Table 1. Shielding materials for collimator characterization.

Material	Neutron attenuation coefficient	Gamma attenuation coefficient	Description
5% BPF	2.35	0.14	Neutron shielding
5% BPE	2.36	0.14	
30% BPE	2.68	0.17	
HDPE	0.63	0.15	
Lead	0.26	3.00	Gamma shielding
Bismuth	0.19	2.98	
Sapphire	0.24	0.76	Fixed

Table 2. Material geometry of neutron diffractometer collimator.

No. of collimator	Material
A	5% borated paraffin + lead + sapphire
B	5% borated paraffin + bismuth + sapphire
C	High density polyethylene + lead + sapphire
D	High density polyethylene + bismuth + sapphire
E	5% borated polyethylene + lead + sapphire
F	5% borated polyethylene + bismuth + sapphire
G	30% borated polyethylene + lead + sapphire
H	30% borated polyethylene + bismuth + sapphire

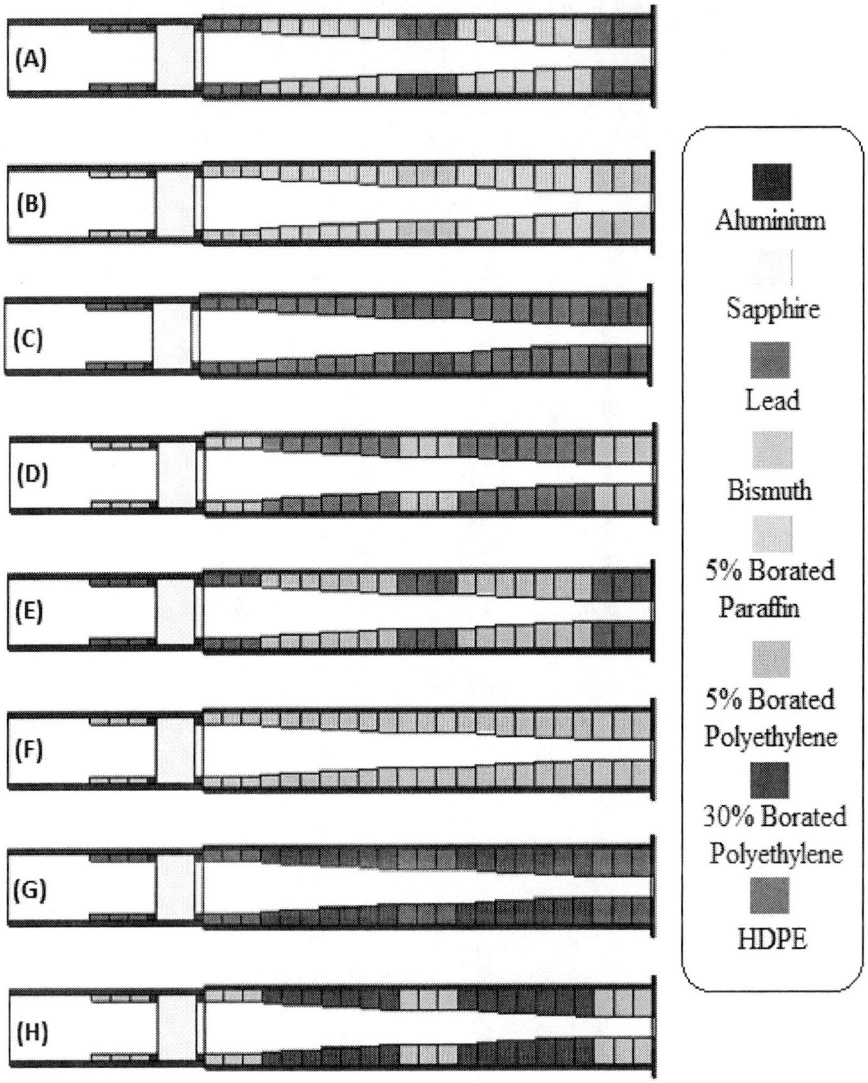

Figure 3. Design of collimators with difference geometries.

3. Results and discussions

The results obtained from MCNPX calculations at the centre and outer part of collimator tube were shown in Figure 4 and Figure 5.

Figure 4 shows that average thermal neutron flux produced at the centre of the collimator was $6.23x10^6$ neutron $cm^{-2}s^{-1}$, fast neutron flux was $2.03x10^5$ neutron$cm^{-2}s^{-1}$ and gamma flux was $1.20x10^1$ γ $cm^{-2}s^{-1}$. The highest thermal neutron flux was obtained at centre part of collimator C with flux of $6.32x10^6$ neutron $cm^{-2}s^{-1}$ which is consisted of HDPE and lead material. On the other hand, the lowest thermal neutron flux was measured at collimator B which consisted of 5% BPF and bismuth material. Thermal neutron flux was high when lead is used instead of bismuth. This is because lead can also moderate fast neutron which indirectly increases intensity of thermal neutrons in the collimator. In addition, lead also has high scattering cross section properties that cause the neutrons to scatter inside of the collimator [10]. In case of fast neutron flux, most of the collimator yields high intensity of fast neutron at the centre region. Therefore, this does not rank so high in the design selection. Furthermore, the fast neutron flux produced in the collimator could be useful in order to increase the intensity of thermal neutron flux by moderation. Besides, based on Figure 3, collimator with bismuth and lead materials give an equivalent result of gamma absorption at the low average of 10^1 γ $cm^{-2}s^{-1}$. The reason is that bismuth and lead has high gamma absorption cross section [11]. Even though the gamma radiations are still present in the collimator, the average gamma flux yielded from the centre region of the collimator were considerably low for bismuth and lead [12].

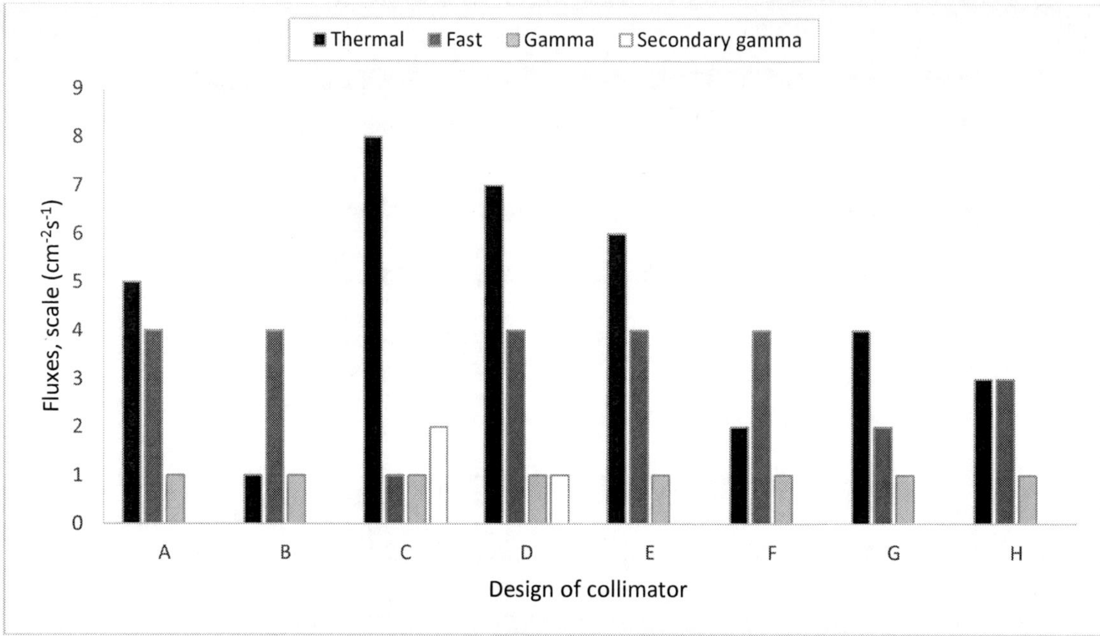

Figure 4. Fluxes at the centre part of collimators

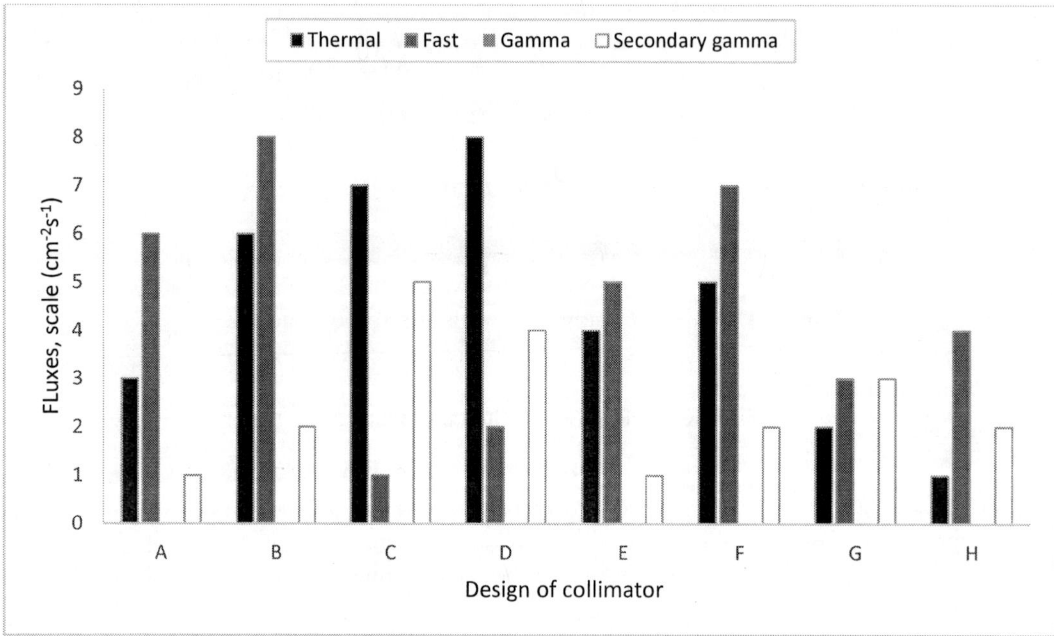

Figure 5. Fluxes at the outer part of collimators.

According to Figure 5, the average neutron flux and gamma flux measured at the end of outer region were lower compared to centre region. This is because the ionising radiations produced from the reactor core directly interacted with the shielding materials at the outer region compared to the centre region of collimator. The average of thermal neutron and fast neutron fluxes yielded at the end window were 3.74×10^3 neutron cm^{-2}s^{-1} and 8.29×10^2 neutron cm^{-2}s^{-1}, respectively. The lowest thermal neutron flux was obtained at collimator H which consist of 30% BPE and bismuth while the lowest fast neutron flux was obtained at collimator C of HDPE and lead. However, the results of neutron fluxes obtained for all collimator design were better which is 99% lower than the fluxes yielded at the centre region. The reason is most of the neutron particles have been absorbed by the neutron shielding materials in the collimator. In case of gamma flux, no flux was simulated at the outer region of the collimator. This implies all of the incident gamma radiations have been totally absorbed by the shielding materials in the collimator. This also proved that both lead and bismuth material are very excellent gamma absorber.

Figure 5 also shows that collimator implemented with bismuth yields more intensity of fast neutrons compared to lead. This shows that while bismuth and lead are very good against gamma shielding, their shielding properties of fast neutron were different. This is because bismuth is more transparent against fast neutron while lead is able to thermalize fast neutron indirectly which substantially reduce intensity of fast neutron. According to Figure 5, direct interaction of neutron particles with these shielding materials causes the production of secondary gamma radiation. The production of secondary gamma radiation was highest at collimator implemented with HDPE material. In case of the other designs which consist of boronated materials, the production of secondary gamma radiation was 87% lower than collimator with HDPE material.

After all the results have been compared, collimator E which composed of 5% BPE and lead material was selected [Figure 6]. This is because of the production of secondary gamma radiation obtained from the neutrons interaction is the lowest compared to the other collimator designs. The emergence of this secondary gamma radiation was not desirable since it may cause harm to the sample and other instrumentations of NDS [2]. Besides, the production of thermal neutron was high in the centre region of the collimator. The high intensity of thermal flux at the window end of the collimator is required since it is one of the most vital criteria of a NDS facility. The fast neutron yielded from the collimator E was also acceptable as the intensity was about average with the other collimator designs. Lastly, there is no gamma radiation at the outer region and some low level of gamma at the centre region of collimator.

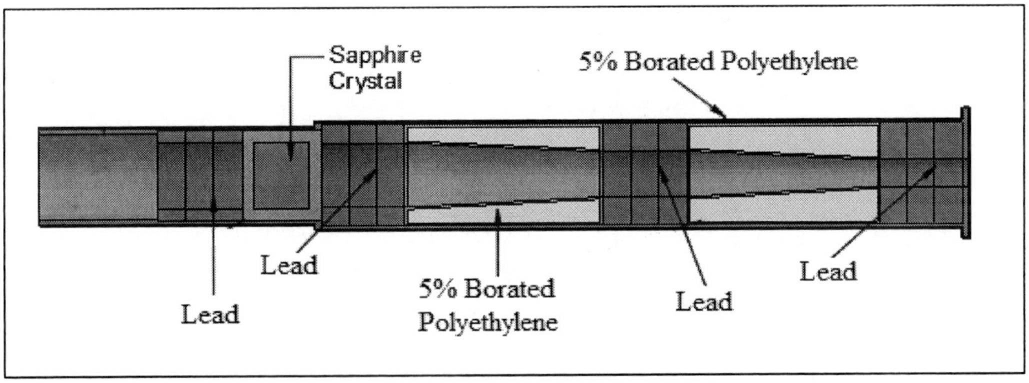

Figure 6. Final design of collimator.

4. Conclusion

The characterization of suitable materials for collimator used in NDS has been simulated using the MCNPX code. Collimator E which composed of 5% borated polyethylene and lead has been selected as the ideal design of collimator for NDS at RTP. The results obtained from this paper may be useful for the construction of new neutron diffraction system facility at TRIGA research reactor.

5. References

[1] Rabir M H 2017 Power and neutron flux calculation for the PUSPATI TRIGA reactor using MCNP. *Jurnal Sains Nuklear Malaysia* **24(1)** p 22-32.

[2] Farhi E, Mohamad Idris F, Mohamed A A, Yazid H B, Jamro R, Harun Al Rashid M and Zin M 2015 Neutron scattering instrumentation at low power reactors for science, engineering and education. *J. of Neutron Research* **18(2-3)** p 61-77.

[3] Solleh M R M, Tajuddin A A, Mohamed A A, Abdel-Munem E, Rabir M H, Karim J A and Yoshiaki K 2011 Collimator and shielding design for boron neutron capture therapy (BNCT)

facility at TRIGA MARK II REACTOR *Journal of Nuclear and Related Technologies* **8(2)** p 41-48.

[4] Dinca M, Pavelescu M, and Iorgulis C 2006 Collimated neutron beam for neutron radiography. Romanian J. of Phys. **51(3-4)** p 435-441.

[5] Lee C H, Kang Y H, and Kuk I H 2000 Present status of neutron beam facilities at the research reactor, HANARO, and its future prospect *KEK Proc.* **1(1)** p 1-9.

[6] Mergia, K., Salevris, A., and Messoloras, S. (2004). The New Neutron Powder Diffractometer at "Demokritos" Research Reactor. *Physica B: Condensed Matter.* 350(1), 162-165.

[7] Xiao-Peng Y, Bo-Xiang Y, Yi-Guo L, Dan P, Jin L, Gao-Long Z, Hang Z, Ai-Wu Z, Chun-Yang, L and Wan-Jin L 2014. Neutron collimator design of neutron radiography based on the BNCT facility. *Chinese Phys. C* **38(2)**, p 1-5.

[8] Joshua A R 2009 *Design, Construction, and Characterization of a Prompt Gamma Activation Analysis Facility at the Oregon State University TRIGA® Reactor* (Oregon State Univ.: M.Sc. thesis).

[9] Solleh M R 2015 *Moderator, Collimator and Shielding Studies for BNCT Research at Malaysian Nuclear Agency* (Universiti Sains Malaysia: PhD Thesis).

[10] Rezaei-Ochbelagh D and Azimkhani S 2012 Investigation of gamma-ray shielding properties of concrete containing different percentages of lead *Appl. Rad. and Isotopes* **70(10)** p2282-2286.

[11] Singh K, Goel A, Mohan S, Arora A and Sharma G 2006 Lead-and bismuth-borate fly-ash glasses as gamma-ray-shielding materials *Nuclear Sci. and Eng.* **154(2)** p 233-240.

[12] Cucinotta F A, Kim M H Y, Chappell L J and Huff J L 2013. How safe is safe enough? Radiation risk for a human mission to Mars (PLoS One) **8(10)** p e74988.

IOP Conf. Series: Materials Science and Engineering **555** (2019) 012017 doi:10.1088/1757-899X/555/1/012017

Optimization and designing collimator for boron neutron capture therapy at the thermal column of RTP

Safwan Shalbi[1], Wan Norharyati Wan Salleh[1], Muhammad Aliff Ashraff Rosdi[1], Faridah Mohamad Idris[2]

[1]Nuclear Engineering Department, School of Chemical and Energy Engineering, Faculty of Engineering, University Teknologi Malaysia, 81310 Skudai, Johor Bahru, Malaysia.
[2]Malaysia Nuclear Agency, 43000 Kajang, Selangor, Malaysia.

wan.norharyati@petroleum.utm.my

Abstract. Boron Neutron Capture Therapy (BNCT) is a promising method to cure cancer by using interaction between boron compound and the slow neutron beam obtained from nuclear power reactor, nuclear research reactor and neutron generator. Based on the previous feasibility study shows that the thermal column can generate higher thermal neutrons for BNCT application at the TRIGA MARK II reactor. Currently, the facility for BNCT are planned to be developed at thermal column. Thus, the main objective was focused on the optimization of thermal neutron and epithermal neutron flux at the thermal column by designing collimator. This paper briefly goes through the characterization of collimator for the ideal neutron collimator design. Several collimators with variable geometries have been designed using Monte Carlo Simulation Codes of MCNPX. The results are then compared in term of thermal neutron, epithermal neutron and gamma fluxes yielded and the ideal collimator then optimize based on aperture size, shielding thickness and collimator condition This research is useful in the selection of neutron collimator design BNCT facility at the TRIGA MARK II reactor.

1. Introduction

BNCT is one of the promising methods in order to cure cancer that used the combination of low energy of neutron (slow neutron) irradiation and the targeting of a tumour site injected with a proper boron containing compound. Basically, the tumour cell was not directly destructed by neutron, but indirectly destroyed by the results of nuclear reaction between neutron and boron [1]. In theory, the BNCT also is one of the forms of radiotherapy which is selectively kills the cancer cells and less effect on other normal cell that used photon that will selectively deposited in tumour cell of Boron carriers such as BPA and BSH [2]. The standard Boron carries compound must be enriched by Boron-10 with about 20% natural abundance. The patient will be irradiated with slow energy of neutron to reach the ratio of high concentration of boron in the tumour cell [3]. The irradiation with slow neutron will cause the nuclear reaction of Boron -10 capture of the thermal neutron and, as a result, of production Boron-10 converts into Boron-11 which then decay into ^{7}Li and emits of an alpha particle [4] that destroy the cancerous cells.

The thermal column of TRIGA MARK II research reactor was believed to supply a sufficient proportion of neutron beam for developing BNCT facility [5]. The thermal neutron flux measurement at the thermal column showed that the thermal neutron flux inside the thermal column was 1.17 x

10^{10} neutron. $cm^{-2}s^{-1}$.which is higher than the IAEA requirement for BNCT (1.0 x 10^9 neutron. $cm^{-2}s^{-1}$) at 1000kW based on previous research [6]. Unfortunately, those thermal neutron fluxes were decreased to 8.58 x 10^8 neutron. $cm^{-2}s^{-1}$ as per measurement at the thermal column door [7]. Thus, the optimization on the thermal neutron flux at the outside section of the thermal column is needed to get the optimize thermal neutron flux inside the thermal column for BNCT facility by designing a neutron collimator.

In order to optimize the neutron beam at RTP for BNCT research facility purpose, the modification for the thermal column must include the designing and installing the neutron collimator [7]. Collimator is well known as a beam limiting device that is widely used in various industries including optical, medical and nuclear [8]. Collimator or beam limiting device acts as a filter of the beam that just allows the beam in a specific direction to travel through. Generally, neutron collimator is composed of the combination of neutron collimating materials, neutron moderator materials and neutron shielding material [8].

2. Design selection

There are several design of collimators with different geometries and materials that have been simulated by using MCNPX Visual Editor as shown in Table 1 and Figure 1. All the collimators have a same total length for the simulation and, only aluminum is fixed as a casing for the collimator because of the low activation energy characteristic as compare to other material and is widely used in nuclear reactor and nuclear applications such as nuclear waste storage and other reactor application [9]. There are two parts of collimator of the thermal column, that is the inner part and outer part. Both parts have different diameter and length, as the inner part has small diameter which is 20 cm with 51.8 cm long while the outer part consist of 24 cm diameter and 54.8 cm of length. In fact, the collimators work based on the principle of radiation interaction with matter, by shifting or filtering methods. The shifting method uses a moderator to lower the neutron energy whereas the filtering method uses materials to absorb neutrons in certain energies [8]. Thermal neutron, fast neutron, and gamma flux were simulated using MCNPX simulation code at the end window of the collimator. The measured flux was divided into two regions which is the centre and outer region. The reason is to identify the difference of the flux values at both regions [10].

Table 1. Summarization of materials used in collimator design.

No. of Collimator Design	Material
D1	HDPE + Aluminium + Lead + 30% BPE
D2	HDPE + Aluminium + Lead + Paraffin
D3	HDPE + Aluminium + Lead
D4	HDPE + Aluminium + Lead + Cadmium
D5	HDPE + Aluminium + Lead + 5% BPE
D6	Paraffin + Aluminium + Lead
D7	Paraffin + Aluminium + Lead + HDPE
D8	Paraffin + Aluminium + Lead + Cadmium
D9	Paraffin + Aluminium + Lead + 5% BPE
D10	Paraffin + Aluminium + Lead + 30% BPE

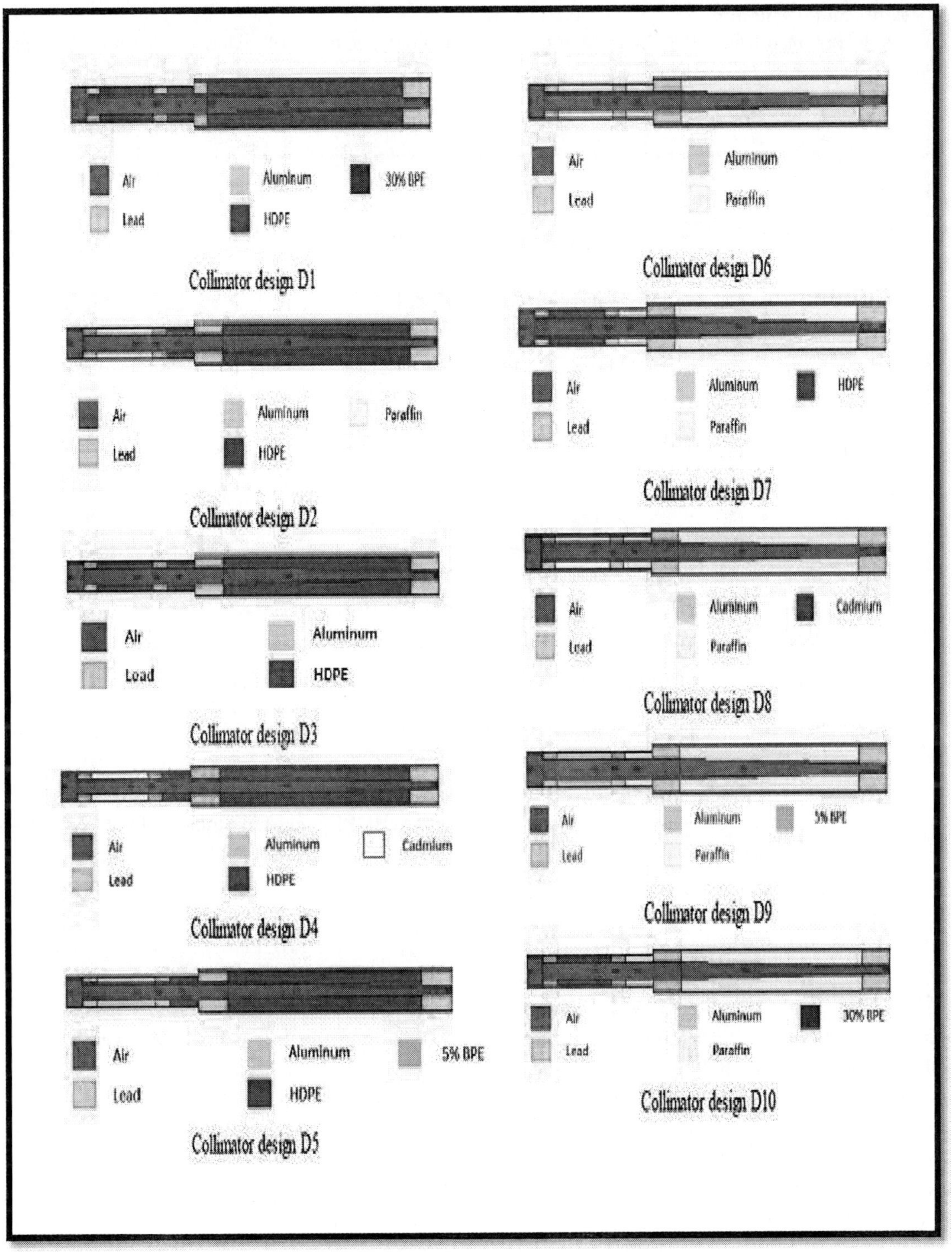

Figure 1. The collimator design D1-D10.

3. Results and discussion

The results obtained from MCNPX code at the centre and outer part of collimator tube were displayed in Figure 2.

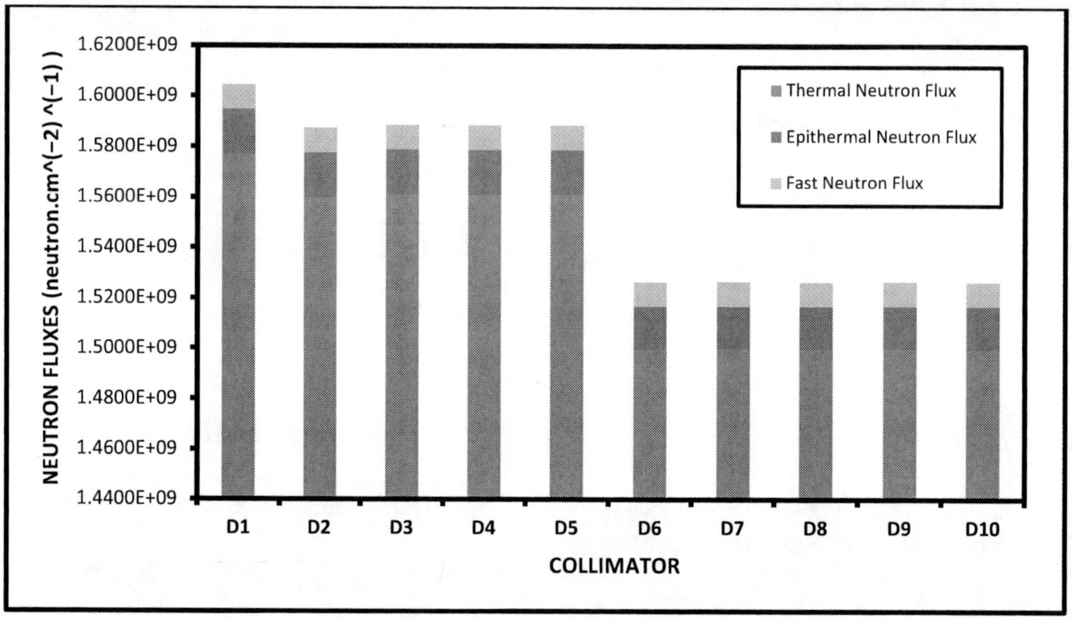

Figure 2. Graph of neutron flux across the beam line of collimator with different material.

Based on the geometries presented in Figure 1, the simulation results show that most of the collimator designs have the same range of neutron flux (thermal, epithermal and fast). For the thermal neutron flux, average flux obtained was 3.75 x 10^9neutron. $cm^{-2}s^{-1}$ and the highest thermal neutron produced was from design D1 that comprise of lead, 30% borated polyethylene, HDPE and aluminium. All the design used HDPE as collimate material (D1, D2, D3, D4 and D5) give simulated result with higher thermal neutron compared to the paraffin-based design (D6, D7, D8, D9 and D10) with 7.8% of relative different between the averages. The presence of boron as the neutron absorber could control the nuclear reactions for all the materials contain boron in the design [11]. The design of collimator using HDPE obtained higher epithermal neutron flux compared to paraffin with 7% of different percentage. For epithermal neutron flux, the used of HDPE on D3 had resulted 1.77 x 10^7neutron. $cm^{-2}s^{-1}$. The strong moderation characteristic from paraffin lead to most of the design used paraffin as neutron collimate material, obtained lower epithermal flux compared to the HDPE. Besides that, the effectiveness of HDPE as fast neutron shielding and moderator was proved on the fast neutron flux measurement. All the design from D1-D5 simulated low fast neutron flux compared to paraffin (D6-D10). In this conjunction, the design D1 was qualified for BNCT collimator based on the MCNPX resulted on highest thermal neutron flux and epithermal neutron flux compared to other designs even though there are less than 1% different of fast neutron with the highest design. The measurement of gamma flux on D1, D2, D3 D4, and D5 have same value of 5.9493 x 10^2gy. $cm^{-2}s^{-1}$ at the outer and 1.4762 x 10^7gy. $cm^{-2}s^{-1}$ for center of collimator. The value of gamma flux for paraffin-based design also all the same from design D6-D10 had obtained highest the gamma flux on outer with 6.1646 x 10^2gy. $cm^{-2}s^{-1}$ and higher for the center of collimator compared to HDPE (D1, D2, D3 D4, and D5) which is negligible.

4. Collimator optimization

In order to select an ideal collimator for BNCT facility at the thermal column, there are several optimizations that were studied in this stage, based on design D1. The optimization of collimator included the aperture of collimator size and the lead thickness. Based on the Table 2, the D1 design still was an ideal collimator for BNCT with the sufficient number of thermal neutron and highly absorbing gamma flux compared to the 5 cm and 10 +5 cm. The 10 cm lead have sufficient quality in shielding gamma but do not give an optimal thermal neutron, as the function of collimator for BNCT is to maximize the thermal neutron. The result is given in Figure 3.

Table 2. The lead thickness result on gamma and thermal from centre and outer.

Lead Thickness (cm)	Thermal Neutron, neuton. $cm^{-2}s^{-1}$		Gamma, $\gamma cm^{-2}s^{-1}$	
	Outer	Centre	Outer	Centre
D1 (5+10)	5.42×10^7	1.58×10^9	5.95×10^2	1.48×10^7
5	3.19×10^7	1.56×10^9	7.81×10^2	1.49×10^7
10	6.47×10^6	3.71×10^8	5.95×10^2	1.48×10^7
10+5	3.20×10^7	1.56×10^9	7.81×10^2	1.49×10^7

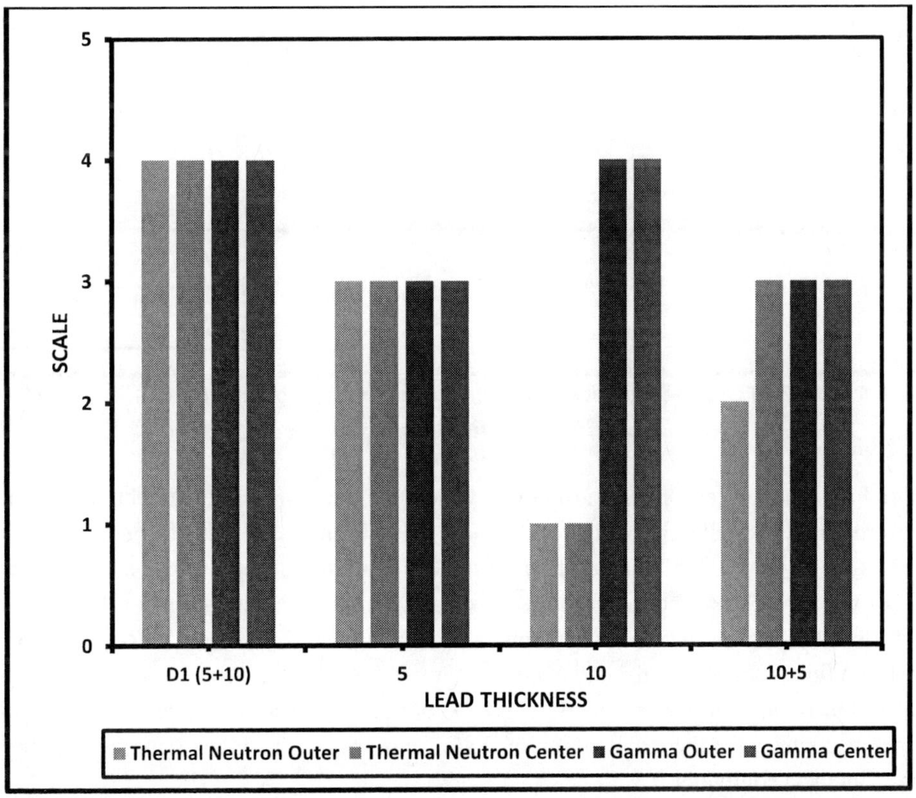

Figure 3. The different lead thickness result on gamma and thermal from centre and outer.

In Figure 3, the scale was 0-4 which means the higher scale, the higher thermal neutron obtained. In contrast, the higher scale for gamma means the higher absorption of gamma for variable thickness. The D1 design still was an ideal collimator for BNCT with the sufficient number of thermal neutron and highly absorbing gamma flux compared to the 5 cm and 10+5 cm. The 10 cm lead have sufficient quality in shielding gamma but not to give an optimal thermal neutron as the function of collimator for BNCT to maximize the thermal neutron production for clinical research study.

In terms of sufficient thermal neutron production, the aperture size was studied in order to see the correlation between the aperture size and thermal neutron production based on simulation result. The result of D1 with 5cm aperture size was compared with different aperture size which is 3 cm, 7 cm and 9 cm. The data was shown on Figure 4.

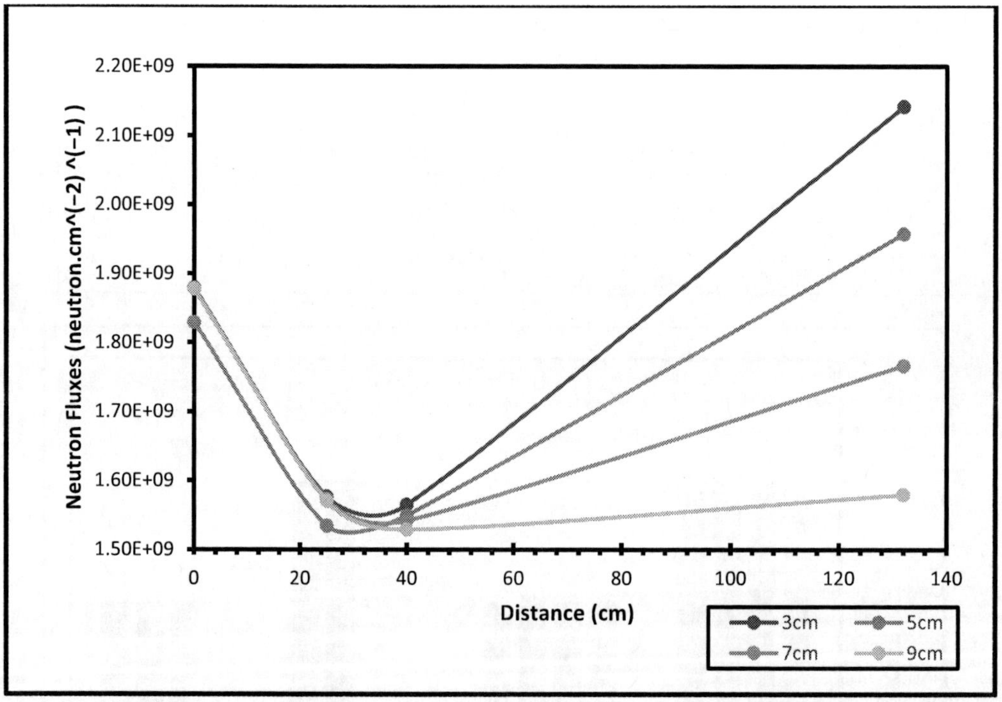

Figure 4. The graphical comparison between thermal neutron flux obtains from different size of aperture along the distance.

Based on Figure 4, neutron fluxes transmitted by the 3 cm of aperture show the higher thermal neutron obtained from MCNP simulation compared to the 5 cm, 7 cm and 9 cm of aperture size for BNCT collimator. At the distance of collimator on 30 cm to 40 cm, all of the aperture size shows the reduction of thermal neutron because of the thermal neutron transmission from the inner collimator (Part 1) to the outside collimator (Part 2). This simulation result shows that the smaller size of collimator aperture, the higher thermal neutron flux produced. At the end of collimator window, all of the thermal neutron was collimating and focusing the beam. The size of aperture gives implication towards of neutron production through collimator. As maximize the thermal neutron for BNCT application, the 3 cm of aperture has been selected to be used as an ideal collimator.

5. Conclusion

The characterization of suitable materials for collimator used in BNCT has been simulated and compared successfully using MCNPX computer code. As result, the collimator D1 which composed of 30% borated polyethylene, aluminum, HDPE and lead materials has been selected as the ideal design of collimator with the optimization of the same thickness and 3 cm of aperture size for BNCT collimator at RTP. The results obtained from this paper are expected to be useful for the construction of the new BNCT facility at TRIGA research reactor around the world.

6. References

[1] Bosko A 2005 General Electric PETtrace cyclotron as a neutron source for boron neutron capture therapy (Texas A&M University).

[2] Gohil C J and Noolvi M N 2015 Selective cancer treatment by Boron Neutron Capture Therapy (BNCT)–a review *Int. J. of Pharmaceutical Chem. and Analysis* **2(3)** p 136-138.

[3] Durisi E, Zanini A, Manfredotti C, Palamara F, Sarotto M, Visca L and Nastasi U 2007 Design of an epithermal column for BNCT based on D–D fusion neutron facility *Nuclear Instr. and Meth. in Phys.Research Section A: Accelerators, Spectrometers, Detectors and Associated Equipment* **574(2)** p 363-369.

[4] Faião-Flores F, Coelho P R P, Muniz R O R, Souza G S, Arruda-Neto J and Maria, D A 2011 Antitumor potential induction and free radicals' production in melanoma cells by Boron Neutron Capture Therapy *Appl. Rad. and Isotopes* **69(12)** p 1748-1751.

[5] Munem, E M E A 2007 Neutron flux measurements with Monte Carlo verification at the thermal column of a Triga mark II reactor: Feasibility study for a BNCT facility.

[6] Solleh M R M, Tajuddin A A, Mohamed A A, Abdel-Munem E, Rabir M H, Karim J A and Yoshiaki K 2011 Collimator and shielding design for boron neutron capture therapy (BNCT) facility at TRIGA MARK II REACTOR *J. of Nuclear and Related Tech.* **8(2**) p41-48.

[7] Solleh M R M 2015 Moderator, collimator and shielding studies for BNCT research at Malaysian Nuclear Agency (Universiti Sains Malaysia: PhD Thesis)

[8] International Atomic Energy Agency 2001 Current status of neutron capture therapy (IAEA: TECDOC-1223).

[9] Chandler 2016 Aluminum used in nuclear reactors and other harsh environments may last longer with new treatment, from https://phys.org/news/2016-03-aluminum-nuclear-reactors-harsh-environments.html (2 Febuary 2017).

[10] Xiang Y, Yi-Guo L, Dan P, Jin L, Gao-Long Z, Hang Z, Ai-Wu Z, Chun-Yang L, and Wan-Jin L 2014 Neutron collimator design of neutron radiography based on the BNCT facility *Chinese Phys. C* **38(2)** p 1-5.

[11] Turgay K A K, Gokhan B, Bunyamin A, Osman G, Aybaba H 2012 Investigation of neutron shielding properties depending on number of boron atoms for colemanite, ulexite and tincal ores by experiments and FLUKA Monte Carlo simulations *Appl. Rad. and Isotopes* **70(1)**. (doi: https://doi.org/10.1016/j.apradiso.2011.09.006) p 341–345.

IOP Conf. Series: Materials Science and Engineering **555** (2019) 012018 doi:10.1088/1757-899X/555/1/012018

Beam characterization of new neutron radiography facility at TRIGA Mark II PUSPATI research reactor

Muhammad Syahir Sarkawi[1, a)], Jasman Zainal[1], Rafhayudi Jamro[2], Muhammad Rawi Mohamed Zain[2], Faridah Mohamad Idris[2], Khair'iah Yazid[2], Arshad Yassin[2]

[1]Department of Energy Engineering, School of Chemical and Energy Engineering, Universiti Teknologi Malaysia, 81310 Johor Bahru, Malaysia
[2]Malaysian Nuclear Agency, 43600 Kajang, Selangor, Malaysia

[a]syahirsarkawi@gmail.com

Abstract. Characterization of neutron beam compositions is important in order to capture great radiographic images. The quality of the neutron beam has significant impacts to the radiographic images produced using that beam. Characterizing of the beam, including determination of the effective value for the thermal neutron content, scattered neutron content, gamma content, and pair production content as well as beam sensitivity, is vital for producing quality radiographic images. This paper provides a characterization of the neutron radiography's beam port (beam port 3) at TRIGA Mark II PUSPATI research reactor (RTP). The experiments to determine the effective beam contents and neutron radiograph sensitivity at the beam are based on American Society for Testing and Materials (ASTM) standards document E545. The Beam Purity Indicator (BPI) and Sensitivity Indicator (SI) are the standard used in these experiments. Neutron radiographic imaging of this standard was performed using direct radiographic method with 0.1 mm thickness Gadolinium (Gd) metal converter and film. Characterization of the beam was carried out for 10, 15, 20, 25 and 30 minutes exposure time at 750 kW reactor operating power. The standard was placed at 110 cm from the beam opening. The current neutron beam performances are compared with previous results.

1. Introduction

Neutron radiography is a non-destructive testing method similar to x-ray, but which uses neutrons as radiation source to expose the film. The quality of the neutron radiographic images and the effective beam contents of the source can be studied using ASTM standards device known as Sensitivity Indicator and Beam Purity Indicator respectively. The ASTM document E545 describes a pair of image quality indicators that have been accepted internationally as a standard for neutron radiography [1].

The BPI device measures the effective beam contents of the source including thermal neutron content, scattered neutron content, gamma content and pair production contribution. It is constructed of four simple materials which the radiographic image of the BPI on the film can give both a qualitative analysis using the human eye for a quick visual check, and a quantitative analysis from measurements using a densitometer.

The SI device has been widely utilized for indicating the sensitivity of details visible on the neutron radiograph [2]. This device has a more complex build as compare to BPI, consisting of four steps of a plastic material, with holes and gaps built in to examine the resolution available. Its values are based on the film reader's ability to see the smallest size hole or gap that can be resolved by the reader.

Both of these devices are used in this study to characterize the new neutron radiography beam at RTP. The image recorder used in this study was AGFA radiographic film together with Gadolinium converter screen of 0.1 mm thickness and aluminium cassette. The characterization of the beam was carried out for 10, 15, 20, 25 and 30 minutes exposure time at 750 kW reactor operating power. The samples were placed at 110 cm from the beam opening which give the L/D ratio at the sample position at 105.

2. Methodology

2.1. Beam Purity Indicator

The purpose of this experiment is to determine the effective beam contents of the collimated beam from beam port #3 at new neutron radiography facility at RTP. This new collimator was design to increase the L/D ratio, decrease gamma content, and has reasonable thermal neutron flux for neutron radiography. The new collimator for this neutron radiography facility is comprehensively discussed in [3]. The BPI is capable to measure the effective beam contents. The percentage of these effective beam contents can be quantitatively determined.

The BPI is made of an 8 mm thick Teflon (26 × 26 mm) plate. It has a central hole of 16 mm in diameter. At the top and bottom of the plate have two pairs of round holes (4 mm in diameter and 2 mm deep) are made to accommodate 2 mm thick boron nitride and lead discs. Square grooves (2 × 2 mm^2) are made to accommodate 12 mm long square (2 × 2 mm^2) cadmium rods. The cadmium rods included in the BPI are used to provide an indication if inherent beam resolution or sharpness. The BPI schematic design is shown in figure 1. The BPI schematic design is explained well in [4].

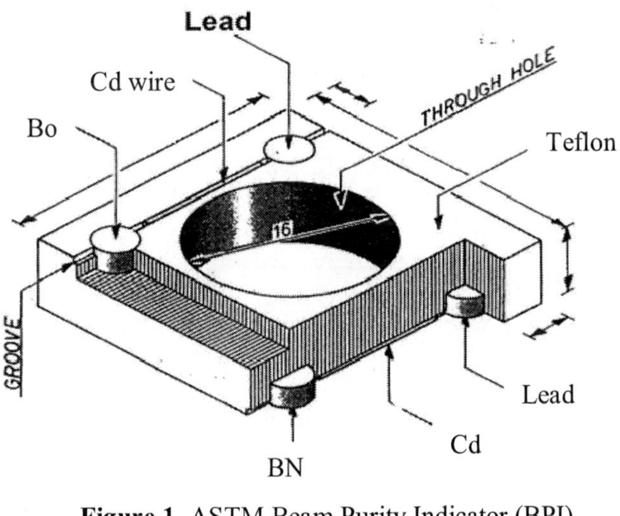

Figure 1. ASTM Beam Purity Indicator (BPI).

The experiments were conducted with exposure time of 10, 15, 20, 25 and 30 minutes. The BPI was attached to the film cassette with aluminium tape and exposed to the radiations beam with AGFA radiographic film embedded with Gadolinium foil. The position of upper and lower Boron Nitrate disc, Lead disk and Cadmium wire was well defined. After finished the exposure, the radiographic film was processed manually in the dark room. Next, the optical density of the BPI images was visualized and measured at six measuring location using X-Rite 361T optical density measurement. The average value of optical density was recorded. The percentage of the effective beam contents and

IOP Conf. Series: Materials Science and Engineering **555** (2019) 012018 doi:10.1088/1757-899X/555/1/012018

interaction characteristic of neutron and gamma recorded on the radiographic film was calculated using the ASTM equations as follows:

$$Themal\ neutron\ contents, C = [\frac{D_5 - (higher\ value\ of\ D_1\ or\ D_2)}{D_5}] \times 100 \quad (1)$$

$$Scattered\ neutron\ contents, S = [\frac{(D_1 - D_2)}{D_5}] \times 100 \quad (2)$$

$$Gamma\ contents, \gamma = [\frac{D_6 - (lower\ value\ of\ D_3\ or\ D_4)}{D_5}] \times 100 \quad (3)$$

$$Pair\ production, P = [\frac{(D_3 - D_4}{D_5}] \times 100 \quad (4)$$

From the neutron radiographs of the BPI, the following film densities are to be measured:

D_1 – Density under the lower boron nitride disc.
D_2 – Density under the upper boron nitride disc.
D_3 – Density under the lower lead disc.
D_4 – Density under the upper lead disc.
D_5 – Density under the centre of the hole.
D_6 – Density under the Teflon body.

2.2. Sensitivity Indicator

The purpose of the sensitivity measurement is to determine the details visible to the neutron radiography system by evaluating the neutron radiographic image of the sensitivity indicator. This device is one of several devices used for qualitative determination of the sensitivity of detail visible on a neutron radiograph.

The SI is a step-wedge device containing gaps and holes of known dimensions. This device has 4 shims and it has 4 holes which marked as A, B, C and D. Shim D is covered with lead and shim A, B and C are covered with cast acrylic resin. The thickness of the shim and its diameter for shim A, B, C and D are 0.13 mm, 0.25 mm, 0.51 mm and 0.25 mm respectively. The cast acrylic resin step separated by aluminum spacers with gap size marked as G. The G values are 0.25 mm, 0.13 mm, 0.10 mm, 0.076 mm, 0.051 mm, 0.025 mm and 0.013 mm. The SI must be enclosed with aluminum cover to protect from dust. The SI schematic design is shown on figure 2. The SI schematic design is comprehensively discussed in [2].

The SI was exposed together with BPI for the same exposure time. This pair of image quality indicators can give a basis for evaluating both the facility at which the film was radiograph, and the resolution of the film itself. The BPI and SI was placed no less than one inch from any edge of the film to avoid the edge effects of captured gamma from the conversion screen. Same as BPI, the orientation of the SI was well defined. The thickest step in the wedge should be placed away from the BPI. The resolution of the film can be subjectively determined by examination of the holes and gaps. The visual examination of the neutron radiograph of the sensitivity indicator, the G and H values was recorded.

The experimental setup for both BPI and SI are shown in figure 3. The exposure factors for this experiment including the diameter of the beam inlet aperture, the distance between inlet aperture to sample, L/D ratio, type of film, exposure time and type of converter are stated in table 1.

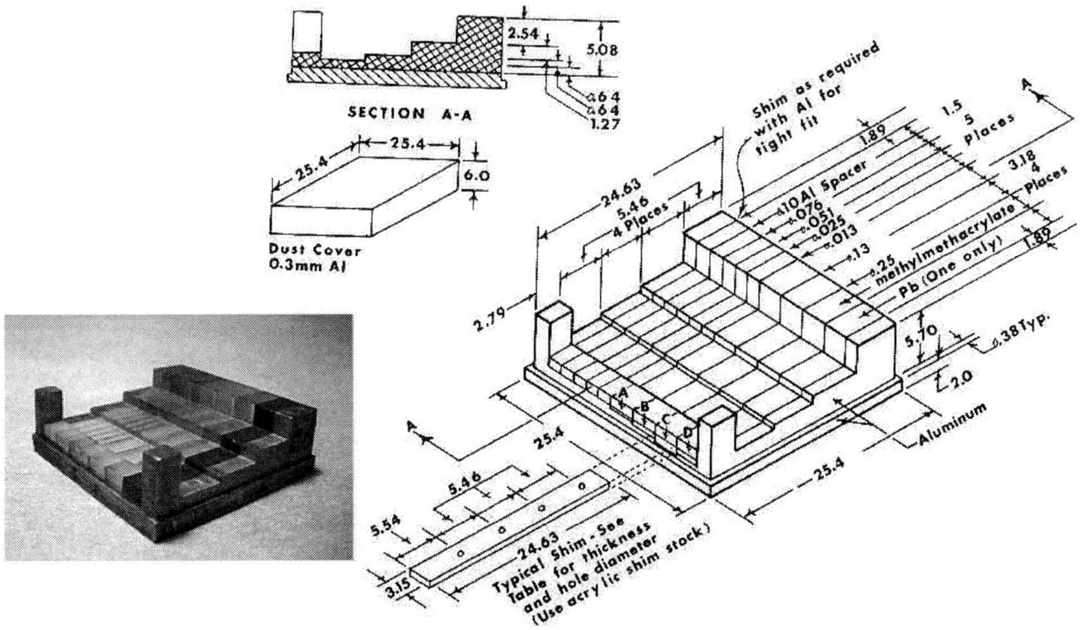

Figure 2. ASTM Sensitivity Indicator (SI).

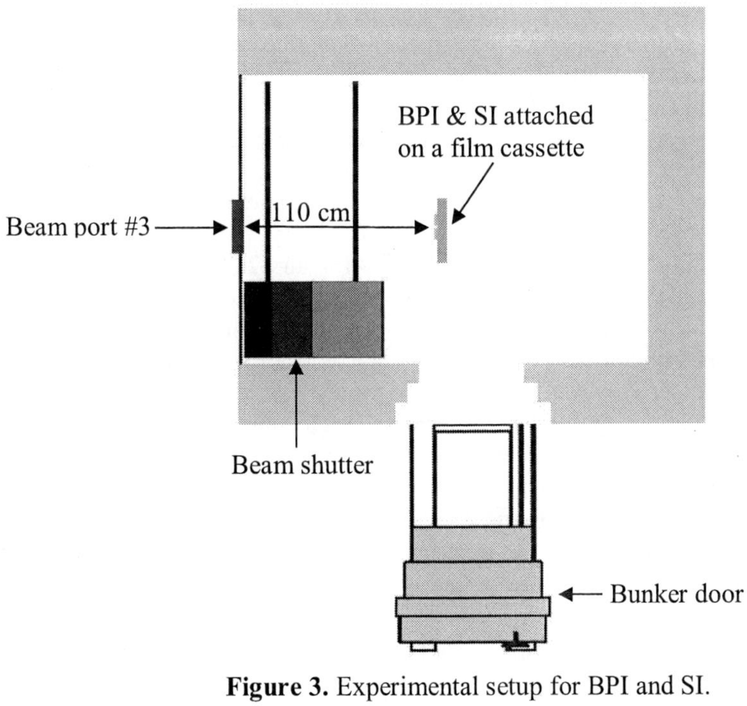

Figure 3. Experimental setup for BPI and SI.

Table 1. The exposure factors for the neutron radiograph.

Factors	Values
Diameter of the beam inlet aperture, D	3 cm
The distance between inlet aperture to sample, L	314.4 cm
The L/D ratio	105
Type of neutron converter	Gadolinium foil with 0.1 mm thickness
Type of film	AGFA radiographic film
Exposure time	10, 15, 20, 25, 30 minutes

3. Results and Discussion

3.1. Beam Radiation Composition

The effective beam contents in the new collimated beam of the neutron radiography facility at RTP are shown in table 2.

Table 2. Radiation contents in the new collimated beam at NR facility at RTP.

Exposure time	10 minutes	15 minutes	20 minutes	25 minutes	30 minutes
Thermal neutron contents, C	22.73%	50.00%	60.23%	61.86%	63.43%
Scattered neutron contents, S	4.55%	4.41%	2.27%	2.06%	2.24%
Gamma contents, γ	0.91%	4.41%	1.14%	1.03%	1.49%
Pair production, P	2.73%	0.00%	2.27%	1.03%	1.49%

From the table 2, it is shown that the scattered neutron contents, gamma contents and pair production (secondary radiation) have lower percentage when the exposure time higher than 20 minutes. Meanwhile, the thermal neutron contents are increased as the exposure time increases because more thermal neutron can be converted and captured by the film when longer exposure time was used. The thermal neutron content is within in an acceptable range between 50.00% to 63.43% as per ASTM [1] confirming that new neutron radiography facility at RTP is in good working conditions. Besides, the 10 minutes exposure time shows it does not have sufficient thermal neutron contents for conducting neutron radiography. The highest value of secondary radiation of scattered neutron, gamma rays and pair production are found to be 4.55%, 4.41% and 2.73% respectively. This results showed that, the secondary radiation is in acceptable range as recommended by ASTM [1].

Figure 4 and figure 5 shows the comparison of the effective beam contents obtained from previous NR facility [5] with the results obtained from new NR facility. From figure 4, the graph showed that the thermal neutron contents from new neutron radiography facility have lower percentage as compared to old neutron radiography facility, especially at 10 minutes exposure time. There are several factors that might contribute in decreasing the thermal neutron contents in new neutron radiography facility. The first factor is smaller beam inlet aperture used in the new collimator [3]. New neutron radiography collimator was design to increase the L/D ratio of the facility, hence the design is made in such a way that it has small beam inlet aperture, D and high length of L. This relation can be expressed with equation below:

$$\frac{Flux\ at\ enterance}{Flux\ at\ exit} = \frac{\phi_0}{\phi} = 16\left(\frac{L}{D}\right)^2 \qquad (5)$$

From the equation, it is clear that neutron flux/neutron contents at the exit of the collimator drops with increasing collimator length L, and rises when the inlet aperture, D is made larger. The other factor that might contribute in decreasing the thermal neutron contents is fuel burnup in the reactor core. However, the thermal neutron contents in new NR facility at RTP still in acceptable range as per ASTM. In the other hand, figure 5 showed that new neutron radiography facility has slightly better performance in managing the secondary radiation especially at 25 minutes exposure time.

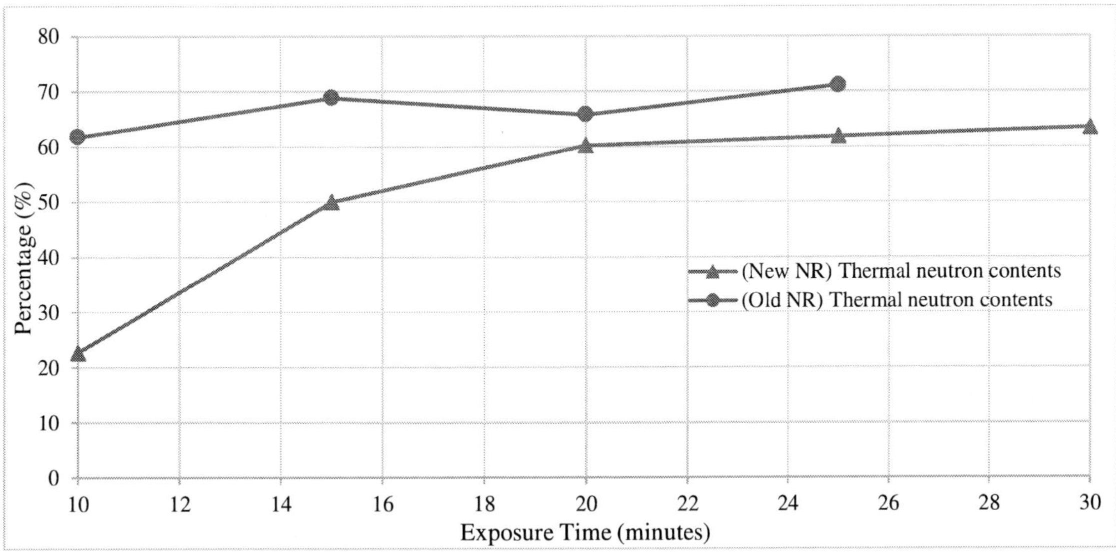

Figure 4. Comparison of thermal neutron contents of new neutron radiography facility with old facility.

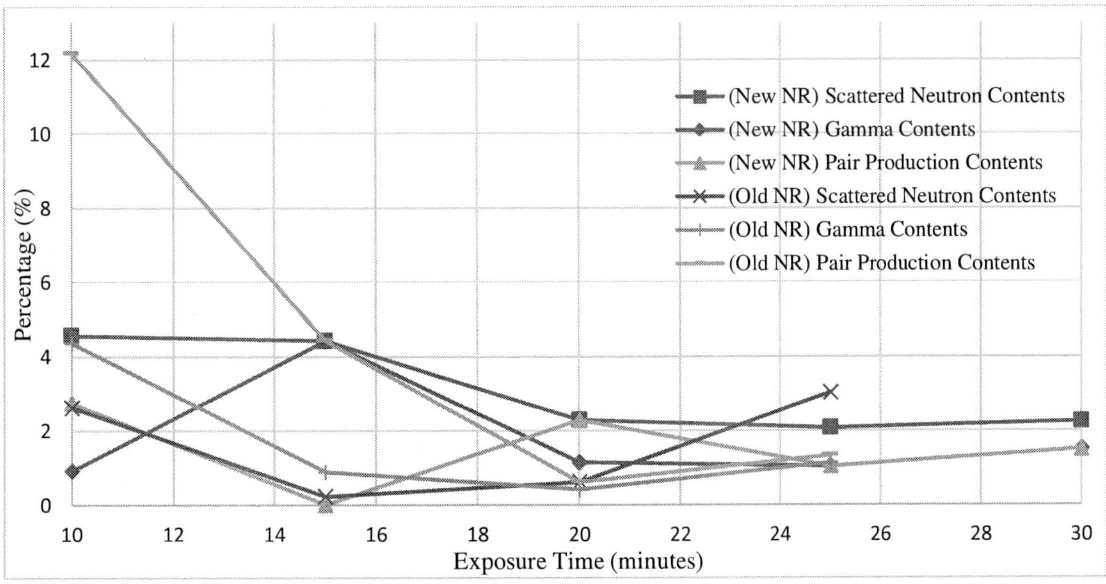

Figure 5. Comparison of secondary radiation composition of new neutron radiography facility with old facility.

3.2. Radiographic Sensitivity

As discussed earlier, the visibility of all holes (H) and gaps (G) on the neutron radiograph of SI device is important criteria to determine whether the system can produce good quality radiograph or not. It was observed that the maximum holes that can be seen in the radiograph was 4 H. The 4 H was visible at exposure times of 20, 25, 30 minutes, while 15 minutes exposure time only have 3 visible H and lastly 1 H for 10 minutes exposure time. These results showed that the radiograph did not achieve the minimum value of H for good quality radiograph. Good quality radiograph should obtain at least 6 H. However, all of the 7 G can be clearly seen on all of the radiograph. These present results of G and H measurement are shown in Table 3 and Table 4 respectively. As compared to previous study by [5], the old NR managed to produce radiograph with 6 H and 7 G at exposure time of 10, 15, 20, and 25 minutes. This results were due to slightly lower thermal neutron contents as compared to previous facility.

Table 3. Radiographic sensitivity (Gap)

Exposure time (minutes)	**Visible gaps (G)** **Spacer** (mm)						
	0.25	**0.125**	**0.0125**	**0.025**	**0.050**	**0.075**	**0.10**
10	yes	yes	yes	yes	yes	yes	yes
15	yes	yes	yes	yes	yes	yes	yes
20	yes	yes	yes	yes	yes	yes	yes
25	yes	yes	yes	yes	yes	yes	yes
30	yes	yes	yes	yes	yes	yes	yes

Table 4. Radiographic sensitivity (Hole)

Exposure time (minutes)	**Number of visible holes (H)**
10	1
15	3
20	4
25	4
30	4

4. Conclusion

The effective beam contents and image quality of the new neutron radiography facility at RTP are obtained from this study. The results have showed that the new neutron radiography facility at beam port #3 at RTP is capable for neutron imaging system in non-destructive application. However, the quality of the radiographic image can be improved in order to meet the minimum requirement for best neutron radiography facility. Based on the results, it showed that this new facility can produce radiographic image with excellent contrast, but has slightly lower sharpness. This experiment can be extended with different L/D ratio for future studies.

5. References

[1] ASTM International 2014 *Standard Test Method for Determining Image Quality in Direct Thermal Neutron Radiographic Examination* (West Conshohocken, PA: E545-14 A).
[2] ASTM 2014 *Standard Practice for Fabrication of Neutron Radiographic Sensitivity Indicators* (E2023-10).

[3] Jamro R *et al.* 2017 *Monte Carlo Simulation for Designing Collimator of the Neutron Radiography Facility in Malaysia.* Phys. Procedia **88** (Supplement C) p 361-368.

[4] ASTM International 2014 *Standard Practice for Fabrication of the Neutron Radiographic Beam Purity Indicators* (West Conshohocken, PA: E2003-10 A).

[5] Jamro R *et al.* 2008 Characteristic of radiation beam composition of neutron radiography facility at TRIGA Mark II.

Acknowledgements

The experiment and instruments used in this paper were performed at Malaysian Nuclear Agency, Bangi, Malaysia. The author would like to acknowledge all the contributors in completing this paper.

IOP Conf. Series: Materials Science and Engineering **555** (2019) 012019 doi:10.1088/1757-899X/555/1/012019

The design of a Hot Cell with interlocking concrete wall

Muhammad Hannan Bahrin[1,a)], Hasni Hasan[1], Anwar Abdul Rahman[1], Mohd Arif Hamzah[1], Mohd Zaid Hassan[1], Mohd Rizal Mamat@Ibrahim[1], Mohamad Hairie Rabir[2], Julia Abdul Karim[2]

[1]Prototype and Plant Development Centre (PDC), Technical Services Division (BST), Malaysian Nuclear Agency, 43000 Kajang, Selangor, Malaysia
[2]Reactor Technology Centre, Technical Services Division (BST), Malaysian Nuclear Agency, 43000 Kajang, Selangor, Malaysia

[a)]hannan@nm.gov.my

Abstract. The Prototype and Plant Development Center (PDC) of Malaysian Nuclear Agency is developing a hot cell facility which is meant for research activities such as spent fuel inspection, fuel fabrication and post irradiation material examination/ inspection. The hot cell is of alpha-gamma shielding hot cell type. The size of the hot cell is 4.4 x 3.9 x 3.825 meter. The 1-meter thickness hot cell wall consists of 89 pieces of interlocking curve design concrete. The 1-meter hot cell roof consists of 41 pieces of square shape concrete. The density of all concrete blocks is 2350kg/m^3. The hot cell is designed to handle up to 227.5 Ci ^{60}Co activity. This paper will discuss the detailed design of the shielding wall, as well as the shielding evaluation/analysis against gamma and fission products of 50% burnup RTP 20wt% fuel type using MCNP.

1. Introduction

Isolation of radioactive material is the best method to protect the operator against hazardous radiation. This isolation is achieved by containing the radioactive materials in an enclosure, which can be in the form of a fume cupboard, glove box, shielded box or hot cell. A shielded enclosure fitted with suitable manipulation systems to allow the performance of operations without subjecting the operator to radiation beyond permissible dose rate is defined as a hot cell [1].

There are many types of hot cell such as alpha–gamma cells, beta–gamma cells, and alpha-beta-gamma cells. These hot cell types are identified by its biological shielding against alpha, beta, gamma and neutron radiation. In addition, the infrastructure of hot cells renders them capable to handle safely large quantities and types of hazardous radio-isotopes inside the enclosed space.

In order to design a hot cell, several specifications are required to be defined such as the type of hot cell, the types of radioactive material to be handled, the expected level of activity to be handled, types of operations to be performed, the viewing system, shielding material selection, size of hot cell and etc.

At Malaysian Nuclear Agency, the existing semi-permanent hot cell is used for radioisotope production such as Technetium-99m (^{99m}Tc), Iodine-131 (^{131}I) and also for research works. Lead with chevron brick-built wall design is used as biological shielding [2]. On the other hand, Malaysian Mobile Hot Cell (MHC) is used to handle and manage high activity disused sealed radioactive sources

(DSRS) such as irradiators and teletherapy heads [3]. River sand with a thickness of 1.5meter is used as biological shielding as it is easily available anywhere and can be procured nearby in the locality of MHC.

In general, a permanent hot cell facility is built next to the nuclear research reactor facility such as hot cell facility in Japan Material Testing Reactor (JMTR), Chalk River Laboratories (CRL) and HANARO (KAERI). These hot cells have been used for research activities such as spent fuel inspection, post irradiation material examination (PIE) and fuel fabrication. However, in Malaysia there is no hot cell facility that is capable of doing such activities. Thus, this paper will study the suitable design of hot cell to support research activities in Malaysian TRIGA PUSPATI Reactor (RTP).

The semi-permanent concept is selected due to the ease/ possibility of assembling and dismantling work within the Nuclear Malaysia site. Currently, no specific site has been selected yet but the possible site is Reactor Hall or "Khidmat Pembangunan" Building. The hot cell also can be assembled near the Thorium Lab if required depending on the user's requirement. The basic design used is the Chevron brick-built wall and curved brick-built wall, the typical interlocking system in designing a semi-permanent hot cell.

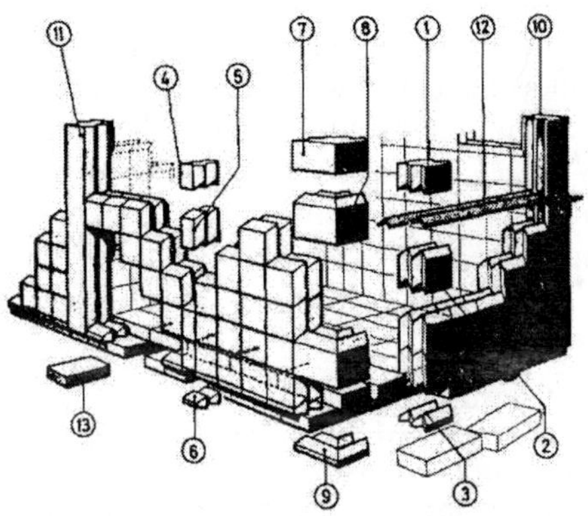

Figure 1. Chevron brick-built wall [1].

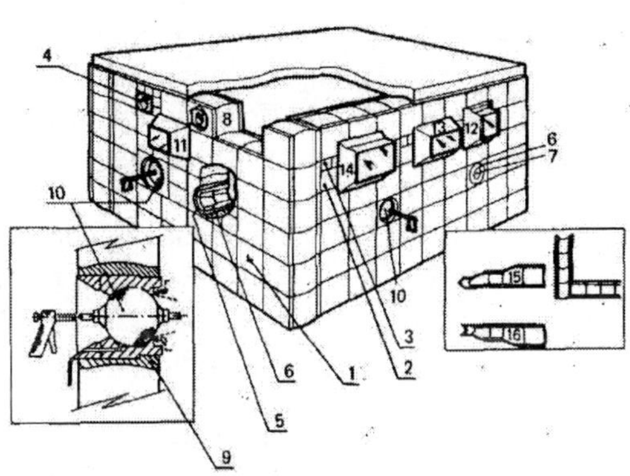

Figure 2. Curved brick-built wall [1].

The main purpose for developing the hot cell is to provide the fuel fabrication preparation for thorium fuel research. The typical type of radiation emitted by thorium isotope are alpha particles and weak gamma rays [4]. Candidates of thorium fuel can be as Th-MOX pellets, Thorium oxide (ThO2) pellets, fuel in a molten state such as solvent mixtures of fluoride salts such as LiF-NaF-BeF2, LiF-NaFRbF,and LiF-BeF2 and etc [5].

According to H. M Glen [6], alpha-gamma hot cell is an enclosed space, in which radioisotopes are handled safely. The enclosing structure is a gamma-ray shield and is sealed to protect the exterior area from alpha contamination [6]. Master slave manipulator will be used to handle the processing or experimental activities remotely. The equipment setup and maintenance are done partially and manually by personnel entering into the hot cell once it is confirmed that hazardous/ radioactive material has been removed. In this study, Alpha-gamma shielding hot cell type with interlocking system design is selected.

2. Methodology

2.1. Radioactive handling capacity

The expected radioactive handling capacity value of hot cell is defined based on the literature of other hot cell designs. For example, the Process Mechanical Cell (PMC) and General-Purpose Cell (GPC) at West Valley Demonstration nuclear fuel reprocessing plant recorded a general dose rates of 100 to 300

IOP Conf. Series: Materials Science and Engineering **555** (2019) 012019 doi:10.1088/1757-899X/555/1/012019

R/hr up to 2000 R/hr hot spots [7]. In addition, other values from studies of RTP spent fuel are also taken into consideration as a future planning on research of RTP spent fuel inspection [8][9]. Thus, the value of 300 R/hr (equivalent to 227.5 Ci of Cobalt-60 (^{60}Co)) and fission products of 50% burnup RTP 20wt% fuel type are used to simulate the hot cell shielding design. The target operating hour for the hot cell is 2000 hour per year. The design of hot cell will ensure that the hot cell operator is safe and the operator will not receive dose higher than 20mSv per year, following the Radiation Protection (Basic Safety Standards) Regulation 1988.

2.2. *Shielding material*

It is very important to select a suitable shielding material for hot cell walls, viewing window and door as it will determine the hot cell capability in handling high activity sources. The common shielding materials of hot cell wall are concrete, lead and steel. In JMTR hot laboratory, cells with different shielding materials are used to conduct different post irradiation examination (PIE) activities, accordingly [10]. The dismantling of irradiated capsules and examination of fuels are conducted in concrete cells. Therefore, concrete is selected as a shielding material for the hot cell wall as it is cheaper and easy to construct. The normal concrete density is about 2300 ~ 2400kg/m^3. The high density concrete will provide better radiation shielding based on the concrete radiation linear attenuation coefficient (μ) value.

In order to design an interlock concrete wall, the suitable concrete grade should be selected according to the concrete grade application. It is important as the concrete does not just act as hot cell wall shielding, but it must also be able to withstand the load in interlocking concrete design of the hot cell. Thus, concrete grade 40 with density of 2350kg/m^3 is selected as shielding material for the hot cell wall as it provides a good radiation shielding and is suitable for interlocking concrete design. Table 1 shows the concrete grade in construction in Malaysia [11]. Table 2 shows the linear attenuation coefficient of different concrete grade [12].

Table 1. Concrete Grade in Construction.

Grade No.	Application
7~10	Concrete without reinforcement
15	Lightweight concrete
20~25	Normal concrete
30	Post stressed concrete
40~60	Pre-stressed concrete
70~100	High performance concrete

Table 2. Linear attenuation coefficient (μ) of different concrete grade.

Radioactive material	Linear attenuation coefficient (cm^{-1})					
	Grade 15	Grade 20	Grade G25	Grade G30	Grade G35	Grade G40
Co 60	7.78	7.62	6.79	6.73	7.22	7.0
Cs 137	25.87	25.31	22.57	22.36	23.99	23.26

The shielding window provides a short observer - to - object distance which permits good depth perception. There are two types of transparent viewing windows - solid type and liquid filled type [13]. The solid type window such as lead glass and phosphate glass are susceptible to discoloration due to irradiation. In addition, solid type is heavy and brittle, causing difficulties on mobility. Thus, liquid shielding is selected for hot cell with interlocking wall. Liquid shielding provides advantages such as easy to be replaced, recovered and transported. However, extra care is needed to prevent leakages. After considering the cost and related problems on handling liquid shielding material, colourless zinc bromide ($ZnBr_2$) 77% w/w Nuclear Optical Grade with density of 2.55 g/ml @ 20°C is chosen as the most practical shielding liquid for viewing window.

Lead door is selected for the hot cell interlocking wall design because of its easiness of handling and the size of the door. The door is designed with lead and layered with boronated HDPE which act as gamma and neutron shielding, respectively. The lead density is 11.35g/cm^3 and boronated HDPE density is 1.08g/cm^3. Boronated HDPE is used in the design of hot cell door for future upgrading as neutron shielding in the hot cell. The attenuated intensity of the gamma radiation that passes through a shielding material can be quantified according to the Lambert's law:

$$I = I_0 e^{-\mu r} \tag{1}$$

where I_o is the incident radiation intensity, I is emerged radiation intensity that passes through the shielding material of thickness r and μ is the linear attenuation coefficient. Based on the calculation, the minimum required concrete wall thickness for 227.5 Ci of Cobalt-60 (^{60}Co) source activity is 1-meter thickness. The door thickness is 250mm with 200mm lead thickness and 50mm boronated HDPE.

2.3. Hot cell design and limitation

The containment size of the hot cell is 2.4 meter x 1.9 meter x 2.825 meter (height). The size is suitable for research activities of the designed hot cell. The space available in the reactor hall is about 6-meter square area. The overhead crane available in the RTP hall is 3 tonnes. By considering safety factor of lifting: 2.0, the maximum weight of each interlocking concrete shall not be more than 1.5 tonne. In addition, lifting plug at each interlock concrete shall be able to withstand the concrete weight during lifting.

During fabrication of the interlock concrete, the lifting plug shall be embedded between reinforced steel bars (rebar) in the concrete. The rebar position shall not be in line with shielding thickness as it will allow radiation streaming through the gap between rebar and concrete. Thus, deformed type of reinforcement steel bar is advised.

In order to limit the numbers of interlock concrete, the concrete is designed as big as possible. Thus, 1-meter concrete thickness is selected in designing the interlock concrete. Two layers of interlocking concrete wall are not advised as it will lead to a bigger number of interlock concrete. Curved brick-built wall design is selected as interlocking concrete design as it is much easier to produce concrete formwork using steel compared to chevron brick-built wall design. Overlapping layers of concrete is selected in designing the hot cell concrete roof. At the concrete roof, there is a top loading chute for source loading from transfer cask into the hot cell. This chute is designed based on the existing transfer cask in RTP. Figure 3 shows the conceptual hot cell design with interlocking concrete walls.

2.4. Monte Carlo N-Particle Transport Code (MCNP)

Detail radiation simulation is performed by using the MCNP5 Monte Carlo code. All shielding inputs are taken into account in the simulation as well as using ENDF/B-VII libraries. These inputs are used for all interactions in simulation. The 227.5 Ci ^{60}Co has 2 high energy gammas, with a total of 1.6835x10^{13} gammas per second. The gamma intensity of 20wt% fuel type with burnup around 50% is 1.56x10^{14} gammas per second. This corresponds to the low energy gamma of 0.5MeV and below [14]. The doses are calculated by means of cell and mesh tally, with flux-to-dose conversion factors, using the data sets of ICRP-21. The defined model and source are shown in Figure 4. The dose-rates inside and outside of the concrete wall are calculated using two mesh tally cases. The first one is axial

IOP Conf. Series: Materials Science and Engineering **555** (2019) 012019 doi:10.1088/1757-899X/555/1/012019

direction (XZ view) and the second one is radial direction (XY view). The MCNP mesh tally results are normalized to the gamma intensity or neutron intensity from the source.

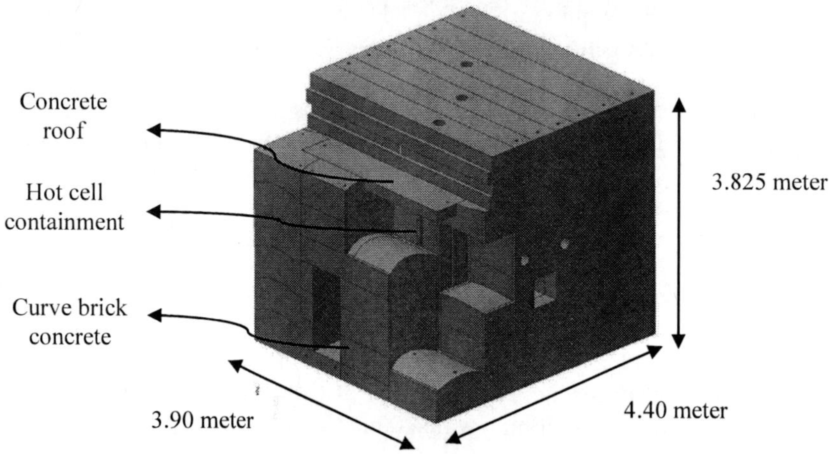

Figure 3. The conceptual hot cell design with interlocking concrete wall.

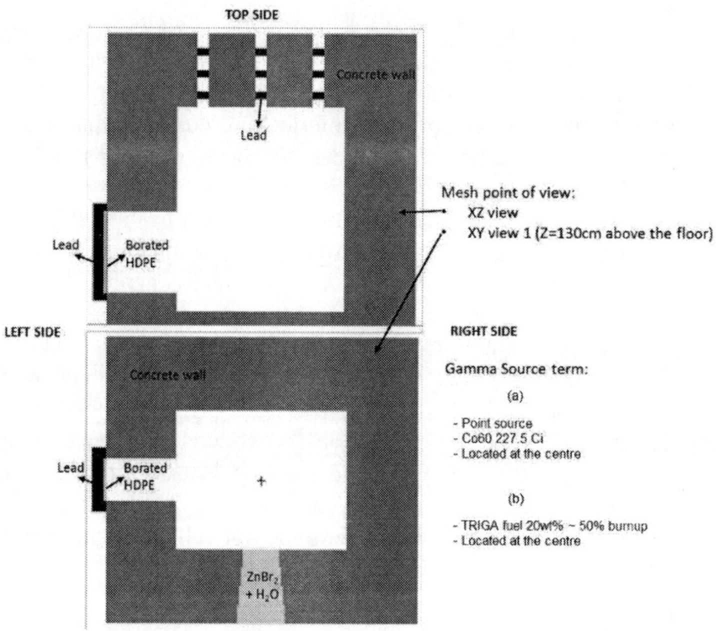

Figure 4. The MCNP model (a) The defined 227.5Ci of Co60 gamma used as source-term, (b) The defined fission product's gamma used as source-term from RTP fuel; and the mesh tally specification.

3. Result and discussion

The overall size of hot cell is 4.4 meter x 3.9 meter x 3.825 meter (height). There are 89 pieces of curve brick concrete with 25 different types of shape, designed for the hot cell wall. The wall thickness is 1 meter, with curve interlocking concrete design as shown in Figure 5. In addition, there are 41 pieces of 24 different types square shape concretes, designed for the hot cell roof. The maximum weight of one piece of concrete is 1.25 tonnes. Thus, each concrete lifting plug is designed for lifting load of up to 2.25 tonnes weight.

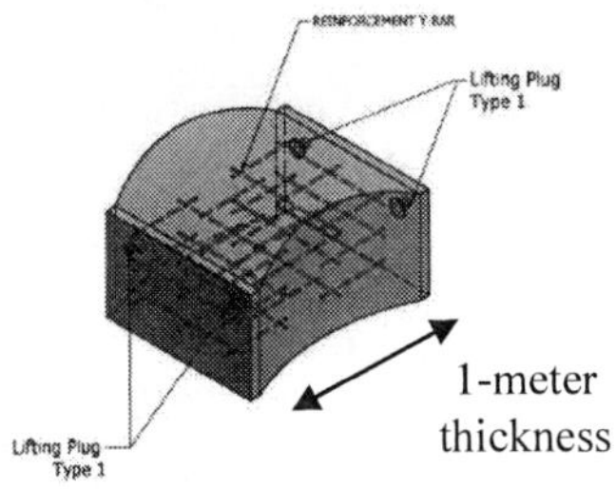

Figure 5. Curve brick interlock concrete.

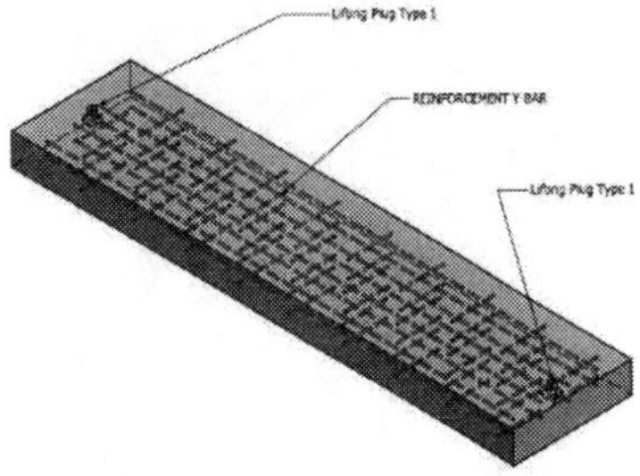

Figure 6. Concrete roof.

The results of the radiation simulations are shown in Figure 7 to Figure 10, the 1-meter concrete shield shows that it is capable of reducing the dose-rate from the inside the wall to the outside surface. The dose-rate at the outer surface of the concrete wall is reduced approximately more than 10 million times when compared to the inner wall surface, the reduced values are 12.1 μSv/hr for gamma source and 0.65 μSv/hr for RTP fuel fission product source. Similar reduction of dose rate is observed at the $ZnBr_2$ windows, both for gamma and RTP fuel fission products. The highest dose rate value in front of $ZnBr_2$ window for gamma and RTP fuel fission product are 10.1μSv/hr and 2.04μSv/hr, respectively. However, the result shows that a relatively high radiation leaked through the door. The highest dose rate, located at lead door are 1570 μSv/hr and 489 μSv/hr for 227.5 ^{60}Ci Co source and RTP fuel fission product, respectively.

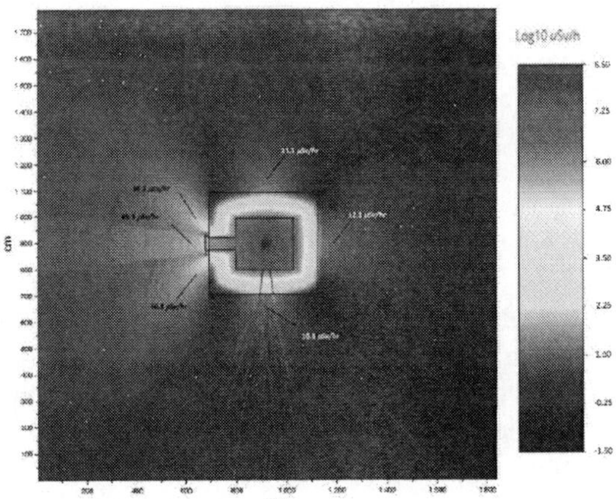

Figure 7. XY axis view, dose-rate map results for 227.5Ci Co60 gamma source.

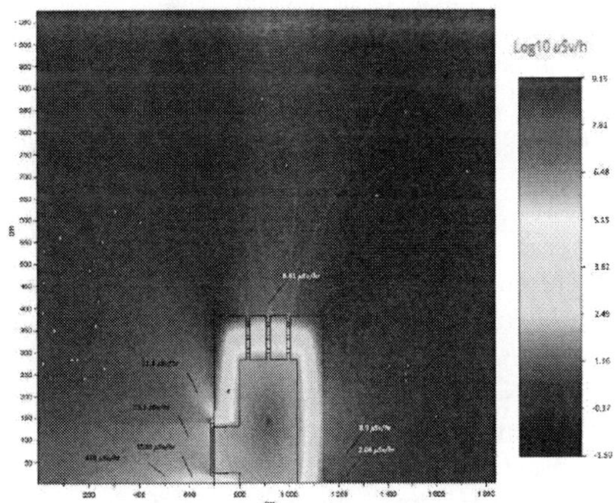

Figure 8. XZ axis view, dose-rate map results for 227.5Ci Co60 gamma source.

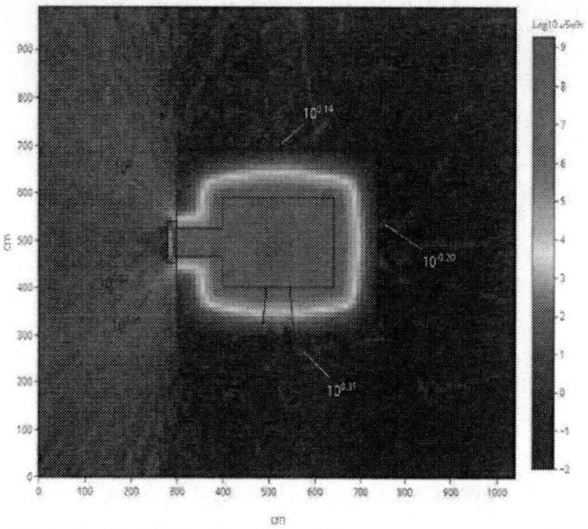

Figure 9. XY axis view, dose-rate map results for fission product's gamma used as source-term from RTP fuel.

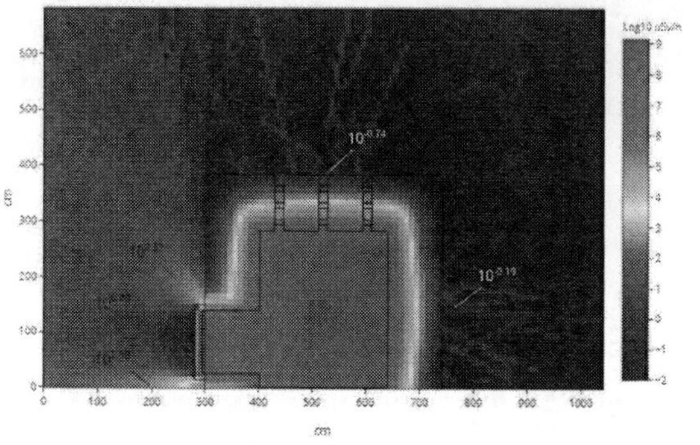

Figure 10. XZ axis view, dose-rate map results for fission product's gamma used as source-term from RTP fuel.

Scattered radiation are observed at the left and right side and also at the upper and bottom side of the door. The low energy gamma, dominant from fission products seems to be stopped by the door. However, the scattered gammas, even the low energy gammas could also escape through the left, right, top or bottom side. Therefore, additional shielding or modification is needed in the future especially at the gap between interlock concrete and the lead door in order to further reduce the dose rate.

4. Conclusion

Through the evaluation, the design of interlocking concrete wall with 1-meter thickness are able to act as a shielding for 227.5Ci ^{60}Co and fission products of 50% burnup RTP 20wt% fuel. However, a set of measurement data is still needed to verify the percentage of discrepancy for the simulation in future. This study proves that the hot cell with interlocking alpha-gamma shielding design, installed in controlled area or RTP Hall is fit to handle research activities.

5. References

[1] International Atomic Energy Agency (IAEA) 1981 *Manual on Safety Aspects of The Design and Equipment of Hot Laboratories* Procedures and Data, Safety Series **30**.

[2] M.N. Radioisotope Production/Supply, Available from: https://www.nuclearmalaysia.gov.my/new/PnS/services/medical/radioIsoProd.php

[3] Bahrin M H 2016 Stress analysis for wall structure in mobile hot cell design *AIP Conf. Proc.* **1704**, https://doi.org/10.1063/1.4940113 p 050017

[4] Radionuclide Basics: Thorium (U.S. Environmental Protection) Available from: https://www.epa.gov/radiation/radionuclide-basics-thorium

[5] Sipaun S 2017 Thorium fueled reactor *AIP Conf. Proc.* **1799** https://doi.org/10.1063/1.4972946 p 050012

[6] Glen H M 1964 A General Purpose Alpha Gamma Hot Laboratory, Nuclear Structural Engineering 1 (ed 1965) (North-Holland Publishing Comp., Amsterdam) p 98-107.

[7] Drake J 2000 Cleanup of the Head End Cells and the Refurbishment of Shield Windows at the West Valley Demonstration Project *WM'00 Conference* February (Tucson, AZ: 27 – March 2, 2000)

[8] Ab Rahim A Nabil 2012 Structural Strength Analysis of Lug Component for RTP Fuel Transfer Cask, *Seminar RND Nuklear Malaysia.*

[9] Usang M D 2014 Estimated Reactivity Increase of Spent Fuel Pool Criticality during Transfer of Fuel Element *Seminar RND Nuklear Malaysia*.

[10] Shibata A 2014 Current status and developments of post irradiation examination techniques in JMTR Hot laboratory *The 7th Int. Symp. Material Testing Reactors (ISMTR-7)* (Otwock-Świerk, Poland: 20-23, October, 2014).

[11] Ismail M P 2012 Concrete mix design for X-and gamma shielding *Seminar RND Nuklear Malaysia.*

[12] Masenwat N A 2010 Effect of different concrete grade on radiation linear attenuation coefficient (μ) *Seminar RND Nuklear Malaysia*.

[13] Bahrin M H 2017 Characterization of $ZnBr_2$ solution as a liquid radiation shield for mobile hot cell window *AIP Conf. Proc.* **1799** https://doi.org/10.1063/1.4972927 p040003.

[14] Rabir M H 2011 Analysis of Gamma Dose Rate for RTP 2MW Core Configuration Using MCNP (IAEA: INIS-MY--2013-183).

IOP Conf. Series: Materials Science and Engineering **555** (2019) 012020 doi:10.1088/1757-899X/555/1/012020

Monte Carlo studies for the radiation shielding in the bunker of the electron beam accelerator using PHITS

Mohd Faiz Mohd Zin, Mohd Rizal Md Chulan, Leo Kwee Wah, Abd Halim Baijan, Rokiah Mohd Sabri, Mohd Azhar Ahmad, Mukhlis Mokhtar, Khaidawaton Abd Malik

Accelerator Development Center, Technical Support Division, Agensi Nuklear Malaysia, Bangi, 43000 Kajang, Selangor, Malaysia

faizzin@nm.gov.my

Abstract. Particle and Heavy Ion Transport Code System (PHITS) is a Monte Carlo simulation program used for transport and collision of nearly all particles over wide energy range. One of its useful features for accelerator facility design is the calculation of dose distribution. This paper describes how the geometry of the bunker for low electron beam accelerator in Nuklear Malaysia (NM) was modelled and used to calculate the equivalent dose caused by the radiation for specific region in the bunker by using PHITS. The dose distribution and energy deposition were obtained. Finally, the simulation results are compared to the measured value from the dose measurement testing inside the bunker facility and found both results are in a good agreement.

1. Introduction

The low energy electron beam accelerator in Nuklear Malaysia (NM) was developed. Although the target energy of 300 keV is relatively low, the radiation shielding design is necessary to protect the workers during the accelerator operation from the secondary radiation. Previously, shielding design has been done analytically to construct a facility to accommodate this electron beam machine and an experiment to measure the dose at certain region in the room has been done during the final phase of the construction. Monte Carlo method simulation has become an essential tool to calculate radiation shielding design. Although in our case, the analytical calculations already suffice in the shielding design calculation, the use of Monte Carlo simulation as our principal design tool for our accelerator facility is beneficial due to its better accuracy and abilities to calculate a much more complex geometry. Furthermore, an extensive dose distribution around the room and at specific regions is important for our next steps in development and improvement of our accelerator. Therefore, the Monte Carlo simulation has been done in the irradiation room of our facility by using Particle and Heavy Ion Transport Code System (PHITS) to provide dose distribution in the irradiation room and at certain important regions for radiation shielding. This calculation is compared with the analytical and the experimental measurement.

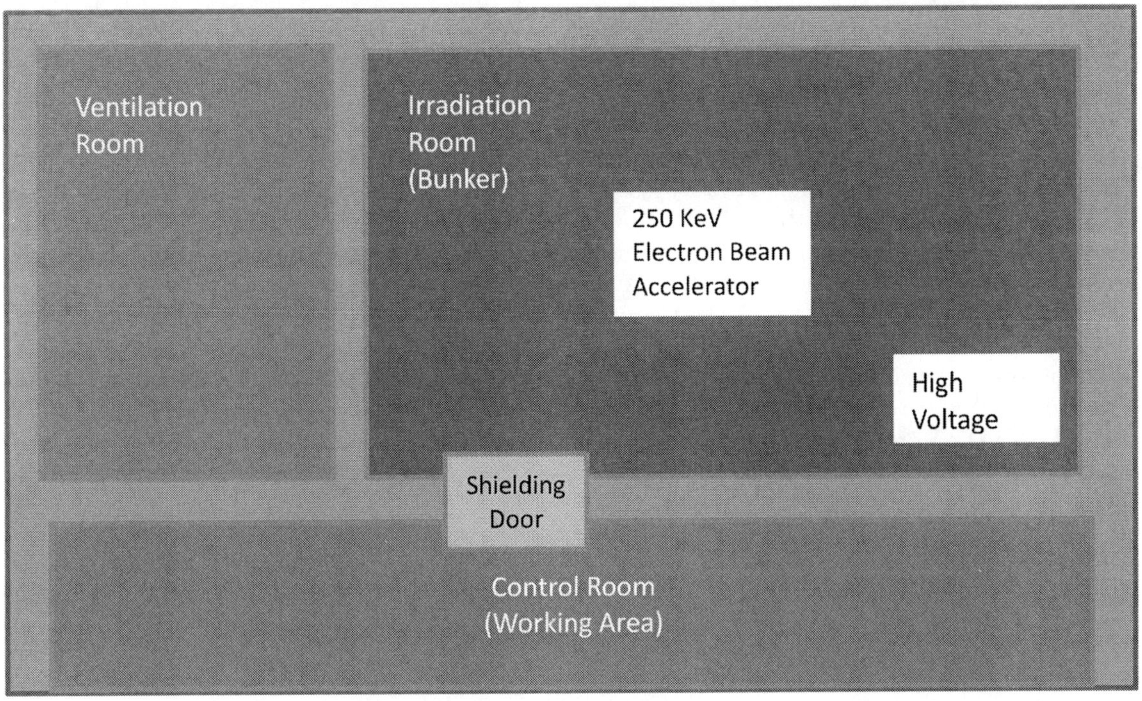

Figure 1. Top-view layout of the low energy electron beam accelerator facility in Nuklear Malaysia comprised of three areas.

Our electron beam accelerator facility can be divided into three areas: working area, ventilation room, and irradiation room (bunker). The working area is the control room which accommodates the control panel monitoring and the open spaces for related accelerator operations. The working area is limited to the radiation worker and is classified as clean area. It is our radiation protection goal for the dose around here to be below 20 μSv/hr [1].

The ventilation room is where the blower and exhaust motor are placed, and they are connected to the irradiation room by the ventilation duct through the wall. The irradiation room housed the electron beam accelerator, high voltage transformers and ventilation system. The walls enclosing the room was constructed using high density concrete with thickness of 85 cm and the height of the room is five meter high. The shielding door of the room was made by same material as the walls, but it is designed as a 'plug' shape to reduce the radiation leaking through the door [2].

2. Monte Carlo Simulation of the Irradiation Room

This paper is to describe the Monte Carlo simulation around the irradiation room and produced the dose distributions around the room and inside the concrete of the specific regions of interest. Monte Carlo method has been developed intensively to solve the practically unmanageable calculational problems [3]. Almost all major physical processes can be calculated accurately, and this can be done for any target with specific shapes and thickness. It is also considered the most competent method for calculating the distribution of photon track-length [4].

A general-purpose Monte Carlo particle transport simulation code, PHITS (Particle and Heavy Ion Transport Code System) is developed under collaboration between JAEA, RIST, KEK and several other institutes [3].

2.1. Geometry Definition

The first step is to design our geometry of the irradiation room. Geometry of the room is shown as in Figure 2. The regions of interest are numbered as followed: - #1 is inside the room which contains the air and the area is 500 cm across the room both horizontal and vertical. #2 is made up by the high-density concrete which enclosed the room. #3, #4, #5, #6 represents the concrete shielding door and designed as plug shaped to minimize the radiation leakage. The multi-layers design is for safety issue [2]. #7 is the maze, a concrete block with 100 cm width and height to minimize the radiation to the hole for ventilation system which we can see in region #8. #8 is the ventilation system that comprised a hole made up for blower from ventilation room to the irradiation room. #9, #10, #100 is the elliptic cylindrical hole that will be used for power supply and other related things to connect the equipment from ventilation area. #12, #13, #14 is the elliptic cylindrical hole to connect the equipment from the control room. The six holes that connected the cables are designed elliptic cylinder from the higher wall to lower region of the wall near the floor to prevent the radiation from leaking through the hole. An elliptic cylinder is a cylinder with an elliptical cross-section.

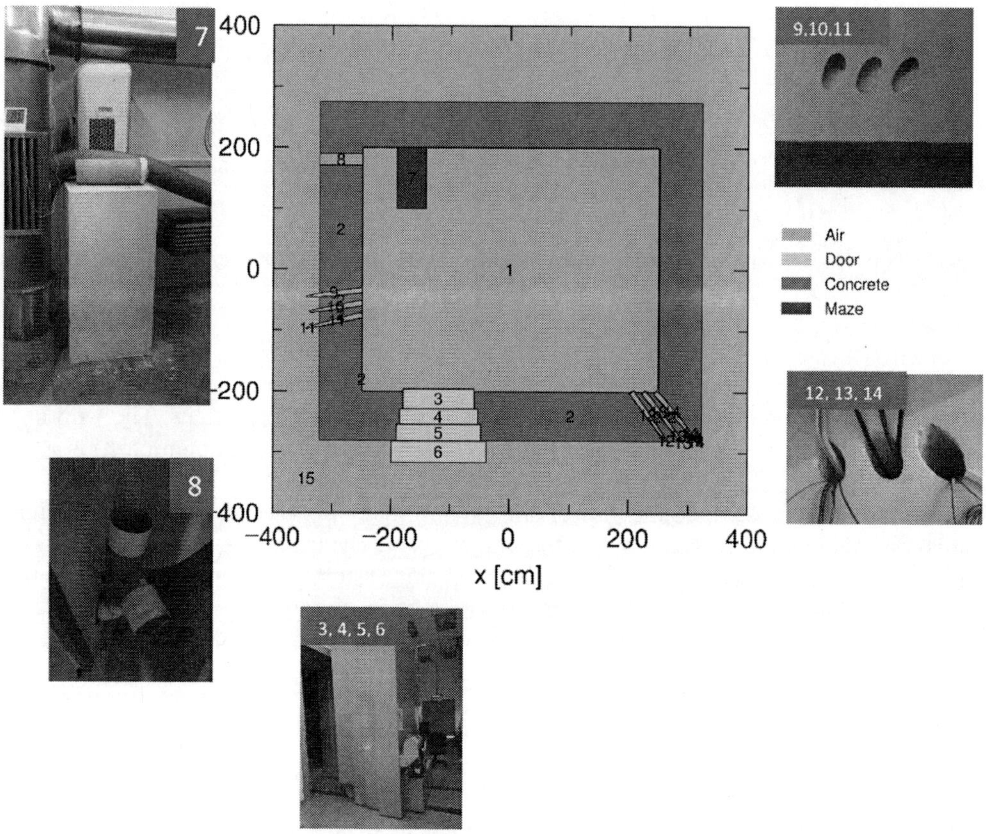

Figure 2. Geometry of the irradiation room from the top view. Each region is marked with a number.

2.2. Source Definition

The parameters that are concerned in radiation shielding in term of electron beam accelerator is electron beam energy E0 and average beam power P [5]. In this case, electron beam energy is considered small and the types of radiations concerned in the bunker is not the primary electron beams but the secondary beams [6]. The bremsstrahlung that results the photon beams are our main concern for radiation shielding. The previous measurement was done by using Iridium 192 (Ir-192), a gamma ray source to simulate the secondary beams from the interaction of our electron beam with the target material. We define Ir-192 like the experimental source. The activity is 1.25 TBq. The direction of the source as isotropic and is situated at the middle of the room and, 77 cm above the floor. Ir-192 is chosen for its strong gamma ray source as it can produce a gamma dose of 5 mSv/h at one meter. The energy spectrum will range from 200 keV to 600 keV, which includes the energy range of our electron beam accelerator.

The method used in gamma-ray dose conversion factor is by applying the parameter sets built-in PHITS for photon. We used the conversion coefficient for the ambient dose equivalent H*(10) from photon fluence [7]. The unit of the dose conversion factors is (μSv/h)/(n/sec/cm2) and these conversion factors were estimated with a condition of Antero-Posterior geometry (AP) irradiation [8].

2.3. Variance Reduced Technique

Using the geometry in Figure 2, there is inaccuracy of the calculation. In Figure 3 (a), the value of dose rate stopped short inside the wall. in Figure 3 (b), clearly dose rate calculation failed at 102 μSv/h. This is statistically unacceptable as we want to ensure the dose is less than 20 μSv/h with an acceptable accuracy and precision. The efficiency of the Monte Carlo simulation is improved using a variance reduction method which will artificially increase the probability of rare event occurrences is used and this will set out the importance values by cells. This method required the splitting of the geometry of our desired target of region of our calculation. In our case, the wall is split into many layers. Each of the layer has value of Importance, I. By setting the value of importance, the particle passing through the layer are split at the same number with the value we have set. This splitting particles are useful to calculate a much more accurate results for the calculation inside the shielding region. The splitting particles are used for calculation as it will increase the reaction inside the region and so will enhance our calculation to get a much more accurate dose or flux. Figure 3 (b) is an example of the differences between setting the importance for the shielding area. The dose rate for without setting the importance will be dropped to zero sharply after the dose only at around 10^2 μSv/h while with the importance, there are still doses calculated even at the outside of the shielding area.

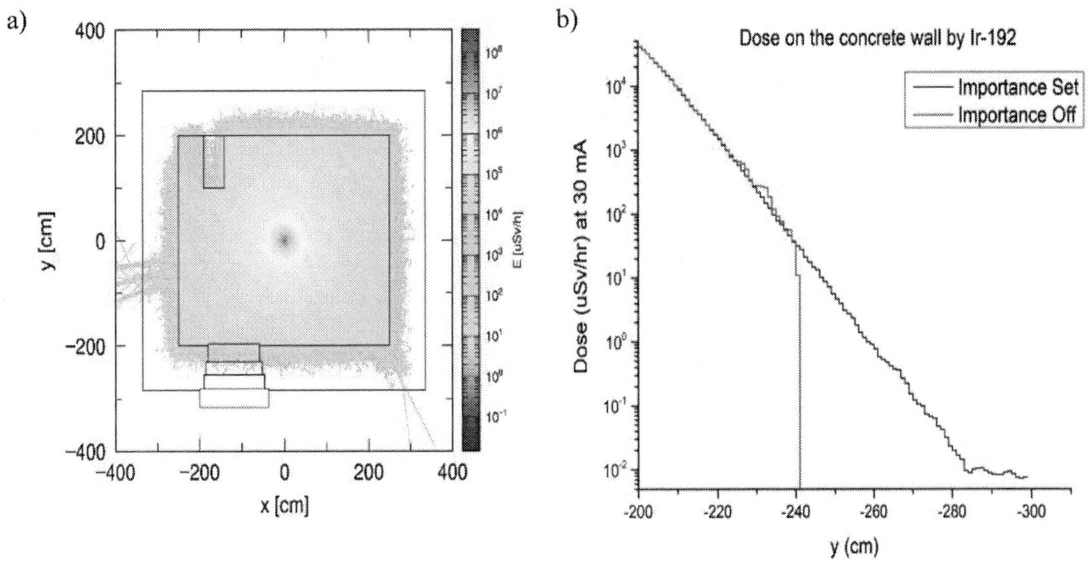

Figure 3. Dose distribution on the wall does not extend to a much higher degree of accuracy in dose rate. In (b), there is a significant improvement in term of accuracy with the variance reduced technique.

IOP Conf. Series: Materials Science and Engineering **555** (2019) 012020 doi:10.1088/1757-899X/555/1/012020

In term of statistical error, at 241 - 242 cm of y-axis, the relative error without variance reduction is 66.7% and 7.4 % while using the variance reduction. 242 cm of y-axis is located inside the concrete wall region and that depth is the limit of depth that we can calculate without using variance reduction. Beyond this depth, it will have 100% relative error. Using the variance method, the relative error of particles reaching the air region, which is outside of the bunker is at 34%. Figure 3 shows particle reaching outside but the calculation on that area gives high relative error (almost 100%), which invalidate the output. Therefore, the calculations using variance reduction on the outside region as shown in Figure 4 (b) and (c) are acceptable.

3. Ir-192 Dose Distribution in Irradiation Room

Figure 4 (a) shows the geometry of the room with consideration of the importance for each layer in the concrete wall and door. Figure 4 (b), 4 (c) is the dose distribution inside and outside the room. The dose at the control room is at 10-2 μSv/h and is comparably lower than the dose at the ventilation room. This is acceptable as our region of interest is in the control room while in the ventilation room, there will be no worker during the operation. In addition, the maze has reduced the radiation leaking through the duct hole for blower.

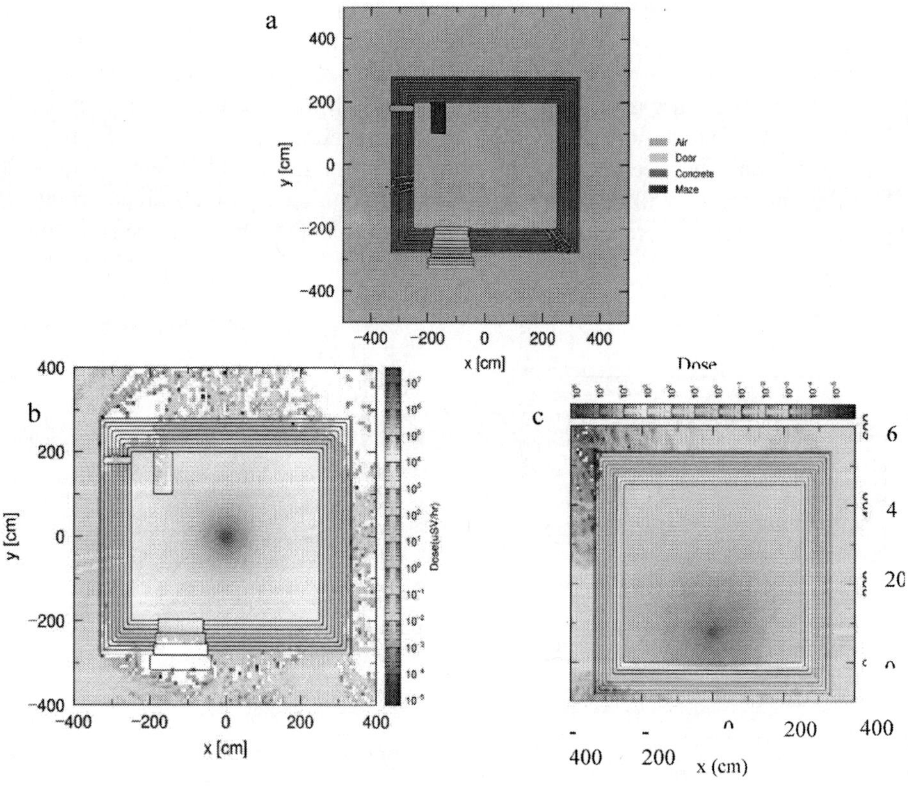

Figure 4. (a) Geometry of the irradiation room. (b), (c). are dose distribution in and outside the room using Ir-192 as source.

Figure 5 (a) is the effective dose inside the room along the width (x-axis) of the room. The dose taken at each point is the average distribution inside the room from the floor to the ceiling and from the front to the back (y-axis). The air attenuates one-tenth of the average effective dose from 104 to 10^3. Therefore, the shielding design reduce the dose by 10^3. Figure 5 (b) and 5 (c) is the dose rate inside the concrete for different regions at the front side of the room. The dose on the concrete wall dropped below 20 μSv/h passing through 45 cm of concrete wall. There is an extra of 40 cm thick concrete for our shielding using this source definition. Through the holes, the dose rate dropped below 20 μSv/h at maximum of 61 cm through concrete. It can be observed that inside the elliptical cylinder holes, the statistics are not good especially hole #6 which might be due to its location at the far corner of the room. However, as reference to the wall, we can accept the dose rate as it is statistically acceptable outside the hole. Figure 5 (d) shows the attenuation factor by one-tenth of the dose.

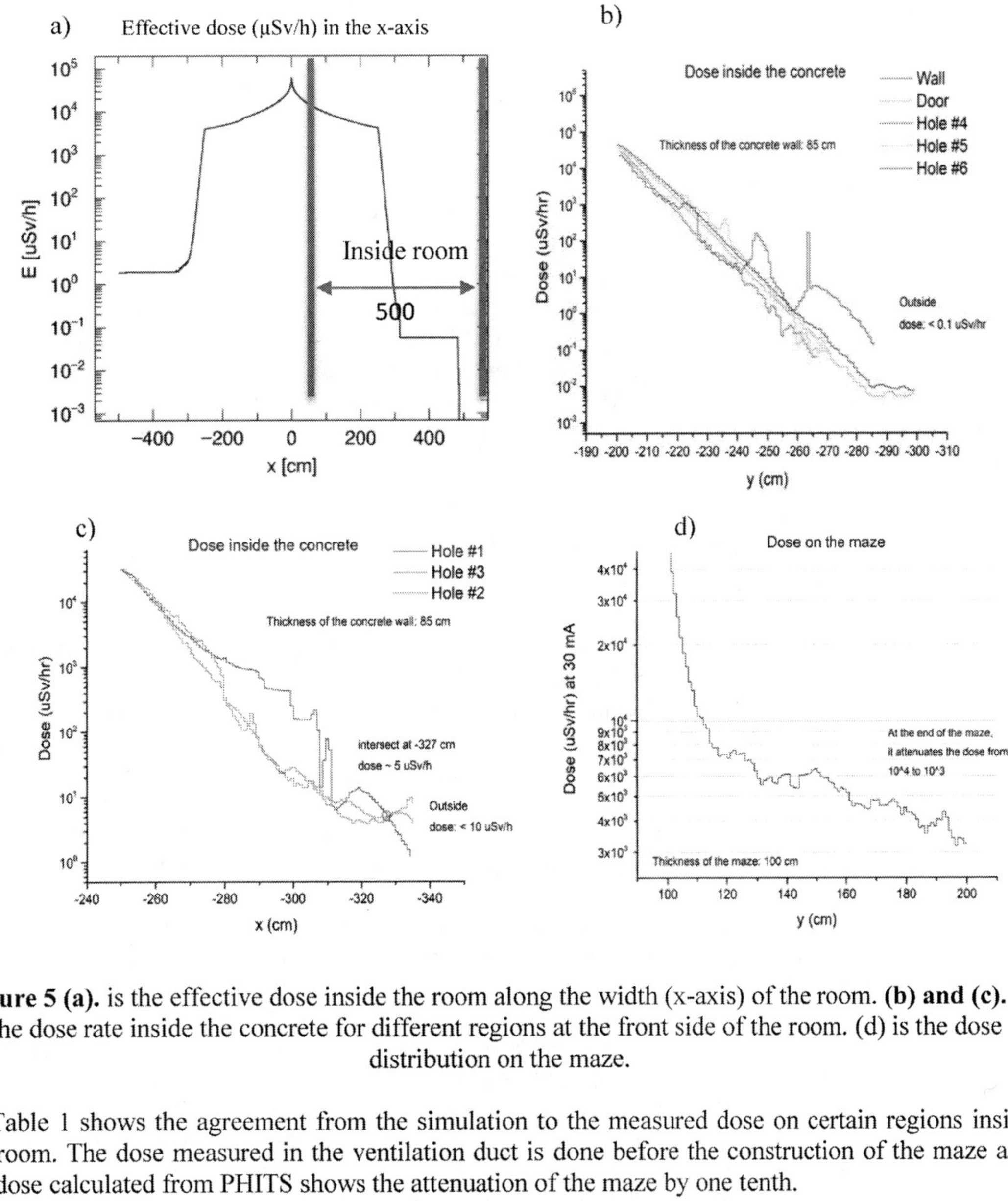

Figure 5 (a). is the effective dose inside the room along the width (x-axis) of the room. **(b) and (c).** is the dose rate inside the concrete for different regions at the front side of the room. (d) is the dose distribution on the maze.

Table 1 shows the agreement from the simulation to the measured dose on certain regions inside the room. The dose measured in the ventilation duct is done before the construction of the maze and the dose calculated from PHITS shows the attenuation of the maze by one tenth.

Table 1. Comparison between measured dose and calculated dose.

Region	Measured Dose (μSv/h)	Calculated Dose (μSv/h)
Wall	0.03 – 0.05	0.08 – 0.09
Door	0.03	0.05
Ventilation duct	20.0 (without maze)	0.3
Hole #1 (left-side)	2.0	1.19
Hole #2 (left-side)	2.0	4.58
Hole #3 (left-side)	2.0	3.86
Hole #4 (front side)	0.3	0.29
Hole #5 (front side)	2.0	1.55
Hole #6 (front side)	2.0	1.21

4. Conclusion

Dose distribution around the bunker has been calculated by using Monte Carlo method from PHITS and has been compared to the measured dose. The methods and results are shown in the paper and for future works, there is a need to improve the calculation inside the six elliptical cylinder duct holes on the concrete wall. Then, the next step is to do simulation using our current electron beam accelerator parameter as source definition with different target materials and for shielding design, a simulation using different shielding material should be done in order to provide a much better and cost-effective shielding design for the facility.

5. References

[1] Kutcher G 1978 Radiation Protection Design Guidelines for 0.1 to 100 MeV Particle Accelerator Facilities Medical Physics 5 73-73

[2] Leo K, Chulan R, Hashim S, Azhar M, Dalim Y, Mokhtar M, Baijan A, Sabri R, Faiz M, Azaman A and Rosli R 2018 Study on the window cooling system of 300 keV Electron Accelerator IOP Conference Series: Materials Science and Engineering 298 012045

[3] Sato T, Niita K, Matsuda N, Hashimoto S, Iwamoto Y, Noda S, Ogawa T, Iwase H, Nakashima H, Fukahori T, Okumura K, Kai T, Chiba S, Furuta T and Sihver L 2013 Particle and Heavy Ion Transport code System, PHITS, version 2.52 Journal of Nuclear Science and Technology 50 913-923

[4] Iida K, Kohama A and Oyamatsu K 2007 Formula for Proton–Nucleus Reaction Cross Section at Intermediate Energies and Its Application Journal of the Physical Society of Japan 76 044201

[5] Leo K, Chulan R, Hashim S, Baijan A, Sabri R, Mohtar M, Glam H, Lojius L, Zahidee M, Azman A and Zaid M 2016 Study on the parameters of the scanning system for the 300 keV electron accelerator

[6] National Council on Radiation Protection and Measurement 1976 Structural Shielding Design and Evaluation For Medical Use of X-Rays and Gamma Rays Up to 10 MeV (National Council on Radiation Protection and Measurement)

[7] International Commission on Radiological Protection 1996 Conversion Coefficients for use in Radiological Protection against External Radiation (International Commission on Radiological Protection)

[8] Anon PHITS Ver. 3.02 User's Manual English version

Acknowledgements
We extend gratitude to Dr. Hiroshi Iwase, from the KEK, Japan, for his guidance and willingness to conduct a workshop NM-KEK PHITS Course 2018 in Nuklear Malaysia in which he contributed in-depth on this subject.

IOP Conf. Series: Materials Science and Engineering **555** (2019) 012021 doi:10.1088/1757-899X/555/1/012021

Design and development of real–time welded metal defect classification automated ultrasonic testing system

Suhairy Sani[1,2,a)], Mohd Hanif Md Saad[2], Nordin Jamaludin[2], Nor Salim Muhammad[3], Ismail Mustapha[1,2], Ishak Mansor[1] and Tengku Sarah Tengku Amran[1]

[1]Industrial Technology Division, Malaysian Nuclear Agency, Bangi 43000 Kajang, Selangor, Malaysia.
[2]Faculty of Engineering and Built Environment, National University of Malaysia (UKM), Bangi 43000 Kajang, Selangor, Malaysia.
[3]Faculty of Mechanical Engineering, Universiti Teknikal Malaysia Melaka (UTeM), Malaysia.

[a]suhairy@nm.gov.my

Abstract. This paper discusses the design and development of a portable PC-based Automated Ultrasonic Testing (AUT) system, which is capable of classifying defect types on a welded metal components in real-time. Three types of welding defects such as a crack, slag inclusion and cluster porosity plus the normal condition (no defect) were used as the test states for the classification process. The system utilizes an ultrasonic pulse-receiver card model attached to computer notebook for signal display. A new specific software package (RUSS) was developed to control the operation of the scanner, displaying the data, extract the signal defect and classify type of defect. The proposed algorithm and complete system were implemented in a computer software developed using Microsoft Visual BASIC 6.0. The classification process NNTool in MATLAB component was manipulated by using the ActiveX. Some result shows the successful of the calibration module and real – time extraction for classification module type of defect

1. Introduction

Non-Destructive Testing (NDT) is important for controlling the quality of structural and engineering materials. This technology is very important for human security and the capacity of industrial plants. The Automated Ultrasonic Testing System (AUT) is an advanced ultrasonic test technique that uses computerized techniques in evaluating and interpreting the reflected signal from any sound reflector including deficiencies in the material. One of the most commonly used imaging techniques is the C-scan system that can provide an image of defects in the welds examined.

However, the ability to interpret the type of disability from interaction with ultrasound can also be important. For example, some ultrasonic reflectors (eg planar crack) are more likely to lead to ultimate failure than volumetric weakness such as slag or porosity. While this development is important, the recognition of ultrasonic patterns is not widely used in practical examination in the industry, since the application of this approach requires the operator's understanding of all the details related to this method [1].

Figure 1 (a) and (b) show the hardware and software in previous development of AUT system for weld flaw analysis to allow real-time measurement of ultrasonic flaw signature features. The system includes an ultrasonic transducer that scans in an x-y plane and is connected to an electronic pulser/receiver, a dual timing gate, a peak detector, and a universal timer. At each position x-y along the scan, the computer estimates the maximum amplitude of any feature that reflects ultrasound, using back wall pulse-echo (also called "time of flight") of the ultrasonic signal. After calibration, the peak detector operates with a time gate chosen so that its output indicates the amplitude of the signal reflected from the feature of interest at x, y. Once the scan has been completed, the computer processes the x-y scanning-position data and the associated amplitude data into a C-scan plotting. The value of the data is plotted using colors to produce detailed images of the internal features in welded plate. The area of interest which contains the maximum amplitude can be chosen automatically from the image and the scanner is set to move the transducer to the target area for flaw signature analysis. The software obtains the A-scan at the point and performs both time-domain and frequency domain analysis to obtain unique characteristics which can be used to classify the type of flaw at the selected point.

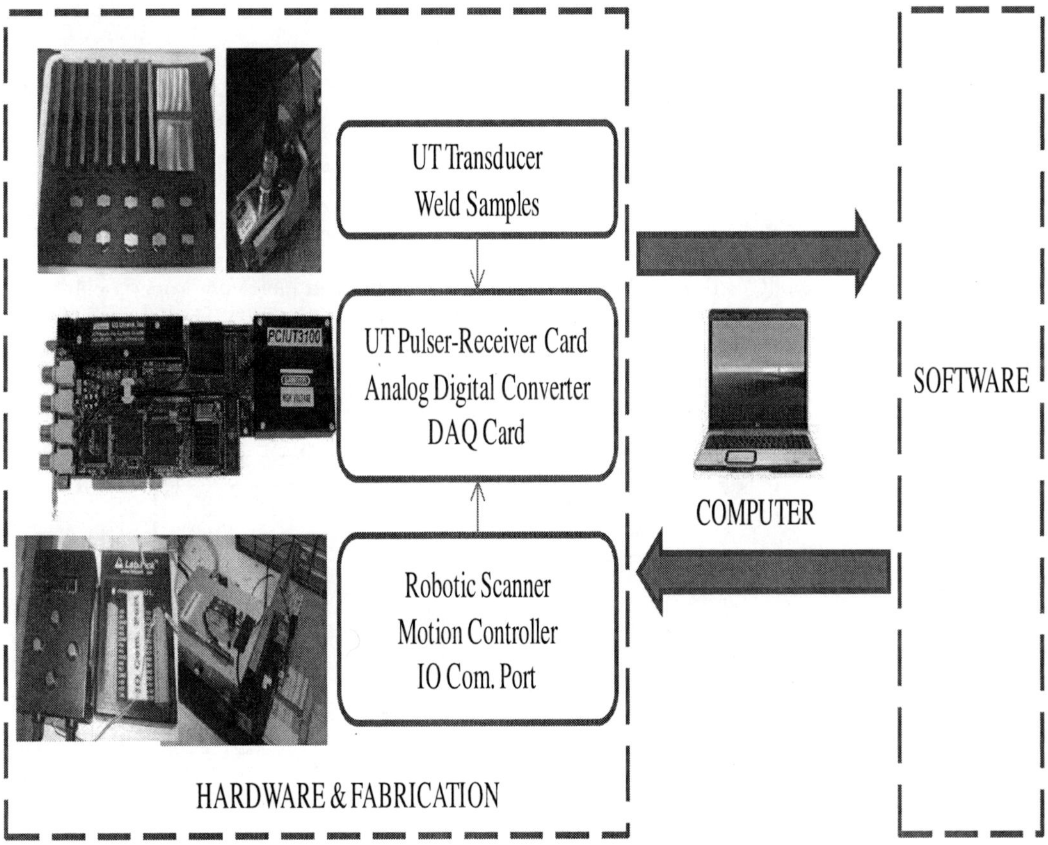

Figure 1. Hardware fabrication for RUSS System.

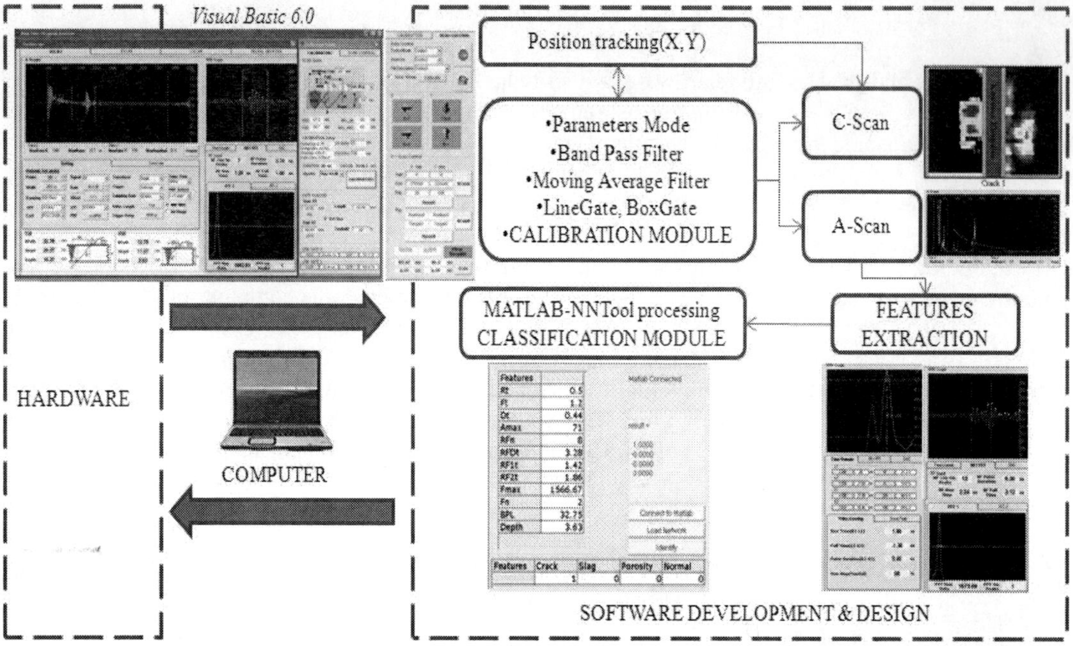

Figure 2. Software modules for RUSS System.

2. Material and Method

The defect signals can be obtained from the echo signal if the equipment has been properly calibrated. In this study, an angle-beam probe 60° with 5MHz and the standard calibration V2 block were used for the equipment calibration. The zero point on the scale is directly connected to the surface of the object being tested.

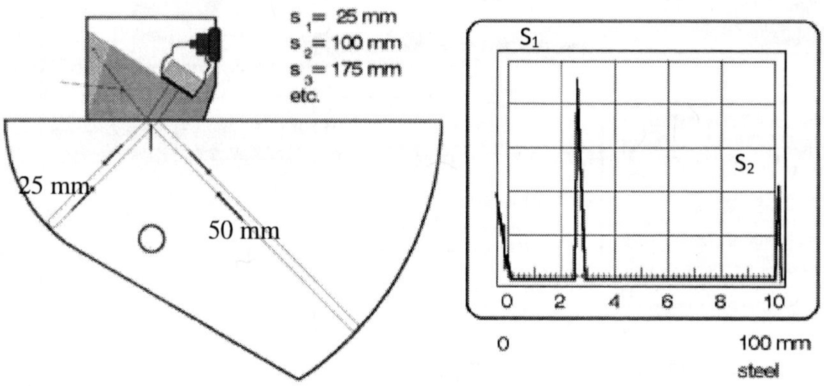

Figure 3. Equipment calibration using V2 block.

When specifying the calibration range, the naming of the material is also important because the displayed distance of the echo, sound path, s (BPL, mm), is always deduced from the time of flight, t of the pulse and the sound velocity, c (km/s) according to the equation [2]:

$$\text{Sound path, } s = \frac{c \times t}{2} \tag{1}$$

Referring to Figure 4, if the signal obtained from the deflection of defect, the depth, T and the actual depth, T_s can be calculated from the equation 4 as below.

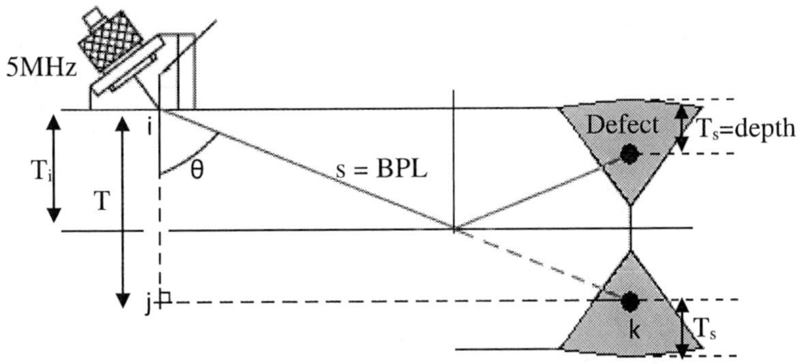

Figure 4. Beam path length (BPL),s from reflected signal.

If sound path, BPL, s = ik then real depth, T_s;

$$\text{Full skip distance, j-k, FSD} = \text{BPL x Sin}\theta \quad (2)$$

$$\text{where, i-j, T} = \text{BPL x Cos}\theta \quad (3)$$

$$\text{Real depth, } T_s = 2T_i - T = 2T_i - (\text{BPL x Cos}\theta) \quad (4)$$

The examination was performed on different welded samples (dimension 145mm x 70mm) with thickness of 10mm. Three types of welding defects such as cracks, slag and petal cans are used as test and normal conditions (no defect) welding as a reference for the next classification process. Figure 5 shows the area of interest which contains the maximum amplitude that can be chosen automatically from the C-Scan image and the scanner can set to move the transducer to the target area for the features extraction analysis.

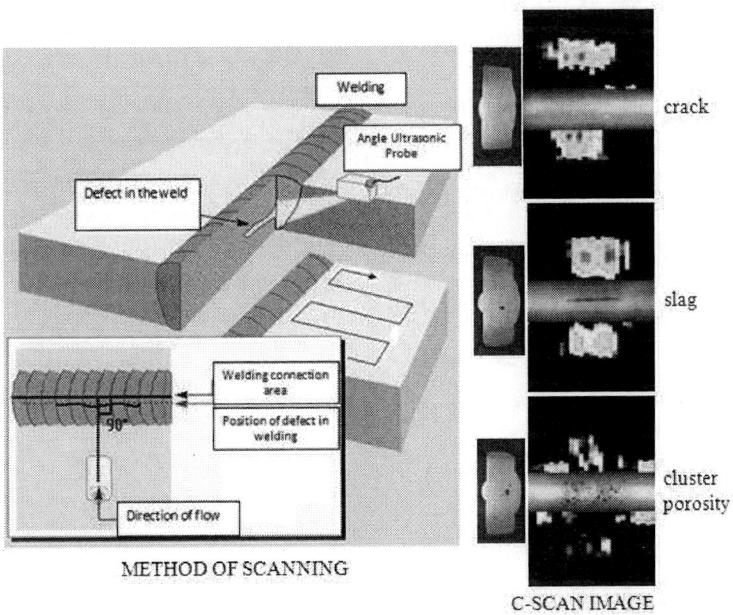

Figure 5. C-scan and DIR image of crack, slag and cluster porosity.

3. Results and Discussion

Figure 6 shows that the calibration process has been successfully performed where the first peak value at the box gate for the beam path length, BPL = 25 mm, skip distance, FSD = 21.65 mm and depth, T = 12.5 mm. The calibration panel shows shear wave velocity for steel calibration blockV2 equivalent to 3266.1 m/s. The distance between the two echo peaks is 75 mm which is equivalent to 2392 sample data and for each 1 mm equals to 31.9 sample data. The zero point and maximum peak are equal to 354.6 and 1120. Examples of calculations for obtaining shear velocity values, BPL, FSD and T are shown as below.

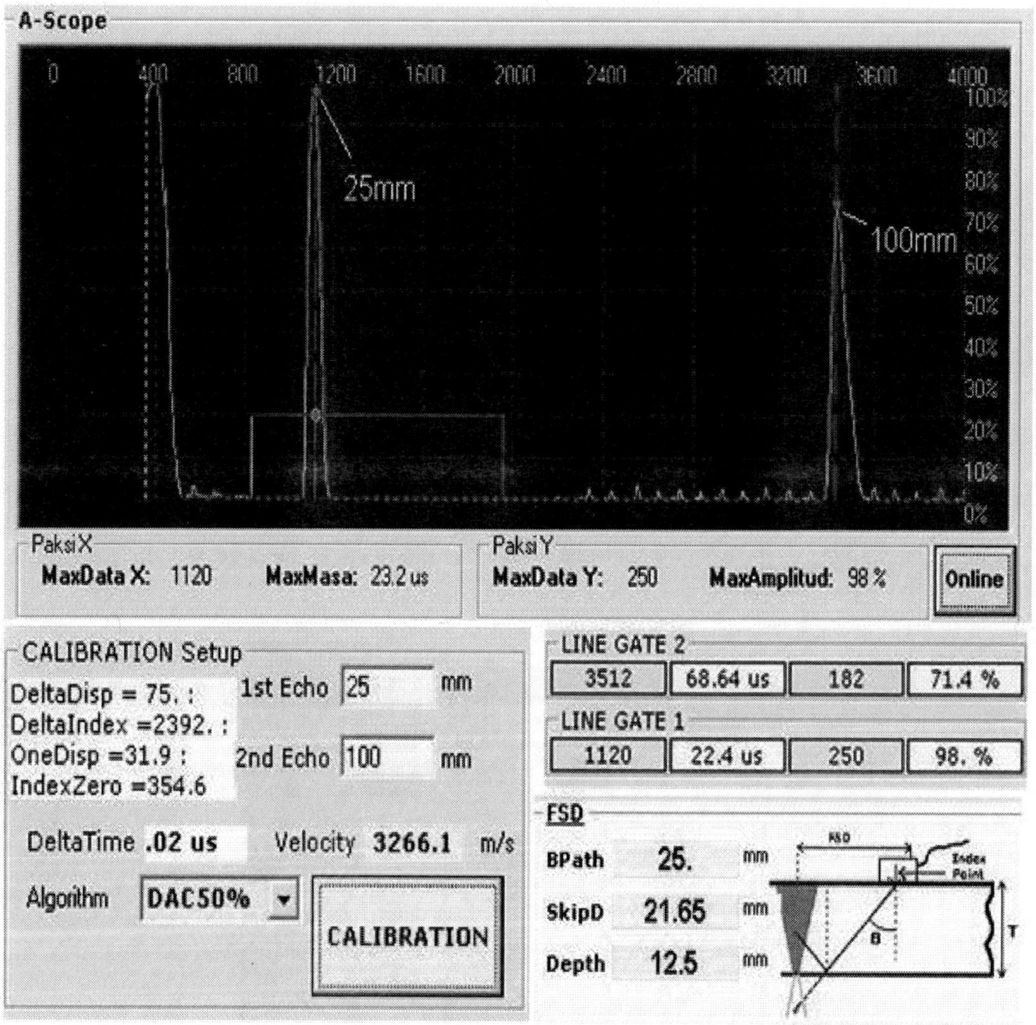

Figure 6. Calibration module using V2 block.

From equation 1, sound path i.e. $S = \frac{c \times t}{2}$:

$$\text{shear velocity} = \frac{2 \times \text{first peak}}{(\text{linegate1} - \text{zeroline}) \times \frac{1}{\text{sampling rate}} \times 1000}$$
$$= \frac{2 \times 25}{(1120 - 354.6) \times \frac{1}{50000000} \times 1000}$$

$$= 3266.3\ {}^{\text{m}}/_{\text{s}} \tag{5}$$

Beam Path Length, BPL is given by:

$$\text{BPL} = \frac{\text{shear velocity} \times (\text{maximum peak} - \text{zeroline})}{2 \times \frac{1}{\text{sampling rate}} \times 1000}$$
$$= \frac{3266.3 \times (1120 - 354.6)}{2 \times \frac{1}{50000000} \times 1000}$$

$$= 25 \text{ mm} \tag{6}$$

From equation 2, Full Skip Distance FSD = BPL x Sinθ:

$$\text{FSD} = \text{BPL} \times \sin\left(\frac{60 \times 3.142}{180}\right)$$
$$= 25 \times 0.8661$$
$$= 21.65 \text{ mm}$$

From equation 3, Depth T = BPL x Cosθ:

$$\text{T} = \text{BPL} \times \cos\left(\frac{60 \times 3.142}{180}\right)$$
$$= 25 \times 0.4999$$
$$= 12.5 \text{ mm} \tag{7}$$

After the noise has been removed, time and frequency domain analysis were executed automatically on the filtered data. The first analysis was done in full wavelength-time domain where the value of the maximum amplitude, pulse duration, rise time and fall time is calculated. The second analysis was done in RF wavelength-time domain where the same features value of number of peaks, pulse duration, rise time and fall time are calculated. Then the software transforms the frequency domain signal via the application of discrete Fourier transform to extract maximum frequency spectrum and number of peak. Figure 7 shows example extraction of the time and frequency domain for slag weld defects. All parameters setting at the same condition for others defect signal.

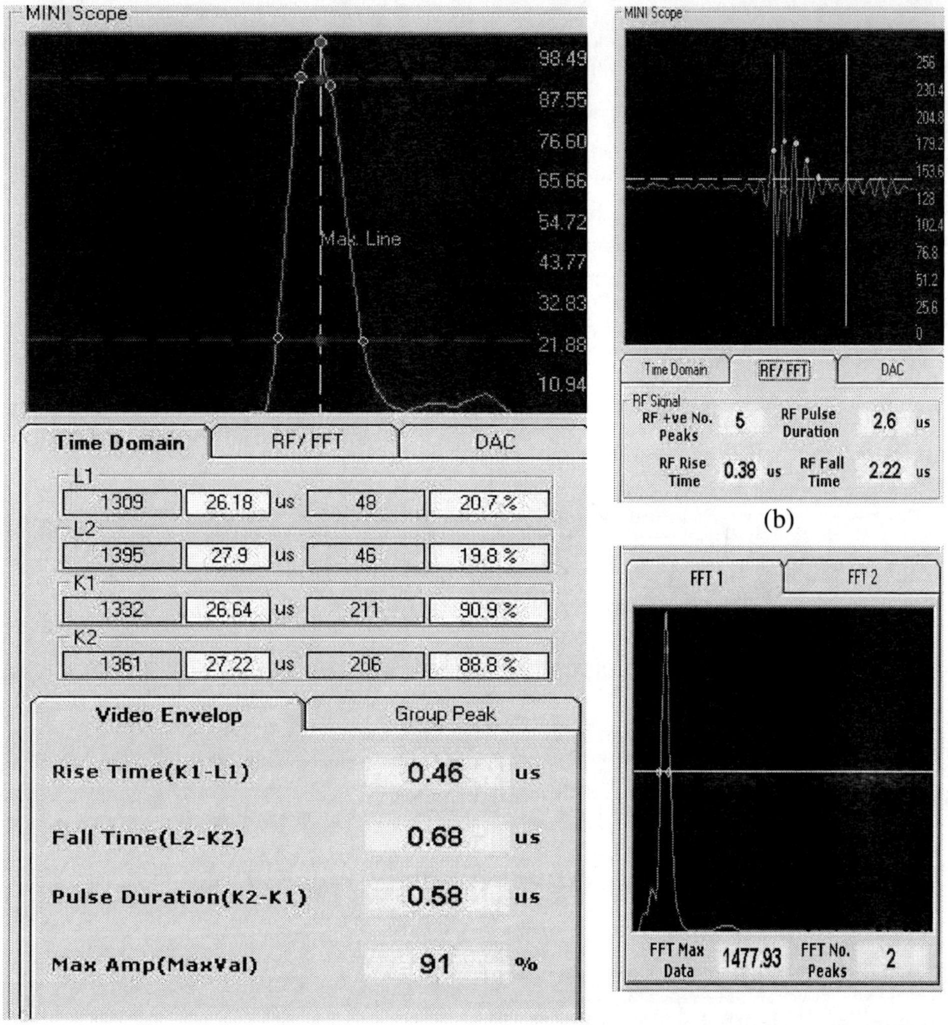

(b)

(a) (c)

Figure 7. Features extraction for slag weld defect (a) full wavelength, (b) RF wavelength and (c) frequency domain.

For early stage, back propagation algorithm is used to train the neural network. The neural network architecture has three layers consists of input, hidden and output layer. Figure 8 shows the twelve real time input of neurons as input data. The hidden layer consists of 3 activation functions (*tansig*, *logsig* and *purelin*) which are determined by trial and error and the output consists of 4 neurons representing type of defect classification.

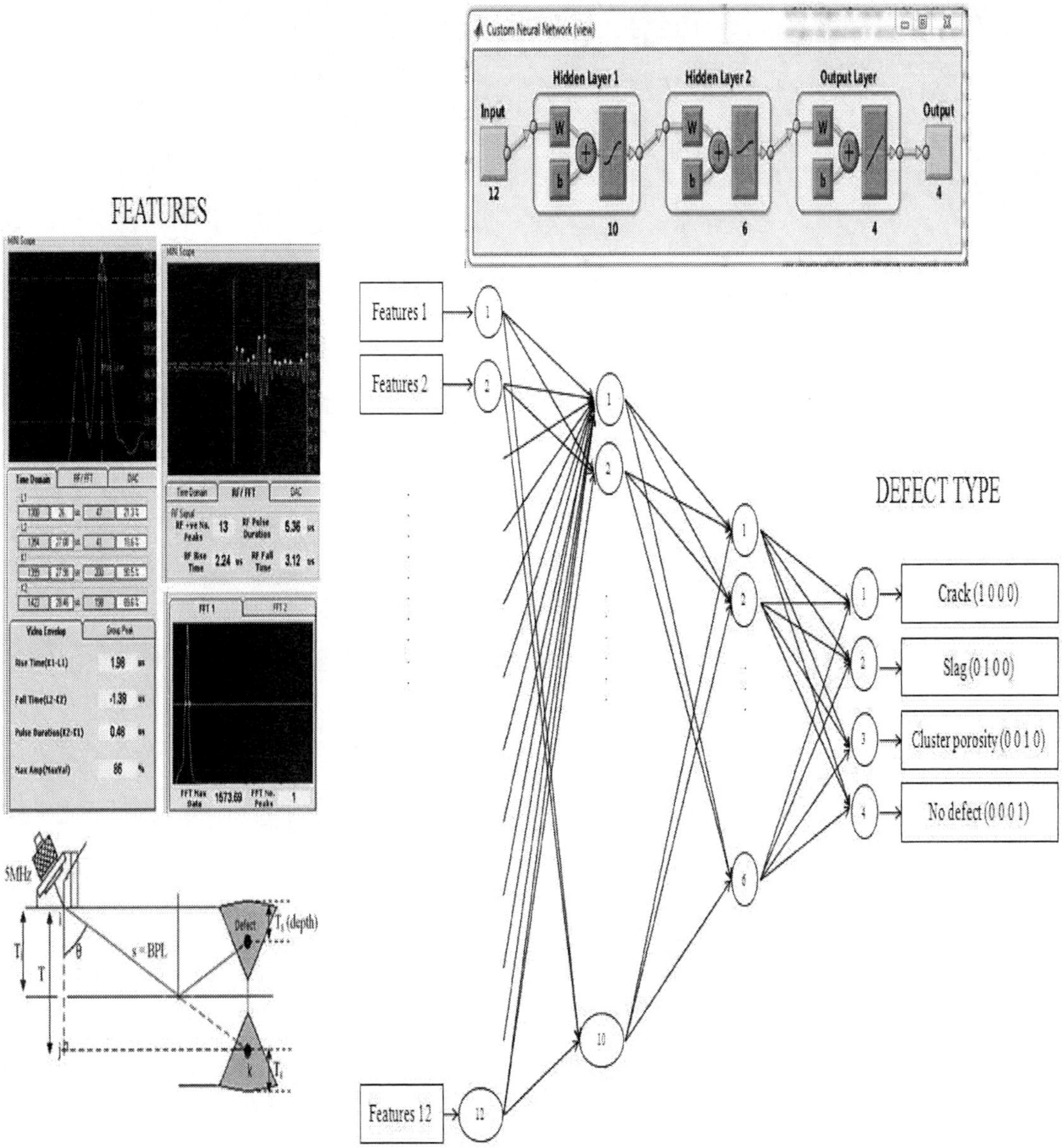

Figure 8. Neural network architecture

Figure 9 shows the results of the classification signal to the specimens crack, slag, cluster porosity and normal. By referring to the feature values in each diagram, this software has successfully automated the classification of each data obtained with a value of '1' which represents the type of defect.

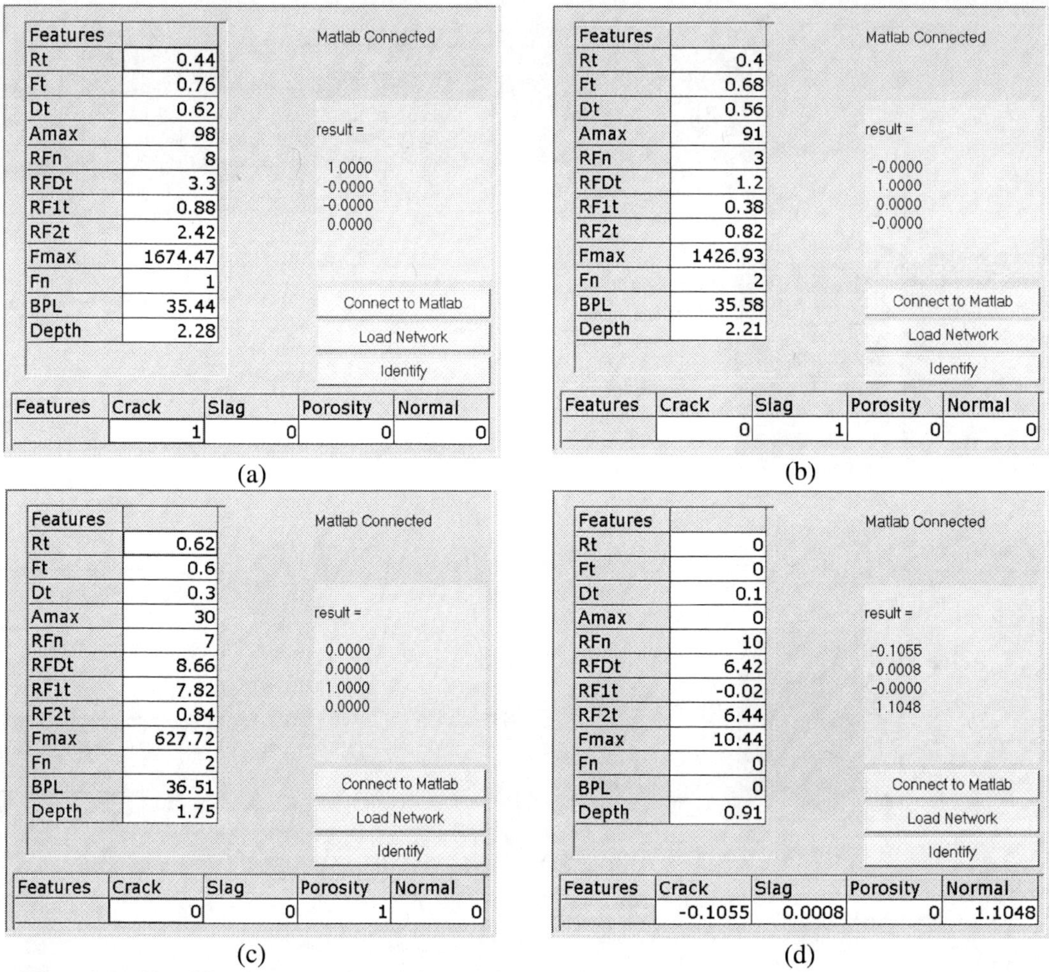

Figure 9. Classification modules of signals for defects (a) crack, (b) slag inclusion, (c) cluster porosity and (d) normal.

4. Conclusion

This paper shows that the combination of time domain, defect location and frequency domain can be used for features extraction of weld defect. Simple neural network models have been developed and successfully classified the types of defect. For future improvement, we recommend confirming results with more specially designed test specimens and others artificial intelligent to provide a better learning environment.

5. References

[1] Song S J, Kim H J and Cho H 2002 Development of an intelligent system for ultrasonic flaw classification in weldments *Nuclear Eng. and Des*ign **212(1-3)** p 307–320.

[2] Berke M 2000 Nondestructive Material Testing with Ultrasonics: Introduction to the Basic Principles **5(09)** (NDT.net: September 2000) [http://www.ndt.net/article/v05n09/berke/berke1.htm].

[3] Szilard J 1982 *Ultrasonic Testing* (USA: John Wiley & Sons).

[4] Birks A S, Green Jr R E, Mcintire P 1991 Waveform and data analysis technique Nondestructive Testing Handbook **7** p 149.

Acknowledgement

The authors would like to express his appreciation to Ministry of Energy, Science, Technology, Environment and Climate Change (MESTECC) also to Malaysian Nuclear Agency for their support to this project.

Study on the filament power supply of the 300 keV electron accelerator

Leo Kwee Wah[1], M. Azhar[1], R.M. Chulan[1], S.A. Hashim[1], M. Mokhtar[1], Khaidawaton[1], H. Baijan[1], R.M. Sabri[1]

[1]Malaysian Nuclear Agency, Bangi 43000 Kajang. Selangor. Malaysia.

leo77@nuclearmalaysia.gov.my

Abstract. This paper describes the study and development of the filament power supply. This locally designed low energy electron accelerator with the present energy of 140 keV will be upgraded to 300 keV. The filaments are used to heating-up and produce the electron. The produced electron will be accelerated through a potential different by a DC High Voltage Power Supply (DCHVPS). Whereby this filament power supply is connected with the DCHVPS, therefore the isolation of high voltage is required. For such purpose, the specifications and the isolation mechanisms of the filament power supply have been identified. As the result, the filament power supply by using an air core transformer has been selected and developed with the required heating power.

1. Introduction

Accelerator is a promising and advanced technology with the capability to support various applications such as flue gases treatment and cable cross linking [1], medical applications such as cancer therapy and isotope production [2][3], as a collider for high energy physic experiment and study [4][5] and agriculture for mutation breeding [6]. In Nuclear Malaysia, a localized, own developed accelerator technology, the so-called the low energy electron accelerator [7][8][9] has been upgraded from 140 keV to 300 keV as shown in figure 1. For such improvement, other auxiliary systems must been upgraded as well such as insulating gas system [8], scanning power supply [9], window cooling system [10], filament power supply and etc. In this paper, the principle and the development of the filament power supply will be described and discussed.

2. Specifications of filament

Basically, an electron accelerator is a machine used to accelerate the electron beam with the required energy and beam current which are mainly determined by the applications. The electron is produced by heating-up the filament or cathode. The heated filament will produce the electron on the surface and with a small applied electric field, the electron is extracted and propagated to the accelerating tube for further acceleration. In this low energy electron accelerator, the filament is embedded at the top of the accelerating tube as shown in figure 2, and the specifications of the filament is shown in table 1.

IOP Conf. Series: Materials Science and Engineering **555** (2019) 012022 doi:10.1088/1757-899X/555/1/012022

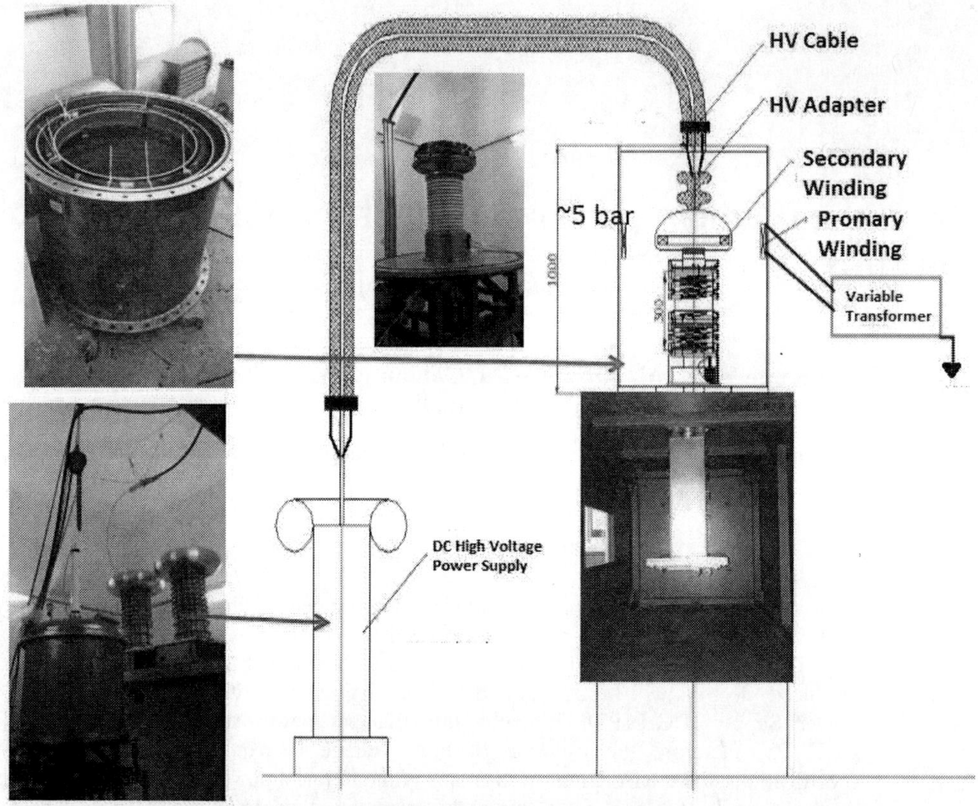

Figure 1. Low energy electron accelerator.

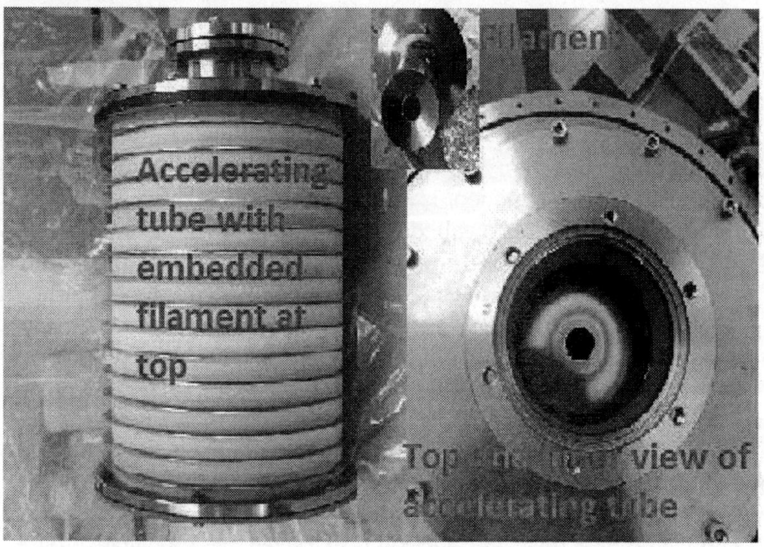

Figure 2. Top and inner view of the accelerating tube with the filament embedded position.

IOP Conf. Series: Materials Science and Engineering **555** (2019) 012022 doi:10.1088/1757-899X/555/1/012022

Table 1. Specs of the filament

Parameters	Figure
Life time	5000 hours
Material	Lanthanum Hexa boride, LaB_6
Diameter	10 mm

Basically, the required heating power is determined by the material of the filament. By referring to figure 3, the filament characteristics dependent to the heating power and temperature are well-defined. Therefore, we could identify the optimized heating power, which is ~ 50 watt. With this characteristic, the developed filament power supply must be able to supply 50 watt or higher heated power for electron beam production.

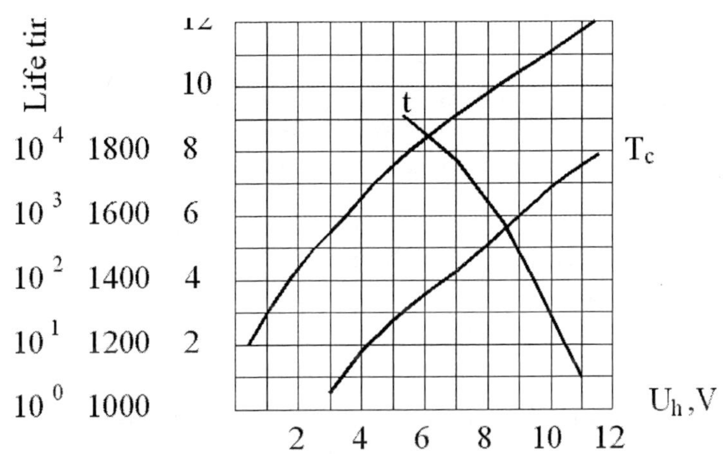

Figure 3. Characteristics of the LaB_6 Filament.

3. Design and requirement of the filament heating system

Basically, the filament heating power supply of the low energy electron accelerator is required to heat-up the filament for electron production. The filament power supply must be isolated from the high voltage source as shown in figure 1. For the high voltage isolation, several techniques are available, such as i) dynamo techniques to convert the mechanical energy to electrical energy for filament heating, ii) by using a high voltage isolation transformer for 300 kV isolation, iii) by using an air core transformer mechanism to transfer the power source indirectly to heat-up the filament. In addition, the design of the filament heating system must consider the physical geometry factor as the dimensions between the accelerating tube and pressure vessel as described in Ref. [8].

Table 2. Comparison of the three filament heating methods

	Dynamo technique	Isolation transformer	Air core transformer
Cost	Θ	◬	△
Required space inside the pressure vessel	◬	△	Θ
Required space outside the pressure vessel	△	◬	Θ
Flexibility	Θ	△	△

(△= Easy/Low, Θ=Medium, ◬ = Difficult/High)

According table 2, the major factors are the space inside the pressure vessel and cost of the system. Therefore, we have selected the air core transformer mechanism as the filament heating system. For the design of the air core transformer mechanism, the concept is rarely simple. The concept is the same as the iron core transformer but instead of iron core, the air core will be applied as a medium to transfer the power from the primary winding to the secondary winding and connected to a matching transformer that connected to the filament. Whereby the filament heating system is located inside the pressure vessel [8], as surrounding by the nitrogen gas as a high voltage insulation medium. For detail parameters optimization, few experiments have been performed as described in next section.

4. Experimental

For the air core transformer mechanism, the parameters such as diameter, number of turns of primary and secondary windings and the power transfer efficiency by a matching transformer must been identified, therefore 3 experiments have been carried out as:

i) To define the physical diameter of primary and secondary windings
For this experimental, the physical diameter optimization for the secondary winding is referred to the magnetic flux density, **B** induced from the primary winding. Whereby the diameter of the primary winding geometry is fixed as 645 mm and 216 turns, which is determined by the diameter of the pressure vessel. As the result as shown in figure 4, the induced B losses ~ 30% at 270 mm distance from the center as indicated with the red line. Moreover the distance between secondary winding and primary winding is crucial to avoid the high voltage discharge because the secondary winding is contacted with the high voltage source. From ref. 8, with respect to the distance of both windings of 5 cm is compatible with the specification of the gas pressure system, therefore the optimized diameter of secondary winding has been fixed as 545 mm as shown in figure 5.

ii) To define the number of turns of primary and secondary windings
With the optimization of each winding's diameters, the number of turns is fixed with 147 turns for primary winding. Whereby this number of turns is based on the winding quantity stored at the laboratory. By considering the voltage increment of ~3 times, the number of turns for secondary winding is selected as ~ 416 turns, which is taking account with the available stock in the laboratory as well. As shown in figure 6: the main power source of variable transformer is set as the primary winding input, the secondary winding is connected with a matching transformer, whereby the matching transformer (60-0-60) is connected filament, LaB_6. As shown in figure 7, the achievable power output power is ~ 24 watt at 150 Vac input and the required heating power is unexpected with this setup even with 240 Vac. Therefore, the specs of the matching transformer must be optimized as described below.

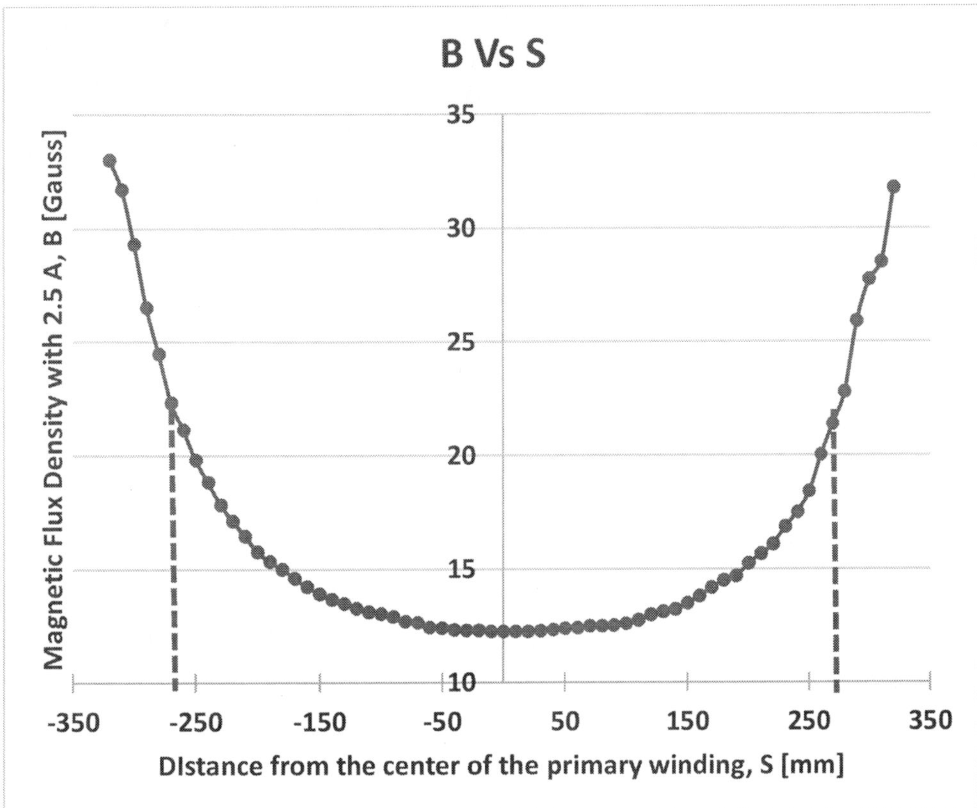

Figure 4. Magnetic flux density induced by the primary winding with the reference position from the center of the diameter as "0"

Figure 5. Overview of the Air Core Transformer Power Supply

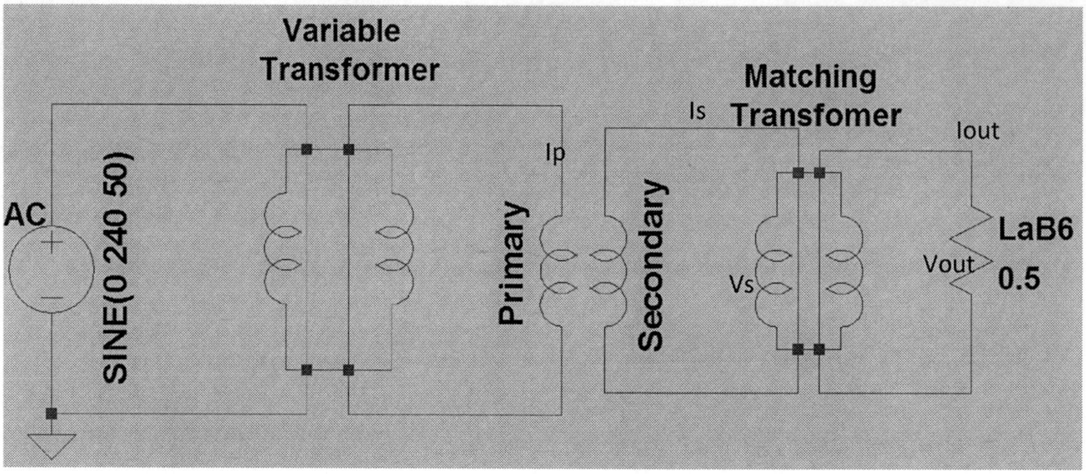

Figure 6. Experiment Setup of the Air Core Transformer Power Supply Circuit.

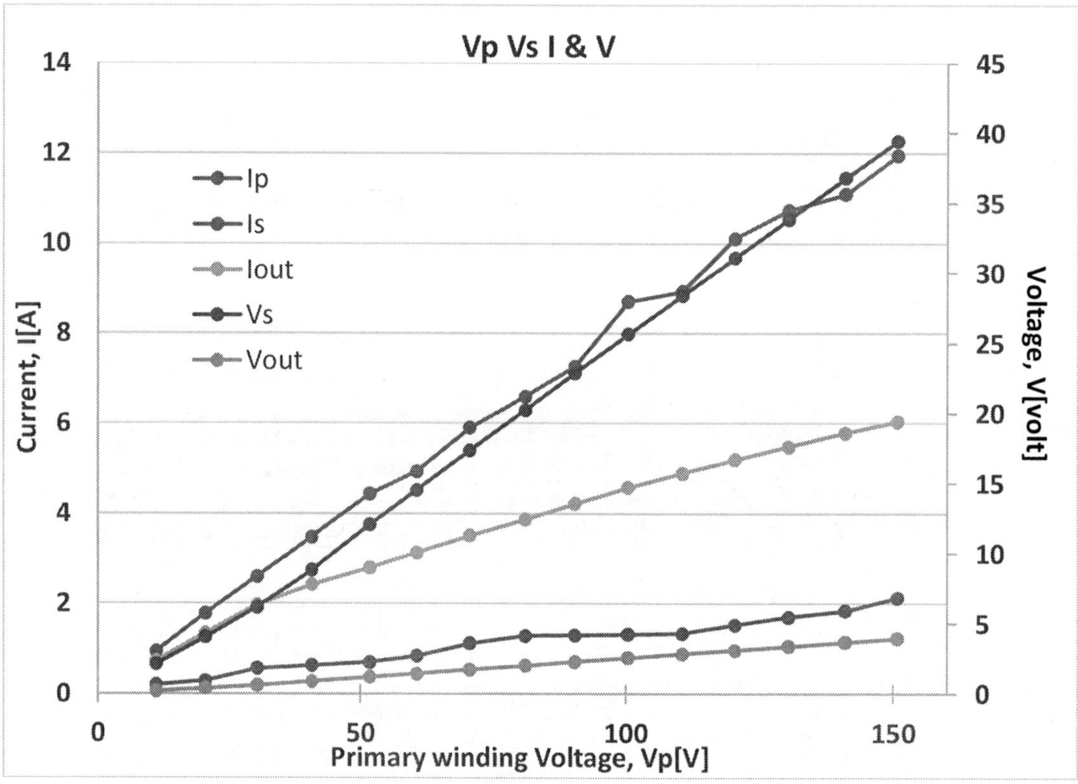

Figure 7. Primary Voltage Vs Current & Voltage of Primary, Secondary and Filament

iii) To optimize the power transfer for filament heating by using a matching transformer According to figure 7, the output voltage is too low, which is incompatible with the output current. Therefore, the matching transformer is required to be changed by the spec of 60-0-60 to 30-0-30. Whereby the setup of the experiment is same as shown in figure 6, only the matching transformer is changed. The result has been improved tremendously by two times as shown in figure 8.

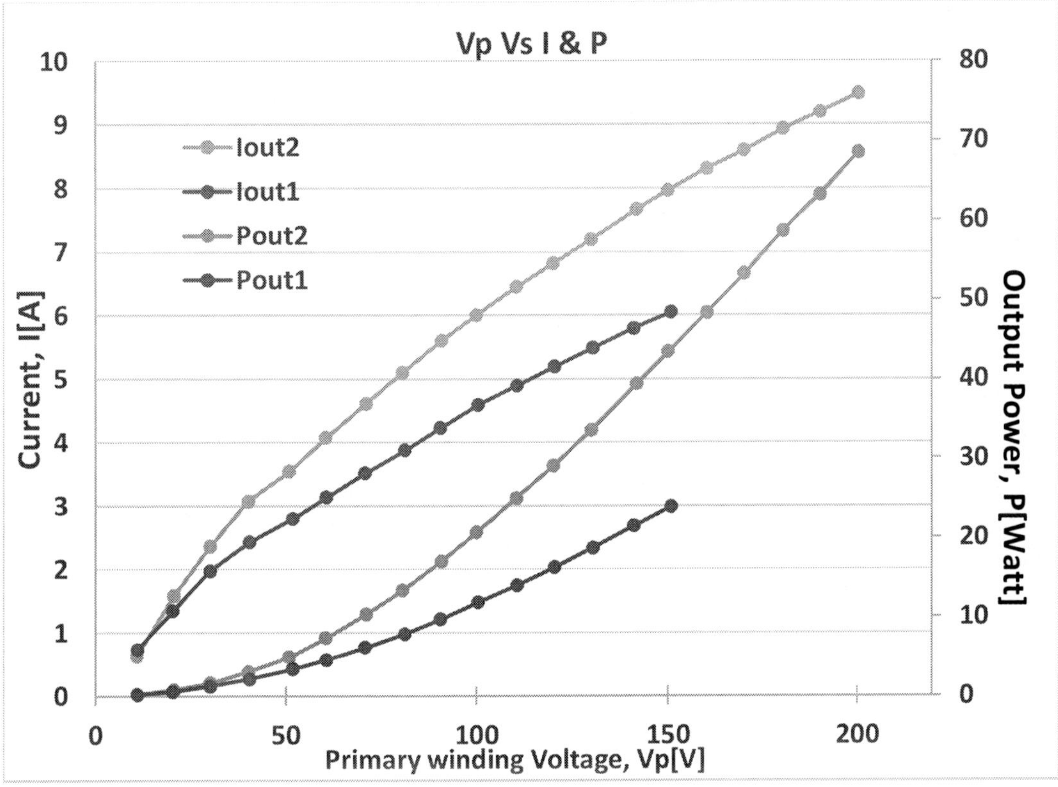

Figure 8. Performance Heating Power with Two Matching Transformer as 60-0-60 and 30-0-30

As the result, the filament power supply with the capability of heating up the filament for 50 watt or higher has been designed and developed. Although, the specs is compatible with our needs but the temperature arise issue during the long time heating must been considered in near future

5. Conclusion

In this paper, parameters of the filament power supply has been identified and developed. Furthermore, the performance of the filament power supply has been tested and the required heating power of 50 watt is achieved. In near future, the filament power supply by using the air core mechanism will be tested using the nitrogen gas inside the pressure vessel and, a motorized control manner will be implemented as well.

6. References

[1] Machi S 2011 Trends for Electron Beam Accelerator Applications in Industry *Reviews of Accelerator Sci. and Tech.* (World Scientific) **4** p 1–10.

[2] Dorman E and Emery R 2013 The University of Washington clinical cyclotron- a summary of current particles and energies used in therapy, isotope production, and clinical research *Proc. of 20th Int. Conf. On Cyclotrons and Their Appl.* (Vancouver, BC, Canada).

[3] Tsutsui H. 2014 *KEK-Nuclear Malaysia-Sokendai Accelerator School in Nuclear Malaysia* 1st-4th December 2014.

[4] Edwards D A and Edwards H T 2008 Particle Colliders for High Energy Physics *Rev. of Accelerator Sci. and Tec.* (World Scientific) **1** p 99–120.

[5] Wilson P B 2008 Electron Linacs for High Energy Physics, *Rev. of Accelerator Sci. and Tech.*, (World Scientific) **1** p 7–41.
[6] Hase Y. 2014 *KEK-Nuclear Malaysia-Sokendai Accelerator School in Nuclear Malaysia* (1st-4th December 2014).
[7] Ghazali A B, Rizal M M, Rosli D, Rizal M C, Lee C H, Leo K W and Zahidee M T 2005 Development of electron beam machine at Nuclear Malaysia *Tech. and Appl. Accelerator Meeting VIII* (P3TM-BATAN, Yogyakarta).
[8] K W Leo *et al.* 2017 Study on the insulation gas system of the 300 keV electron accelerator *Advancing Nuclear Sci. and Eng. for Sustainable Nuclear Energy Knowledge AIP Conf. Proc.* **1799** p 050008-1–050008-8.
[9] K W Leo *et al.* 2016 Study on the parameters of the scanning system for the 300 keV electron accelerator *Advancing Nuclear Sci. and Eng. for Sustainable Nuclear Energy Infrastructure AIP Conf. Proc.* **1704** p 020010-1–020010-12.
[10] K W Leo *et al.* 2018 Study on the window cooling system of 300 keV Electron Accelerator *IOP Conf. Series: Materials Science and Engineering* **298** p 012045.

Acknowledgments
We would like to express our gratitude to Dr. Fadeev, from the Budker Institute of Nuclear Physics, Russia, for his guidance and willingness to join us at the Workshop on Electron Accelerator Technology during which he had shared his experience and knowledge with us theoretically and practically.

IOP Conf. Series: Materials Science and Engineering **555** (2019) 012023 doi:10.1088/1757-899X/555/1/012023

A study on surface contact effect using ground penetrating radar

Mohamad Syafiq Mohd Amin[1, a], Fatasya Izreen Hanim Alias[2], Tengku Sarah Tengku Amran[1], Mohamad Ridzuan Ahmad[1], Ishak Mansor[1], Shaharudin Sayuti[1]

[1]Non-Destructive Testing - Material Structure Integrity group (NDT-MSI), Industrial Technology Division, Malaysian Nuclear Agency, 43000 Kajang, Selangor, Malaysia.
[2]Department of Physics, Faculty of Science, Universiti Teknologi Malaysia (UTM), 81310 Skudai, Johor, Malaysia.

[a]syafiqamin@nm.gov.my

Abstract. In this paper, the effect of contact and non-contact antenna system of ground penetrating radar (GPR) for buried pipes are presented. The frequency for GPR antenna used in this project is 400MHz which is suitable for GPR system to detect buried pipes for the tested surfaces, since the soil involved is sand which has low electrical conductivity. The results of the radargram obtained from this experiment and the limitations occurred are discussed.

1. Introduction

Ground penetrating radar (GPR) is one of Non-Destructive Testing (NDT) method which uses geological method radar pulses to image the subsurface. This NDT method uses electromagnetic radiation in the microwave band of radio spectrum, and detects the reflected signal from subsurface structures. GPR can have application in a variety of field such as geological and hydro-geological investigations, engineering investigations, location and evaluation of buried structures, site inspection and subsurface mapping for underground substructures including pipes. GPR is easily moved to inspection area since the equipment is portable and on-site interpretation due to instant graphic display. The basic GPR technology had aided the development of NDT methods. This can be proved by the growth of GPR since 1970's (Conyers, 1997) and GPR application in various fields.

This project's objectives are to operate basic working principle of GPR and to detect the location of pipes with different sizes, depth and separation by using GPR and the project is attempted at Block 60, Malaysian Nuclear Agency. For this project, antenna with frequency of 400MHz is used since the pipe is below 2 meter from the ground surface and the data obtained is clearer than antenna with higher frequency used. Furthermore, from the data obtained, the limitation of GPR is discovered.

Underground piping is commonly used for utilities in civil engineering. To monitor the exact location and condition of the utilities pipe, engineers normally map the underground layer using various methods, especially GPR. Unfortunately, due to location of the new un-developed site or developed site, most civil site is not smooth and prevent GPR antenna to make a full contact with ground surface.

2. Experimental works

2.1. Pre-data collection

To conduct the GPR, firstly, we need to know the target characteristic which is either the target is metal lines, plastic pipes, a concrete culvert or bedrock. In this project, the objective is to determine the location of pipes with 3 different conditions – depth, separation and size. Besides knowing the target characteristics, the electrical properties of and the pipes and soils should we understand since the GPR's antenna emits electromagnetic impulses. It is related to dielectric constant and conductivity of materials. Then, we should know the capability of antenna used by predicting the survey area such as depth of the pipes and types of soil involved in the survey. In this project, the depth is about 2 meter from the ground surface and type of soil involved is sand. After that, we need to choose the survey design either in 2D or 3D. Since the target is utility line and they are in a known direction, it may to do in 2D transect.

Figure 1. Location of pipes.

2.2. During data collection

When collecting the data, the antenna is dragged along the survey line to identify pipe positions, depths and sizes on the surface. While conducting this project, two different condition of antenna that is the level of air gap between the antenna and the ground. The radargram will appear on the screen at SIR 4000. Also, the depth and position of pipes measured by using cross-hairs cursor. The corresponding X (distance) and Y (depth), coordinates are shown in the bottom right corner.

2.3. Post-data collection

Then, the data is transferred to PC and analysed using the computer monitor for data interpretation. A postprocessing software called RADAN (Figure 2) is used so that the data analysis is more accurate and precise. Some filtrations are done to achieve the objectives of this project.

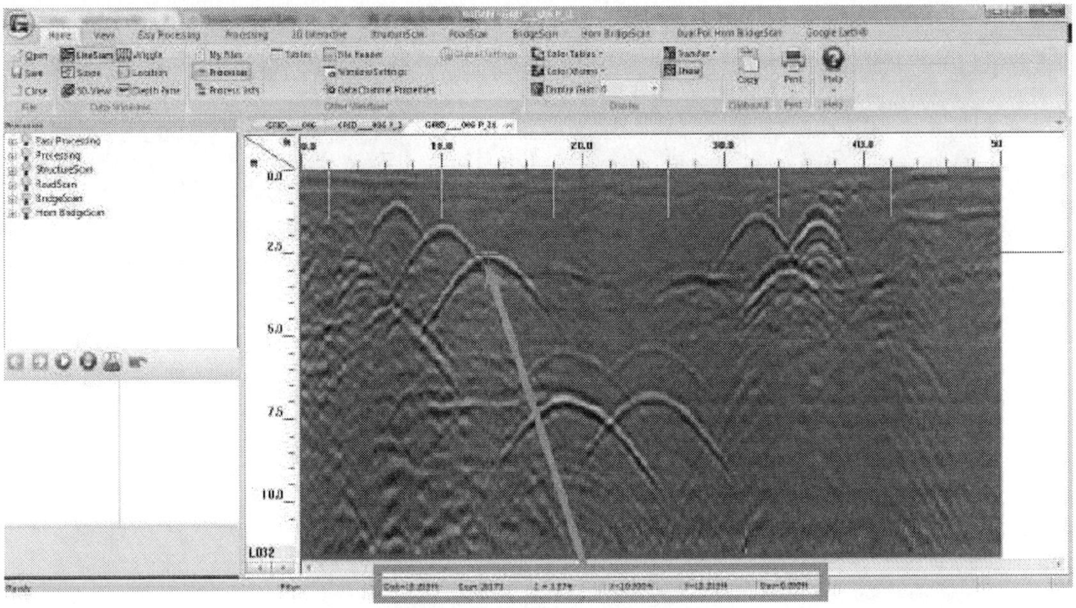

Figure 2. RADAN software.

The test site was filled with dry sand. The metal (iron) pipe was buried into the test area in three conditions. First with different pipe separation gap. Second, pipe with different depth, and the third pipe with different size. The antenna setting for 400 MHz.

3. Result and analysis

The purposes of this project are to demonstrate the basic working principle of GPR and to use this technique to detect the location of pipe with different sizes, depths and separations. The project was done with two conditions of antenna which was attached and detached on the ground. From the Figures 4, 6 and 8 (where the antenna was attached on ground), the hyperbolic shape on radar profile is more complete compared to the Figure 3, 5 and 7 (where the antenna detached on the ground). This changing pattern is due to the electromagnetic impulses are able to penetrate within the soils until the buried pipes. When the antenna is in contact to the ground, there is no gap between them. The air gap contains particles that would absorb the energy from the electromagnetic waves. The energy is not lost when the electromagnetic waves reach the ground surface, which then is able to detect the buried pipes with fewer obstacles. Thus, the hyperbolic shape obtained is more complete and clearer than when the antenna is not in contact with the ground surface.

IOP Conf. Series: Materials Science and Engineering **555** (2019) 012023 doi:10.1088/1757-899X/555/1/012023

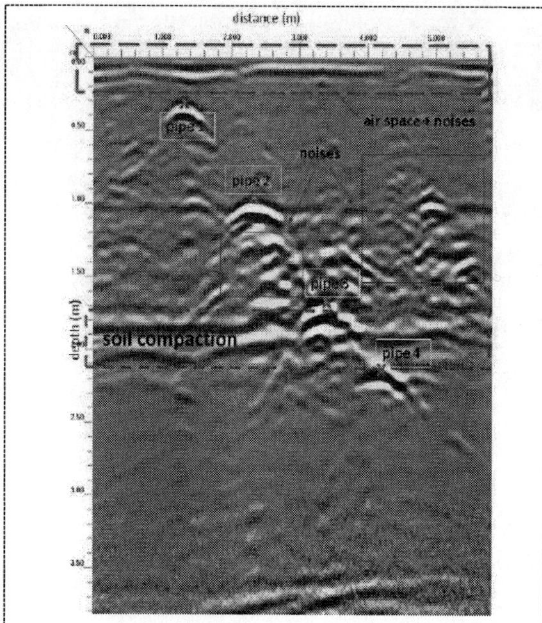

Figure 3. Pipes with different depth: radar profile for antenna detached on ground.

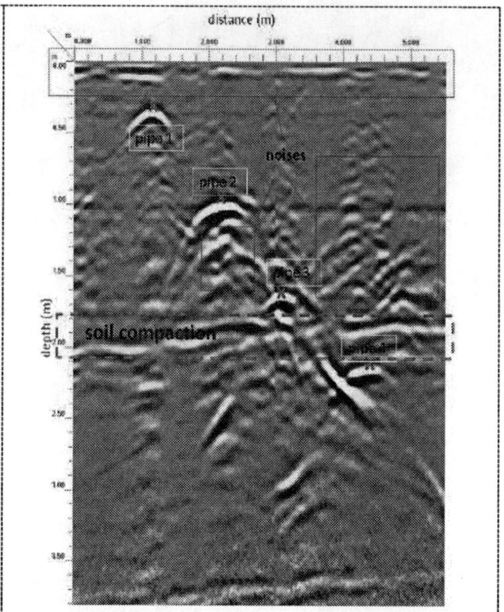

Figure 4. Pipes with different depth: radar profile for antenna attached on ground.

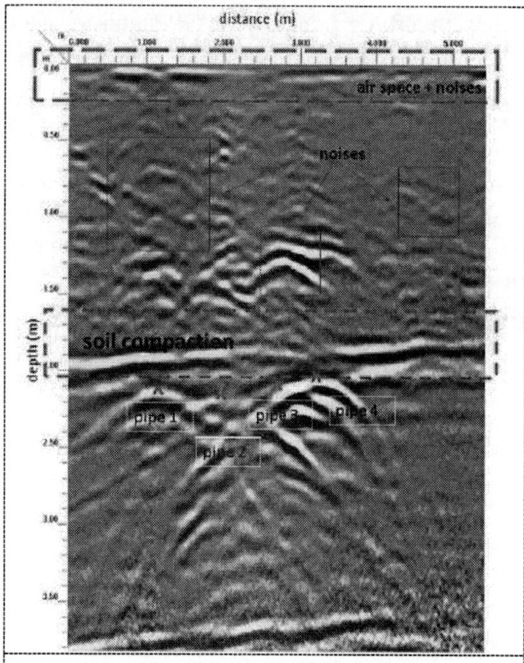

Figure 5. Pipes with different separation: radar profile for antenna detached on ground.

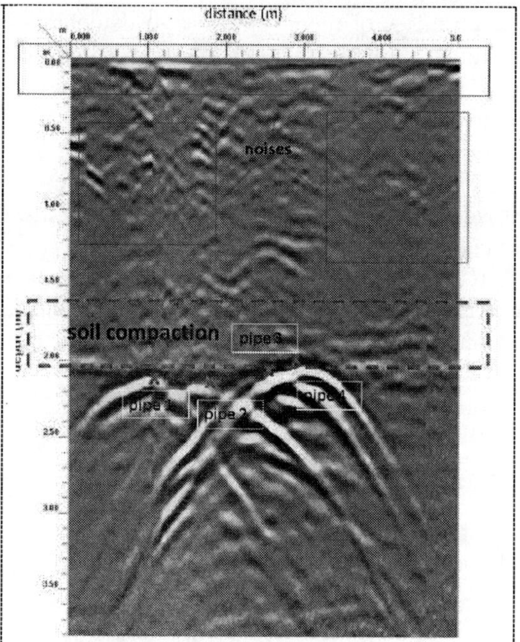

Figure 6. Pipes with different separation: radar profile for antenna attached on ground.

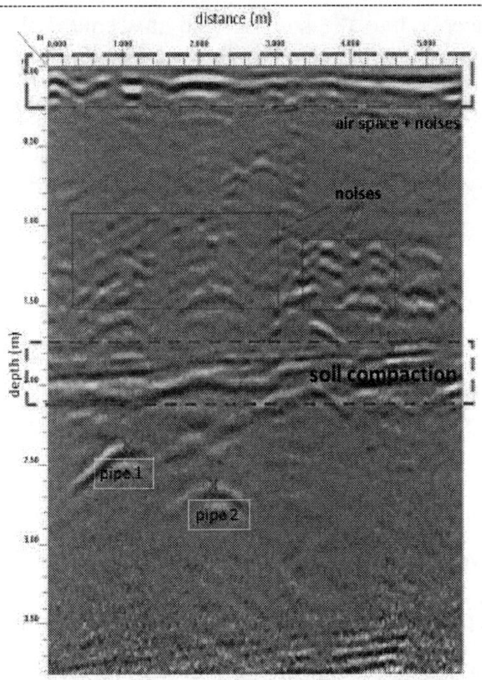

Figure 7. Pipes with different size: radar profile for antenna detached on ground.

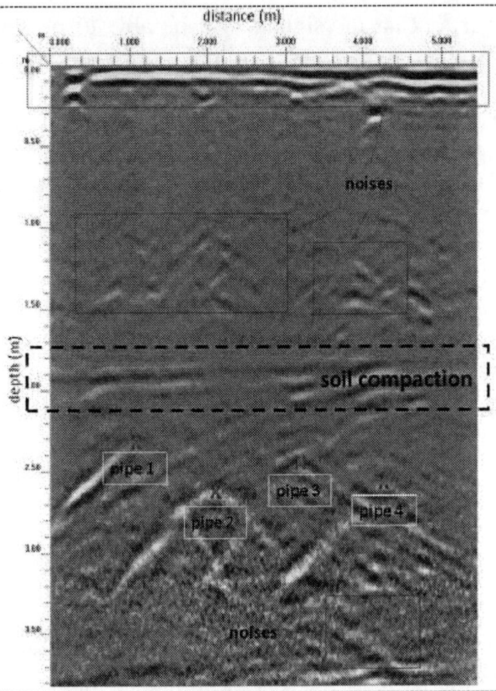

Figure 8. Pipes with different size: radar profile for antenna attached on ground.

The results obtained from all radargrams are in hyperbolic form. This is due to the distance of the antenna and the detected buried pipes. As the GPR was placed far from the buried pipe, the incident electromagnetic waves need to travel with a long path to reach the pipe. Then the reflected waves from the buried pipes received by the antenna travel with a long path as well. From here, it is reasonable to conclude that there was significant energy loss before it reached the antenna. When the GPR was moved closer towards the pipes, the distance travelled by the electromagnetic waves to reach the buried pipes and reflected to the antenna was shorter. Due to these differences in distance and energy loss, the density of the reflected waves received by the antenna was different as well.

The noises from all figures were detected due to the sensitivity of GPR. GPR is sensitive and has a high resolution coverage of the survey area. Therefore it is able to detect small material, including particles of soil and air. However, the noises showed in the radargram were not significant. To overcome this problem, some filtrations can be done during data interpretation by using a software called RADAN. We only focused on the perfect hyperbolic shape that shows the location of buried pipe, and then filtered all the unwanted noises.

The continuous lines on all radar profiles indicate soil compaction. The GPR is also able to detect soil compaction. Compaction of soil contains dense and many particles, so some of the energy from the electromagnetic waves were captured before the waves reached the target. If there is no soil compaction, we can assume that the hyperbolic shape from the radargram is clearer than the results obtained.

There was a limitation of GPR. If the target is too deep and small, the signals would not go through the target because the energy is lost before the signals reach the target. Figure 7 (radar profile of pipes with different sizes with antenna detached on the ground) illustrate this point; the third and fourth pipes were not detected by GPR. To overcome this problem, antenna with high frequency should be used. According to the relation between frequency, wavelength and energy, higher frequency of antenna result in shorter signal wavelength, thus the greater the energy, the higher the penetrating power so that the signals will be able to penetrate deeper into the soil. The signals emitted by the transmitter antenna are also limited in materials with high electrical conductivity such as clays.

4. Conclusion

In conclusion, there are possible techniques to detect all pipes with both conditions - antenna attached and detached on ground. However, the hyperbolic structure of radar profile is more complete when the antenna is attached to the ground.

5. References

[1]. Conyers B 1997 *Ground Penetrating Radar for Archaeology* (Rowman & Littlefield Publishers, Inc).

[2]. Yelf R J 2007 *Application of Ground Penetrating Radar to Civil and Geotechnical Engineering Electromagnetic Phenomena* **7(1)**, p 102-117.

[3]. Hunaidi O, Chu W, Wang A and Guan W 2000 *Detecting Leaks in Plastic Water Distribution Pipes* **92(2)** (Nournal AWWA) p 88-94.

[4]. Davis J L and Annan A P 1989 Ground penetrating radar for high resolution mapping of soil and rock stratigraphy *Geophys. Prospecting*, **37**, p 531-551.

[5]. Harith Z Z T . and Rosli BS 2003 The effectiveness of ground penetrating radar in detecting buried objects (University Sains Malaysia).

[6]. Pain R, Nordiana M M, Sulaiman N, Hisham H 2016 The analysis of EM wave for different media by GPR technique (University Sains Malaysia).

[7]. Lee K F, Wang T K, Kang Y M, Wang C S and L A 2013 Identification of pipelines from the secondary reflect wave travel time of ground penetrating radar waves *J. of Marine Sci. and Tech.*

Acknowledgments

The authors would like to acknowledge the financial support from MESTECC and Malaysian Nuclear Agency as well as the Material and Structure Integrity Group (NDT-MSI) for the support.

IOP Conf. Series: Materials Science and Engineering **555** (2019) 012024 doi:10.1088/1757-899X/555/1/012024

Preliminary analysis of gamma and neutron irradiation on sand and activated carbon by using SEM, XRD and SAXS

Asyraf Arif Abu Bakar[1, a)], Anas Muhamad Pauzi[1, b)], Abdul Aziz Mohamed[1, c)], Faridah Mohamad Idris[2,d)] Hafizal Yazid[2,e)]

[1]College of Engineering, Universiti Tenaga Nasional, Jalan IKRAM-UNITEN, 43000 Kajang, Selangor, Malaysia
[2]Malaysian Nuclear Agency, Bangi, 43000 Kajang, Selangor, Malaysia.

[a]aasyraf@uniten.edu.my
[b]anas@uniten.edu.my
[c]azizm@uniten.edu.my
[d]faridah_idris@nuclearmalaysia.gov.my
[e]hafizal@nuclearmalaysia.gov.my

Abstract. Gamma and neutron irradiation effect on sand and activated carbon were studied in this paper intensively and, were found from various studies to have strong correlation between neutron fluence or gamma energy on physical and mechanical damage of materials. Sand and activated carbon were irradiated with neutron at Reactor TRIGA PUSPATI (RTP) and gamma at Sinagamma Malaysia Nuclear Agency (Nuklear Malaysia). Morphological analysis was carried out on the irradiated elements using the SEM, XRD and SAXS which showed changes on the microstructure of irradiated samples. Irradiation of activated carbon showed increase in the numbers of pores while change of textural profile of the surface take place at gamma irradiated sand. XRD pattern graph did not indicate any changes, however the specific surface area for both irradiation materials decreased.

1. Introduction

The National Security Council (NSC) of Malaysia secretary Datuk Mohamed Thajudeen confirmed that the massive floods which hit Kelantan in 2014 was among the worst in history. In these times of tragedy, people in affected areas usually suffer from shortages of clean water and electricity. Due to the limitations of the transportation network during emergency, only limited amount of clean water supplies could be sent to evacuation centres. As a result, some people who were familiar with water filtration often resorted to building simple home-made water filtration system using empty bottles and filter elements (cotton/coffee filter, sands and gravels)[1]. The rate of producing clean water from the flood water using this method is rather slow as several cycle filtering are required before the water can be used. This can be improved by using better filter materials. One of the ways of bettering the existing filter materials is to modify the properties of the materials. Studies had been done previously where silica (closest form of compound to sands or gravels that are commonly used as filters) was irradiated with neutron and gamma [1][2]. The effect of radiation to the adsorption and mechanical properties of activated carbon was also

IOP Conf. Series: Materials Science and Engineering **555** (2019) 012024 doi:10.1088/1757-899X/555/1/012024

studied [3]. Several authors observed that irradiation could introduce defects which resulted in the formation of micropores and different levels of irradiation level would result in different concentration of defects [4][5]. It is established that higher amount of irradiation results in higher concentration of defects in the irradiated materials[2].

2. Methodology

A Scanning Electron Microscope (SEM) was used to examine the surface structure (micropore) of non-irradiated and irradiated sand and activated carbon. The samples were coated with platinum to enhance their image visibility. X-Ray Diffraction (XRD) analyses were carried out using 6000 Shidmadzu, Japan at Universiti Tenaga Nasional (UNITEN). XRD patterns were measured with the following conditions: Cu K-alpha, (λ = 0.15418 nm), tube voltage of 30 kV, current of 20 mA, scanning speed of 4°/min and 2θ of 20° to 75°. The results were analysed by using X'Pert Highscore Plus software. The surface area of samples was examined through Small Angle Scattering of X-Ray (SAXS) to characterize the porosity of the samples. The SAXS uses X-ray as a beam to penetrate the nanostructured material. The process took three frames where each of frame lasted for five minutes. The scattering pattern was further analysed in results and discussion. Table 1 and Table 2 describe the sample properties and the testing information of the SAXS procedure.

Table 1. Sample properties.

No	Sample	Form	Solid Density	Apparent Density
1	Non-irradiated Activated Carbon	Powder	2.26 g/cm^3	0.58 g/cm^3
2	Activated Carbon (Gamma)	Powder	2.26 g/cm^3	0.58 g/cm^3
3	Activated Carbon (Neutron)	Powder	2.26 g/cm^3	0.58 g/cm^3
4	Non-irradiated Sand	Powder	2.65 g/cm^3	1.09 g/cm^3
5	Sand (Gamma)	Powder	2.65 g/cm^3	1.09 g/cm^3
6	Sand (Neutron)	Powder	2.65 g/cm^3	1.09 g/cm^3

Table 2. Testing information.

Testing Information	Details
Sample	1-6
X-rays	Cu-Kα, λ = 0.1542nm
Power	50W (50 kV, 1 mA)
Sample size	0.5 mm (thickness)
Detector	2D Eiger R Series (Dectris), 75 x 75 μm^2
Calibration	Ag-behenate (d= 58.38Å)
Measured q-range	0.035 nm^{-1} < q < 5 nm^{-1} (SAXS) 0.035 nm^{-1} < q < 23.5 nm^{-1} (WAXS)
Temperature	20°C
Exposure time	180 s

3. Results and discussion

The surface morphology of the non-irradiated and irradiated by gamma and neutron was evaluated by scanning electron microscope (SEM) to confirm the effect of gamma and neutron treatment on the sand and activated carbon. Figure 1 shows SEM images of the non-irradiated activated carbon, gamma activated carbon, and neutron activated carbon in 5000× magnification. These images show the changes on their surface structure after irradiation. It can be observed that there were many pores formed on them. This was due to the ability of gamma-rays and neutrons to cause atomic displacement [13]. They interacted with carbon nucleus mainly through Compton scattering and absorption. The recoiled electrons with sufficiently high energy from the Compton scattering would collide with neighbouring carbon nuclei, shifting them from their lattice position. Then, the knocked-on atoms would continue to collide with their neighbouring atoms, resulted in transferring of energy. If the energy exceeded the threshold of further interactions and displacements, it would result in cascading damage [33]. The average size of pores was taken using dimension tools provided by the SEM where pores were chosen at random. The neutron-radiated activated carbon was average pore diameter of 0.7881 μm. Meanwhile, the non-irradiated activated carbon and gamma-radiated activated carbon showed no significant difference in average pore diameters which were 1.429 μm and 1.603 μm respectively.

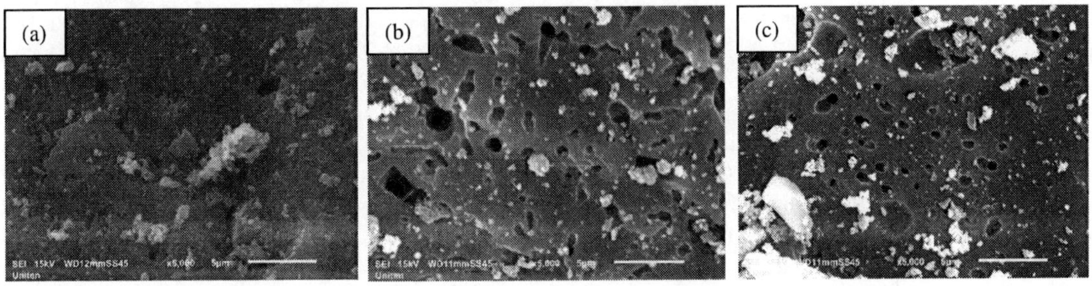

Figure 1. SEM image at (5000×) magnification of: (a) non-irradiated activated carbon, (b) irradiated activated carbon with 50kGy gamma dose, (c) irradiated activated carbon with 2.000×10^{12} cm^{-2} s^{-1} neutron flux.

Figure 2 shows the SEM images of non-irradiated and irradiated sand at 4000× magnification. Generally, sand (mainly build-up from quartz) has a particle shape with higher sphericity index and lower internal porosity [3][4]. The SEM image for gamma-radiated sand in Figure 4.2(b) shows the change in textural profile of its surface resulting valley and hill areas as compared to the non-irradiated sand in Figure 4.2(a) and the neutron-radiated sand in Figure 4.2(c). Pores were only visible on the surface of the non-irradiated sand and the neutron-radiated sand where their average sizes were 1.334 μm and 1.043 μm respectively. The change in topography of the surface and the small average size of pores prove that irradiation of neutron and gamma-ray can induce micropores or defects in the materials [4][5].

IOP Conf. Series: Materials Science and Engineering **555** (2019) 012024 doi:10.1088/1757-899X/555/1/012024

Figure 2. SEM image at (4000×) magnification of: (a) non-irradiated sand, (b) irradiated sand with 50 kGy gamma dose, (c) irradiated sand with 2.000×10^{12} cm^{-2} s^{-1} neutron flux.

Based on the discussion above, the neutron-radiated activated carbon has the smallest average pore size of 0.7881 μm. This might contribute to the highest capability of filtration as the small pores could trap foreign and contaminated substances in the water. The XRD analysis was performed to investigate the structure and crystallinity of the non-irradiated and irradiated sand and activated carbon which are presented in Figure 3 and Figure 4 respectively. It is known that if the radiation has high enough energy, it can cause build-up of defects on the materials. This may collapse the crystalline lattices to the extent of inducing crystalline-amorphous transition due to the existing of vacancies-interstitials pair [3].

Based on the XRD profiles, the non-irradiated sand shows the highest peak which indicates that it is highly crystallized with 011 hkl at 2θ of 26.6586°. This suggests that it has hexagonal structure [4]. Using Debye-Scherrer equation on high intensity peak, the average crystallite size of particles was calculated to be 62.39 nm. Apparently, the XRD pattern for irradiated sand did not show substantial changes, following the same pattern as the non-irradiated sand. However, there were nominal changes in the average crystallite size of gamma-radiated sand and neutron-radiated sand which were 68.46 nm and 64.50 nm. Compared to non-irradiated sand, both of the irradiated sand samples had slightly small crystallite size which might improve their filtration effectiveness [5][6].

Figure 4 illustrates the XRD spectra for the activated carbon samples. The diffractogram pattern of these samples exhibits two distinctive broad reflections in the 2θ angles approximately at 25° and 44°. The peak broadening is an indication of an amorphous structural aspect. The broad reflection in the 2θ at 25° and 44° are associated with (002) and (001) diffraction line respectively. The similar pattern shown concludes that nothing had changed in the amorphous structure of activated carbon after being irradiated by gamma and neutron. This happened because radiation does not carry sufficient energy to cause the change [3].

IOP Conf. Series: Materials Science and Engineering **555** (2019) 012024 doi:10.1088/1757-899X/555/1/012024

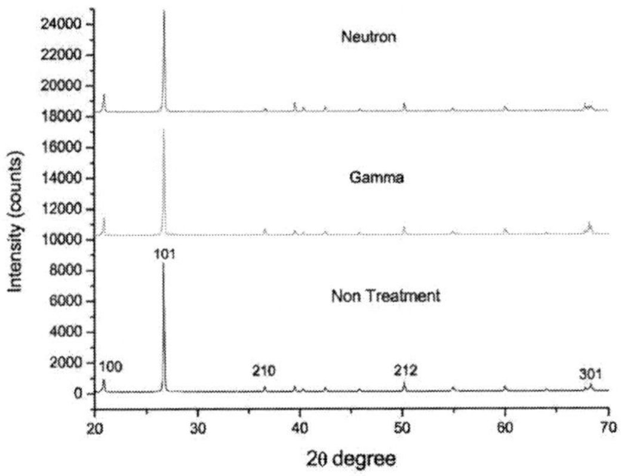

Figure 3.. XRD of Sand.

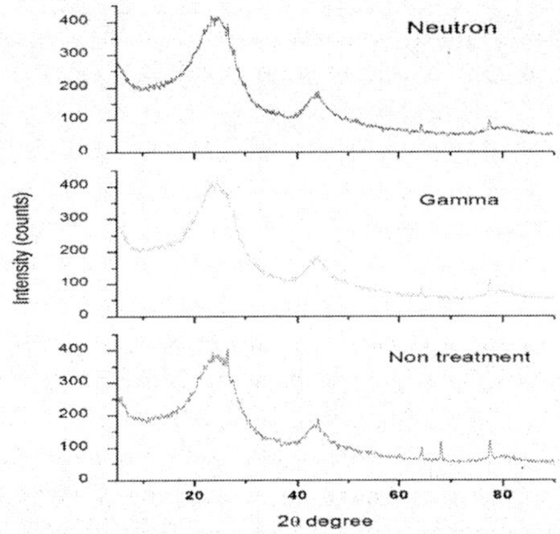

Figure 4. XRD of Activated Carbon.

However, a previous study showed that ionising radiation can cause loss of crystallinity or loss of weight in graphite. This stems from the breakage of the chemical bonds which results in the creation of defects and in the loss of regular structure. In the case of persistent or higher concentration of radiation, a fraction of the defects condenses and may form extended defects by arranging carbon atoms in the basal layers [7].

A sample of non-irradiated sand was evaluated with SAXS to identify the nanosized particle. When X-rays penetrate a nanostructured material, they are scattered on the interfaces of the nanostructures at a certain scattering angle. The resulting scattering pattern is unique to the structure and this is used to characterize the nanomaterial. The data were processed with SAXStreat and SAXSquant softwares. Then, evaluation was performed with the SAXSquant and SGI (Space group indexing). Table 3 shows specific surface area value for each sample.

Table 3. Specific Surface area, Si value.

Sample	Si (m^2/g)
Non-irradiated Activated Carbon	1437.5
Gamma Activated Carbon	863.6
Neutron Activated carbon	679.9
Non-irradiated Sand	122.6
Gamma Sand	104.6
Neutron Sand	81.89

Based on Table 3, the specific surface area of the sample decreased after irradiation. As the area decreased, the particle size became small. Specific surface area can also decrease if the sample has less pores [8]. These results seem to agree with a previous study which mentioned that the gamma-irradiated activated carbon had reduced specific surface area and micropores volume [9]. Gamma-rays interactions with bulk materials are different from that of charged ions and neutrons. It is known that the probability of photoelectric interaction and pair production is insignificant for the case of highly-energetic gamma-rays and absorbing material of low atomic number such as graphite. Therefore, Compton scattering is believed to be the major interaction mechanism of high-energy photons with the carbon atoms in graphite lattice [7]. Another reason for the decreasing specific surface area of sand is due to the energy transferred to the quartz, resulting in the knocking and breaking of some of the porous wall. This blocks the porosity formation, disturbing the surface area of micropores [10]. Table 4 shows the summary of radiation effects of sand and activated carbon

Table 4. Summary of radiation effects.

<table>
<tr><th colspan="2">Analysis
Sample</th><th>Morphology</th><th colspan="2">Structure</th><th colspan="2">Surface Area (m2/g)</th></tr>
<tr><td rowspan="2">Sand</td><td>Neutron</td><td>•small number of pore does visible in SEM image
•have small average pore diameter = 1.043μm compared to non-irradiated (1.334 μm)</td><td rowspan="2">•The XRD shows same pattern throughout all types of samples.
•The abundant element is SiO2 (quartz) indicates the hexagonal structure</td><td rowspan="2">•Nominal change in average crystallite size irradiated sand with gamma is 68.46nm
•irradiated sand with neutron is 64.50nm (non-irradiated =62.39nm</td><td>•81.89 m2/g</td><td rowspan="2">•Specific surface area were decreasing after irradiate with gamma and neutron from non-irradiated is 122.6m2/g</td></tr>
<tr><td>Gamma</td><td>•change of textural profile of the surface compared to non-irradiated and neutron sand
•pore does not visible in SEM image</td><td>•104.6 m2/g</td></tr>
<tr><td rowspan="2">Activated Carbon</td><td>Neutron</td><td>•Increasing number of pores
•have small average pore diameter = 0.7881μm compared to non-irradiated and gamma sample</td><td colspan="2" rowspan="2">•The diffratogram pattern of these samples exhibits 2 distinctive broad reflections in the 2θ angles located approximately at 25° (002) and 44° (001).
•The peak broadening is an indicative of an amorphous structural aspect.</td><td>•679.9 m2/g</td><td rowspan="2">•Specific surface area were decreasing after irradiate with gamma and neutron from non-irradiated is 1437.5 m2/g</td></tr>
<tr><td>Gamma</td><td>•Increasing number of pores
•average pore diameter (1.429 μm)does not have significant change compared to non-irradiated activated carbon (1.603 μm)</td><td>•863.6 m2/g</td></tr>
</table>

4. Conclusions

It was hypothesised, based on previous studies, that the irradiation of sand and activated carbon might defect the structure of materials with sufficient energy and time. In this research, the effect of gamma-rays and neutrons to sand and activated carbon were analysed using SEM, XRD and SAXS. Different radiation treatment produced different effect to the structure of the materials[2]. We observed more irradiation energy is required to obtain further understanding related to structural changes in the samples.

5. References

[1] Anuar N, Pauzi A M, and Bakar A A A 2017, "Methodology of water quality index (WQI) development for filtrated water using irradiated basic filter elements" *AIP Conf. Proc.* **1799** p 040010.

[2] Bakar A A A, Pauzi A M, Mujaini M, Mohamed A A andMohamad Idris F 2018 Study on the irradiation effect towards water filtration element *IOP Conf. Series: Materials Sci. and Eng.* **298** p 012024.

[3] Pedraza D F 1988 Radiation-induced collapse of the crystalline structure *J. of the Less Common Metals* **140** p 219-230.

[4] Bahoria B V, Parbat D K and Nagarnaik P B 2018 XRD Analysis of Natural sand, Quarry dust, waste plastic (ldpe) to be used as a fine aggregate in concrete *Materials Today: Proc.* **5** p 1432-1438.

[5] Kaminski I, Vescan N and Adin A 1997 Particle size distribution and wastewater filter performance *Water Sci. and Tech.* **36** p. 217-224.

[6] Bakar A A A, P. Pauzi A M, Mohamed A, Sharifuddin S S and Mohamad Idris F 2018 Preliminary analysis on the water quality index (WQI) of irradiated basic filter elements *IOP Conf. Series: Materials Sci.and Eng.* **298** p 012005,.

[7] Marsden R K A N J B J 2015 Gamma Radiation Studies on Reactor Graphite Grades.

[8] Wolfrom R L Specifics on surface area (Online).

[9] Ercin D, Eken M, Aktas Z, Cetinkaya S, Sakintuna B and Yurum Y 2004 Effect of gamma-irradiation of activated carbons on the generation of new free radicals and porosity.

[10] Nishijima S, Matsumura S, Asakuma Y and Saptoro A 2017 Promotion of nucleation for nano-particle formation by two-stage microwave irradiation *IOP Conf. Series: Materials Sci. and Eng.* **206** p 012055.

Acknowledgement

Special thanks for Malaysia's Ministry of Higher Education (MOHE) for sponsoring the research project under the Fundamental Research Grant Scheme (FRGS) project number 117289-136338.

Impact of gamma and electron irradiations on superconducting properties of $Bi_{1.6}Pb_{0.4}Sr_2Ca_2Cu_3O_{10}$ (Bi-2223) superconductor

Zaahidah 'Atiqah Mohiju[1], Nasri A. Hamid[1,a)], Yusof Abdullah[2]

[1]Nuclear Engineering and Energy Group, College of Engineering, Universiti Tenaga Nasional, Jalan IKRAM-UNITEN, 43000 Kajang, Selangor, Malaysia

[2]Materials Technology Group, Industrial Technology Division, Malaysian Nuclear Agency, Bangi, 43000 Kajang, Selangor, Malaysia

[a]Nasri@uniten.edu.my

Abstract. In several studies on the impact of radiation towards superconductor, there were evidences that changes may occur in superconducting properties due to alteration in lattice structure of HTS materials. In this work, the solid-state reaction method was employed in preparing the bulk samples of $Bi_{1.6}Pb_{0.4}Sr_2Ca_2Cu_3O_{10}$ (hereafter Bi-2223) superconductor. The samples were then irradiated with gamma and electron sources with dosage of 100 kGray each. X-ray powder diffraction (XRD) patterns were used to determine the phases that present in the samples. Results of XRD patterns for both the non-irradiated and irradiated samples showed well-defined peaks all of which could be indexed based on the Bi-2223 phase structure. Besides, the XRD patterns indicate that gamma and electron irradiations did not affect the Bi-2223 superconducting phase structure. Measurements of the critical temperature, T_C and the critical current density, J_C of all the samples were conducted before and after irradiation using AC susceptibility technique. T_C for non-irradiated, electron irradiated and gamma irradiated sample is 97 K, 95 K and 103 K, respectively. On the other hand the J_C for non-irradiated, electron irradiated and gamma irradiated sample is 1618 A/cm^2, 1729 A/cm^2 and 1549 A/cm^2, respectively. Results of scanning electron microscopy (SEM) micrographs show improvement in the grain growth and disorientation in the texture of all the gamma and electron irradiated samples.

1. Introduction

Potential applications in the electricity distribution industry are more promising. Among the use of superconducting wire is to produce a transmission cable – in urban areas, where large amounts of power must be transmitted by underground cables, high current density is important and thus superconducting wires are an attractive proposition. There is also interest in the development of transformers, where superconducting materials would have several advantages over those using conventional materials. Another application of superconducting materials is the fault current limiter, which uses the resistive transition to reduce the fault current in an electrical circuit.

Radiation effects including electron irradiation and gamma irradiation are caused primarily by alteration in the structure of the lattice, and the study of radiation effects have been mainly an attempt to resolve these alterations into a moderate number of lattice defects, and to elucidate the properties of these defects [1]. Moreover, the large difference between the critical currents achieved in textured materials and random polycrystals in early experiments on high Tc materials indicated that the grain boundaries are responsible for reducing J_C [2]. This was first demonstrated directly by Chaudhari et al. [3], where the critical current inside a grain and across a grain boundary in yttrium-based superconductor (YBCO) film was studied and found that the inter-grain critical current to be more than an order of magnitude lower.

The effect of the grain boundary misorientation angle was studied by Dimos *et al.* [4] who fabricated thin films of YBCO on $SrTiO_3$ (STO) bicrystals. Grain boundaries with three different geometries have been studied; the transport properties of all three types of boundaries are essentially identical, which implies that the poor superconducting coupling between grains is the result of the intrinsic structural disorder at the boundary. The purpose of this research work is to study the impact of electron and gamma irradiations on Bi-2223 superconducting properties by applying the same amount of radiation exposure. All the results are presented in the following section.

2. Experimental

Specimens with nominal composition $Bi_{1.6}Pb_{0.4}Sr_2Ca_2Cu_3O_{10}$ were prepared via conventional solid state method. Samples were made of oxide and carbonate powders, homogenously mixed according to its stoichiometric molar ratio of Bi_2O_3, PbO, Sr_2CO_3, $CaCO_3$ and CuO. The mixture was ground using mortar and pestle, and heated in a tube furnace with free air flow at temperature of 810°C for 30 hours to remove impurities and form the Bi-2223 oxide. Then, the oxide powder was pressed using hydraulic press machine at room temperature with pressure of 7 tonnes. Bulk samples of 13 mm in diameter and 3 mm in thickness were formed and synthesized under the same condition at 850°C for 60 hours, then furnace-cooled to room temperature.

Electron irradiation on one of the sample was carried out using a beam of 3 MeV, current measurement of 10 mA and radiation doses of 100 kGray. The irradiation was conducted using the EPS-3000 electron-beam accelerator. Meanwhile, another sample was exposed to gamma ray irradiation and it has been conducted in SINAGAMA Plant using JS10000 (IR-219) irradiator with radiation doses of 100 kGray. Both facilities are provided by Malaysian Nuclear Agency.

Structural investigations of non-irradiated and irradiated samples were conducted using a Bruker D8 Advanced X-Ray Diffractometer (XRD) to confirm their phase formation. Microstructure investigation was carried out using a Hitachi S3400N Scanning Electron Microscope (SEM). Superconducting properties for each sample, such as transition temperature, T_C and the critical current density, J_C were determined by using AC susceptibility technique where the magnetic field varies sinusoidally.

3. Results and analysis

Figure 1 shows the XRD patterns of Bi-2223 for both non-irradiated and irradiated samples. The XRD patterns for Bi-2223 of non-irradiated sample show sharper and well-defined (0010) and (119) peaks. A few unknown peaks and other impurities are also observed. The position of diffraction peaks of Bi-2223 phase in all patterns are identical, which suggests that no change in lattice parameters occurred within the resolution of XRD. Visible changes at the intensity of peak (1111) after being exposed with 100k Gray of electron and gamma irradiations are observed. This indicates that irradiations could increase the volume fraction of Bi-2223 phase and promotes crystal growth to the superconductor ceramics [5]. Meanwhile, lattice parameters for all prepared samples were calculated from Bragg's law and Miller indices *hkl*. Since the values obtained for a = 5.2662 Å, b = 5.4705 Å and c = 38.3014 Å, so it is characterized as orthorhombic. According to previous research work, partial substitution of Bi by Pb is effectively enhanced the phase formation kinetics as well as increased the yield of Bi-2223 phase [4]. In our study, similar substitution was used in preparing our samples.

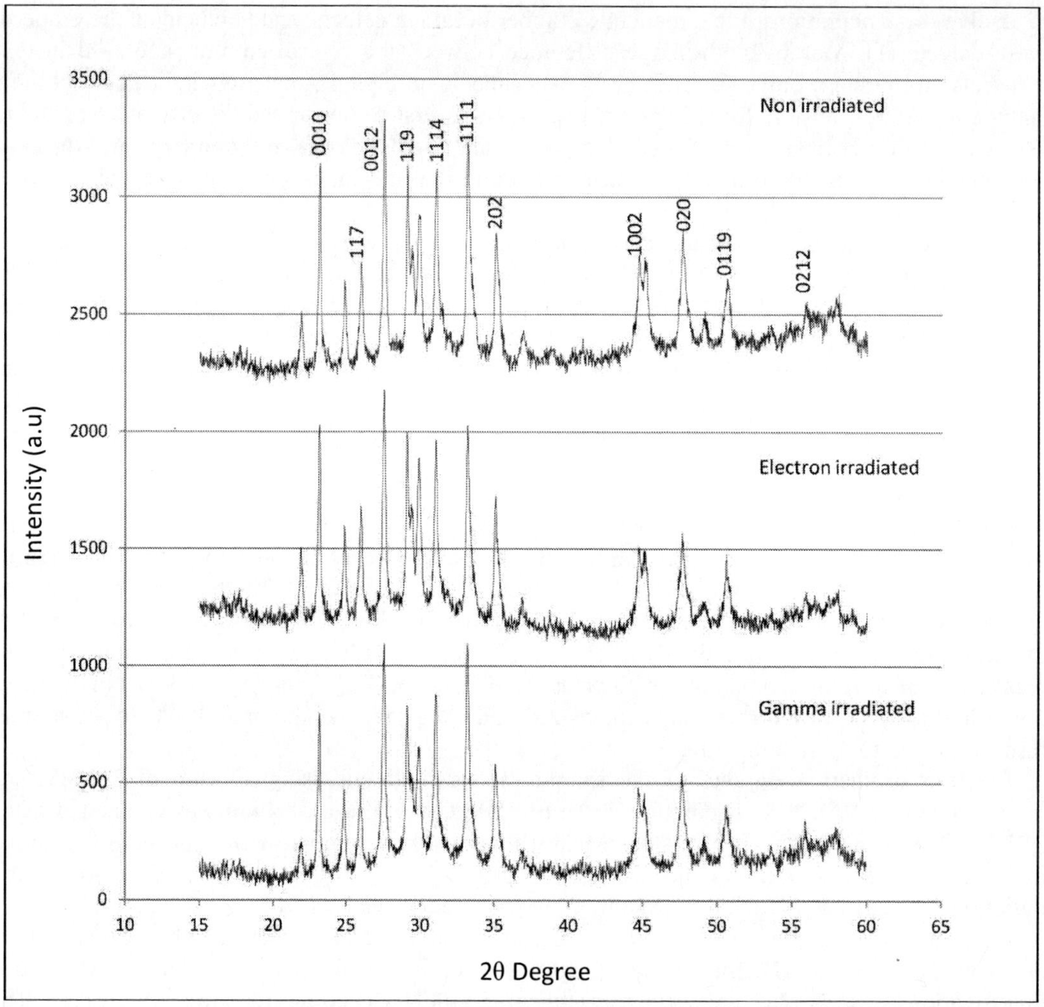

Figure 1. XRD patterns of Bi-2223 non-irradiated, electron ang gamma irradiated samples exposed with 100kGray, respectively.

AC susceptibility has been used to measure the critical temperature and critical current of the samples. Figure 2(a), 2(b) and 2(c) show the real parts of susceptibility (χ') of non-irradiated, electron irradiated and gamma irradiated samples, respectively. In each graph, the position of inflection point on the curve is the critical temperature. This is due to the considerable difference between lower critical magnetic field of the intergranular matrix and the grains, because the flux will penetrate the intergranular matrix first before it reaches the grains [6].

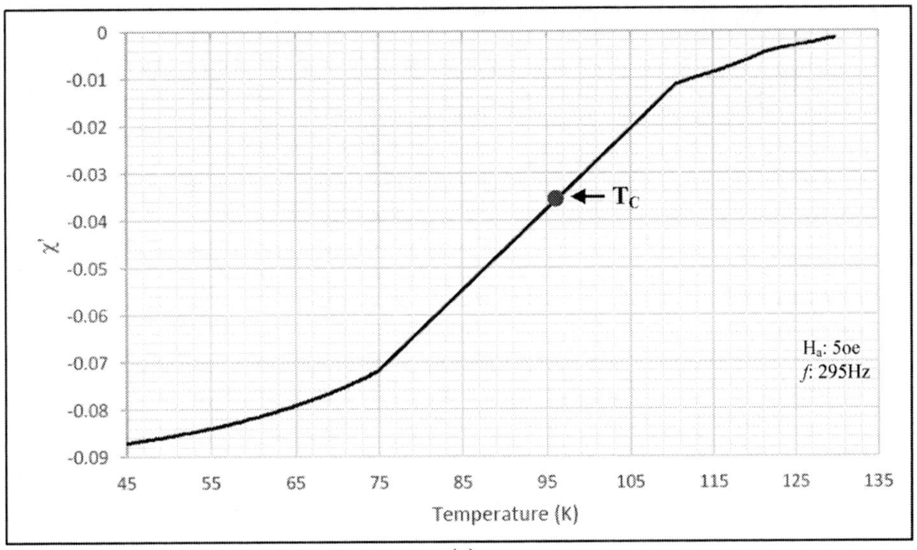
χ'
Temperature (K)
T_C
H_a: 5oe
f: 295Hz

(a)

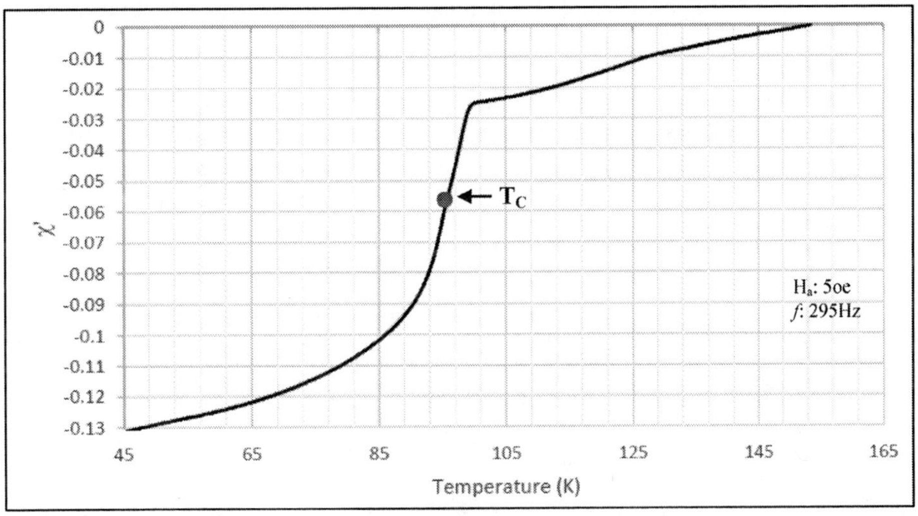
χ'
Temperature (K)
T_C
H_a: 5oe
f: 295Hz

(b)

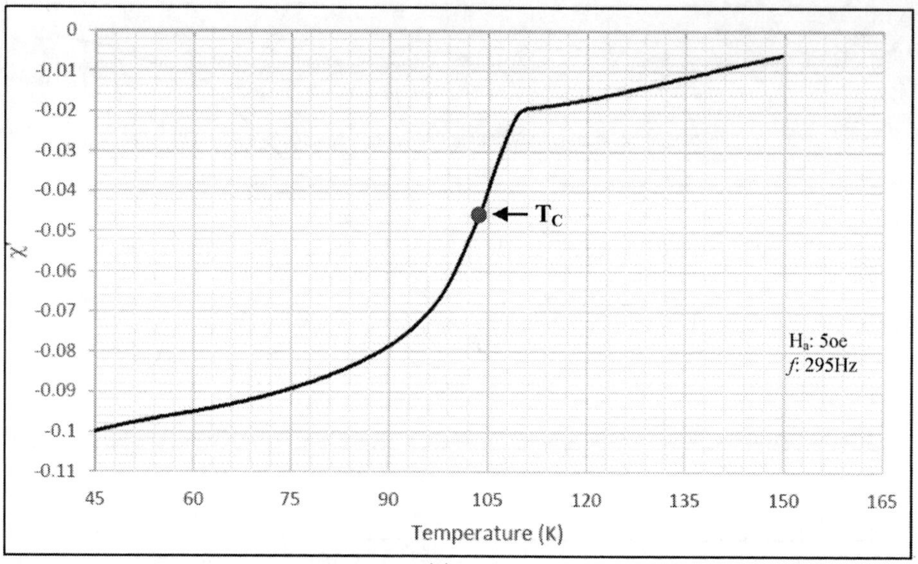

(c)

Figure 2. Real (χ') parts of susceptibility for (a) non-irradiated sample, (b) electron irradiated sample, and (c) gamma irradiated sample for applied magnetic field, H_a=5oe and frequency of 295Hz.

Rullier-Albenque *et al.* [7] stated that electron irradiation is an excellent technique to introduce homogeneously distributed point defects within the superconducting grains that produces higher J_C. Nevertheless, bonding between the adjacent grains within the sample is weakened after irradiation and leads to reduction in T_C. J_C at 70 K for all samples were calculated by using this equation,

$$J_C(T_P) = \frac{400A}{a \times b \ cm^2} \tag{1}$$

where a and b are the width and height of the sample.

Based on the above results and from Table 1, T_C of electron irradiated sample shows reduction of 2 K while its J_C improved by 6.91% from non-irradiated sample. Unlike gamma irradiated sample, T_C value increased to 103 K but J_C value dropped by 4.25%.

Table 1, Transition temperature, T_C, surface area, and critical current density, J_C of samples.

Type of sample	T_C, K	*a*, mm	*b*, mm	$(a \times b)$, cm^2	J_C, A/cm^2
Non- irradiated	97	2.37	2.58	6.11×10^{-2}	1617.62
Electron irradiated	95	2.54	2.11	5.35×10^{-2}	1729.35
Gamma irradiated	103	2.46	2.71	6.67×10^{-2}	1548.81

The SEM micrographs in Figure 3(a), 3(b) and 3(c) show the microstructures of Bi-2223 phase of non-irradiated and irradiated samples. The granular morphology of all Bi-2223 samples consist of flaky layers of large platelet-like structure. Non-irradiated sample shows lower grain alignment disorientation and the platelet grains are well linked when compared to gamma irradiation samples and this is in agreement with the finding reported by Mohiju et al. [8]. Nevertheless the electron irradiation sample show better connectivity between the grain boundaries and this explained the higher J_C in the sample. The disorientation of grains in the structure of irradiated samples were caused by particle collision [1]. There are increments in terms of number of layers in irradiated samples compared to non-irradiated.

The microstructures present in oxide superconductors are expected to play a prominent role in determining the current carrying capacity of these materials but the major bottleneck for the flow of the current is in the structure of grains or grain boundaries that connect these colonies. In case of Bi-2223 phase, it has been claimed that a macroscopic c-axis alignment is enough to give high critical currents [9]. Meanwhile, increases in the number of thin rectangular, indicating the decreases in T_C of the Bi-2223 phase and decreasing in porosity of the microstructure.

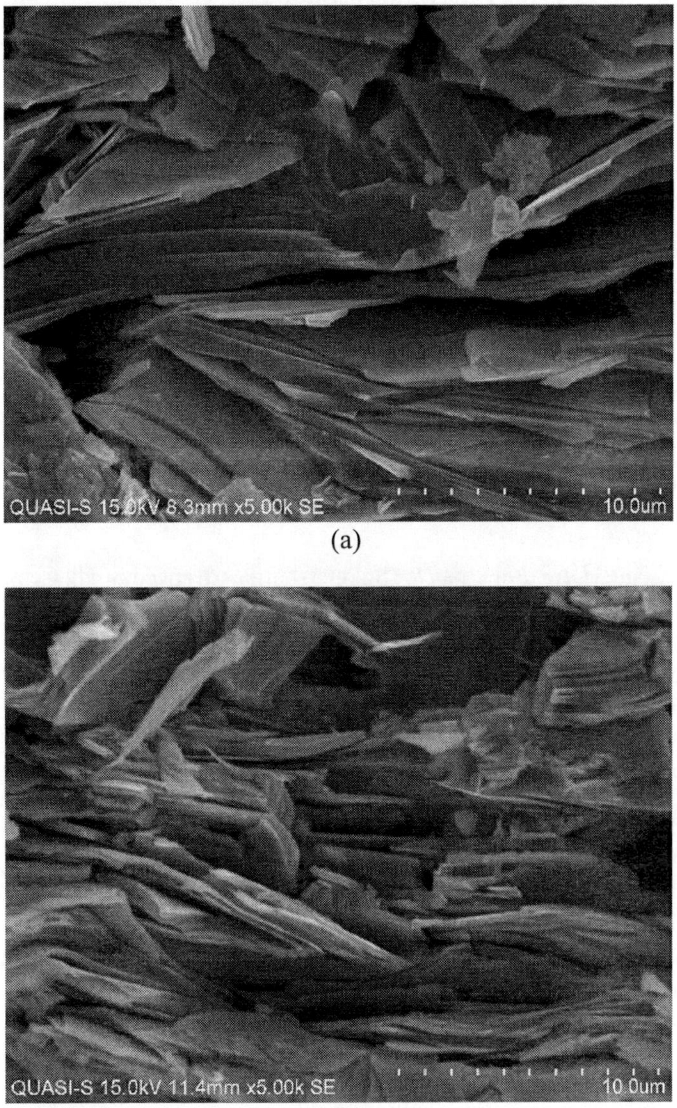

(a)

(b)

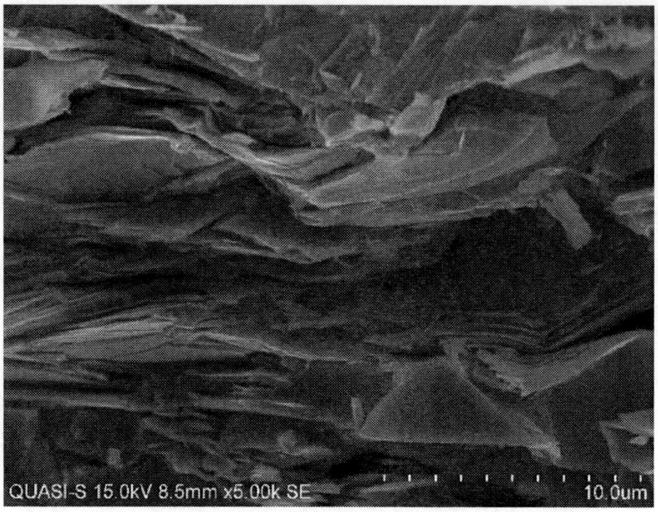

(c)

Figure 3. SEM micrographs of (a) non-irradiated sample, (b) electron irradiated sample, and (c) gamma irradiated sample, respectively.

4. Conclusion

Samples of $Bi_{1.6}$ $Pb_{0.4}Sr_2Ca_2Cu_3O_{10}$ were successfully prepared by conventional solid state method. In the initial stage of the sample preparation, the partial substitution of Bi by Pb plays an important role to improve the structure, as well as facilitating the grain growth. Results show that electron irradiation is a noble method to further enhance the superconducting properties of Bi-2223 phase superconductor. The J_C of electron irradiated sample improved by 6.91% compared to non-irradiated sample without much degradation in T_C.

5. References

[1] Pande C S 2000 *Mater. Phys. Mech.* (St.Petersburg: Advanced Study Center Co. Ltd.) **2** p 1.
[2] Amrein T, Schultz L, Kabius B and Urban K 1995 Phys. Rev. B **51** (College Park: American Physical Society) p 6792.
[3] Chaudhary S, Kashyap S C, Pandya D K, Kulkarni V N 1997 *Physica C* **280** (Amsterdam: North-Holland/Elsevier) p 37.
[4] Dimos J, Chaudhari P and Mannhart J, 1990 *Phys. Rev.* B **41** (College Park: American Physical Society) p 4038.
[5] Knauf N, Harnischmacher J, Müller R, Borowski R, Roden B and Wohlleben D 1991 *Physica C* **173** (Amsterdam: North-Holland/Elsevier) p 414.
[6] Nikolo M 1995 *Am. J. Phys.* **63** (New York: AIP Publishing) p 57.
[7] Albenque F R, Legris A, Berger H and Forro L 1995 *Physica C* **254** (Amsterdam: North-Holland/Elsevier) p 88.
[8] Zaahidah 'Atiqah Mohiju, Natasha Alieya Adnan, Nasri A. Hamid and Yusof Abdullah 2018 *IOP Conf. Series: Mat. Sc. & Eng.* **298** (Bristol: IOP Publishing) p 012021.
[9] Jaroslav K and Jan K 2013 *Materials Science Monographs* **79** (Amsterdam: Elsevier B. V.) p 22.

Acknowledgement

This project is funded by the Malaysian Ministry of Higher Education, reference number FRGS/1/2017/STG02/UNITEN/02/05 and conducted under collaboration with the Malaysian Nuclear Agency. The authors wish to thank Prof. Abd-Shukor from the Universiti Kebangsaan Malaysia for providing his research facilities for this work.

IOP Conf. Series: Materials Science and Engineering **555** (2019) 012026 doi:10.1088/1757-899X/555/1/012026

Radiation dose required for the vulcanization of natural rubber latex via hybrid gamma radiation and peroxide vulcanizations

Sofian Ibrahim[1,2], Khairiah Badri[2], Chantara Thevy Ratnam[1], Noor Hasni M. Ali[1] & Muhammad Abdurrahman Munir[2]

[1]RAYMINTEX Plant, Malaysian Nuclear Agency, 43000 Kajang, Selangor, Malaysia
[2]Faculty of Science and Technology, Universiti Kebangsaan Malaysia, 43600 Bangi, Selangor, Malaysia

sofian_ibrahim@nm.gov.my

Abstract. To enhance the crosslinking of prevulcanized natural rubber latex, combination of Irradiation and peroxide vulcanizations were used. Through this method, hexanediol diacrylate (HDDA) from irradiation vulcanization acted as the main sensitizer, while cumene hydroperoxide (CHPO) from peroxide vulcanization would act as the co-sensitizer. The effects of irradiation doses on the mechanical properties of latex film were investigated. 16 kGy irradiation dose, 2.5 parts perhundred rubber (pphr) of HDDA, 0.1 pphr of CHPO and 2.5 phr of Aquanox LP antioxidant were found to be the optimum conditions for compounding formulation. The rubber film obtained had tensile strength, modulus at 500% and modulus at 700% of 27.7, 3.5 and 12.4 MPa respectively, which is more than 21% increment compared to control film. Besides, the crosslink percentage of the rubber film showed 7 % increment from 90.7% to 97.7%.

1. Introduction

Radiation vulcanization of natural rubber latex (RVNRL) and peroxides vulcanization of natural rubber latex (PVNRL) have several advantages over the conventional vulcanization with sulphur such as less or absence of toxicity, free from nitrosamines and accelerator induced allergies, low in cytotoxicity and cleaner process [1-3]. These properties are important for many products, particularly catheters, surgical gloves and other medical and hospital supplies. For such uses, it is important that the products are free of contaminants, toxic and carcinogenic components to avoid harmful effects in human beings. However, the tensile strength of end products from both processes still failed to achieve a minimum of 24 Mega Pascal (MPa) as required by ASTM D3577-01a; which is the Standard Specification for Rubber Surgical Gloves [4].

Based on our previous studies, hybrid vulcanization method which consisted of radiation and peroxide vulcanizations could help to improve the mechanical properties of the vulcanized natural rubber latex films [5][6]. In this paper, the required radiation doses for vulcanization of natural rubber latex via hybrid irradiation and peroxide vulcanizations systems are determined. The other factor such as effect of molecular structure of peroxide on the mechanical properties of irradiated latex film is also investigated.

2. Materials and Methods

2.1. Materials

The latex utilized in this work was a high ammonia type of latex (HA latex) supplied by Revertex (M) Pt. Ltd., Malaysia. The sensitizer and co-sensitizer used were hexanediol diacrylate (HDDA) supplied by Allnex, China and cumene hydroperoxide (CHPO) supplied by Merck, Germany respectively. The stabilizer used was potassium laurate supplied by Tiarco Chemical (M) Pt. Ltd., Malaysia and the antioxidant used was Aquanox Lp supplied by Aquaspersion (M) Pt. Ltd., Malaysia. These materials were used as received.

2.2. Preparation of RVNRL compounding formulations

A typical latex compounding formulation for RVNRL (control) preparation is given in Table 1. The sensitizer, stabilizer, antioxidant and water were homogenized into an emulsion prior to addition into the latex with gentle stirring [5][6]. Once the addition of the emulsified materials was completed, the latex mixture was left stirred for three hours. It was then transferred into a one litre screw capped plastic container and irradiated with gamma rays from a Cobalt-60 source at a dose of 12 kGy [6][7]. After almost 6 hours of irradiation, the RVNRL was formed into film by coagulant dipping method.

Table 1. Compounding formulation of standard RVNRL (control).

Materials	Part perhundred (phr)
NR Latex (62% Total Solid Content)	100
Stabilizer	0.06
HDDA	2.50
Antioxidant	2.50
Water	Add to 52% Total Solid Content

The experiment was repeated by adding 0.1 phr of CHPO as the co-sensitizer into 9 kg of latex as formulated in Table 2 [6]. It was then transferred into 9 seperated one litre screw capped plastic container and irradiated with gamma rays from a cobalt-60 source at varying doses of 2, 4, 6, 8, 10, 12, 14, 16 and 18 kGy respectively.

Table 2. Compounding formulation of RVNRL with co-sensitizer.

Materials	Part perhundred (phr)
NR Latex (62% Total Solid Content)	100
Stabilizer	0.06
HDDA	2.50
CHPO	0.10
Antioxidant	2.50
Water	Add to 52% Total Solid Content

IOP Conf. Series: Materials Science and Engineering **555** (2019) 012026 doi:10.1088/1757-899X/555/1/012026

2.3. Irradiation
The latex formulations were irradiated using gamma rays from Co-60 isotope at MINTec-Sinagama Plant, Malaysian Nuclear Agency. Currently the activities of Co-60 are 447000 Curie with dose rate of 2.08 kGy/hr.

2.4. Measurement of tensile properties
Specimens for tensile testing were prepared using the coagulant dipping method to simulate the production of gloves in the gloves production line, where a glass plate was immersed in the coagulant and then placed in an oven at 100°C to partially dry the coagulant. It was then immersed in the latex compound for 20 seconds. The wet gel was allowed to consolidate at 100°C for 1 minute, and followed by leaching in distilled water at 60°C for 5 minute to remove hydrophilic materials and excess chemicals from latex dipped produce by washing them in hot water. The latex film was finally dried at 100°C for 30 minute and subjected to tensile test using Universal Testing Machine Instron 5564 in accordance to ASTM D412 [8]. The latex films samples were cut into dumbbell shape test pieces (Figure 1). Five samples were used for tensile test and a median value was taken as the final result.

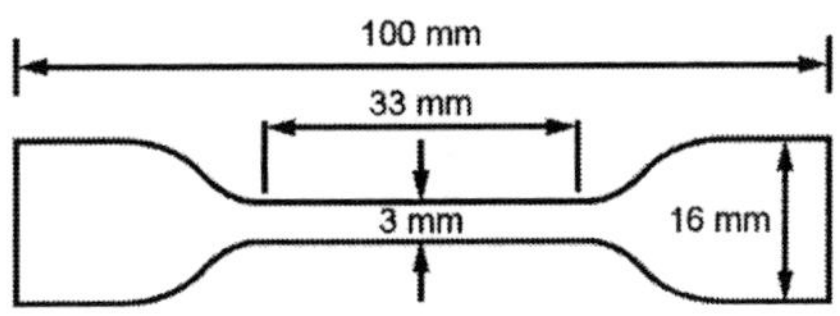

Figure 1. Dimension of dumbbell cut

2.5. ATR spectroscopy
In this study, FTIR spectroscopy analysis was carried out using Bruker's Tensor II Platinum Attenuated total reflection (ATR) spectrophotometer.

2.6. Determination of gel content
The gel content of the crosslinked samples were determined by the extraction of samples in toluene for 8 hours using Soxhlet apparatus [9][14]. The extraction samples were dried in an oven at 70°C until constant weight was achieved. The gel fraction was calculated as Eq. i:

$$\text{Gel content, \%} = \frac{w_1}{w_o} \times 100 \tag{1}$$

where w_0 and w_1 are the weights of the dried samples before and after extraction, respectively.

3. Results and discussion

3.1. Effect of irradiation doses on tensile strength of RVNRL
In gloves production, tensile strength and modulus values are considered as the commercial importance parameter. Tensile strength value are referred to the extent of the film undergo stress, whilst modulus values are always referred as the degree of crosslinking in the films [10]. Thus, the precise amount of irradiation dose is very important in preparation of RVNRL because the irradiation dose gives a major impact on the mechanical properties of RVNRL [2].

In previous studies, aliphatic type peroxide; *t*-butyl hydroperoxide successfully have helped to produce rubber film with tensile strength, modulus at 500% and modulus at 700% of 27.0, 3.0 and

11.0 MPa respectively at irradiation dose of 6 kGy [11]. However, in this study CHPO with aromatic molecular structure was found to give totally different results as compared to peroxide with aliphatic molecular structure. Table 3 gives the mechanical properties of control and hybrid RVNRL-peroxide samples that have been prepared by gamma irradiation at various doses and tested as required by ASTM D412 standard.

Table 3. Mechanical properties of RVNRL (median value from tensile test).

Sample	Irradiation dose (kGy)	Modulus @ 500% (MPa)	Modulus @ 700% (MPa)	Tensile strength (MPa)
RVNRL (Control)	12	2.8	10.5	22.5
Hybrid RVNRL-peroxide (D2)	2	0.9	2.5	9.3
Hybrid RVNRL-peroxide (D4)	4	1.5	4.9	15.6
Hybrid RVNRL-peroxide (D6)	6	2.2	7.5	18.9
Hybrid RVNRL-peroxide (D8)	8	2.4	8.8	22.7
Hybrid RVNRL-peroxide (D10)	10	2.5	9.1	23.8
Hybrid RVNRL-peroxide (D12)	12	2.9	11.3	24.3
Hybrid RVNRL-peroxide (D14)	14	3.0	12.3	26.4
Hybrid RVNRL-peroxide (D16)	16	3.6	12.4	27.3
Hybrid RVNRL-peroxide (D18)	18	3.5	13.5	27.3

It was observed that rubber film obtained from irradiation at 16-18 kGy had tensile strength, modulus at 500% and modulus at 700% of 27.3, 3.5 and 12.4 MPa respectively, which was more than 21% increment as compared to control. In contrast to the use of aliphatic type peroxide, decomposition of peroxide with aromatic molecular structure required high irradiation dose to engender the radical reaction. This is due to the high energy required to break aromatic rings that are very stable, where the electron density is delocalized in the ring so each bond is more similar to 1.5 bonds than either single or double bonds [12]. Besides that, higher irradiation dose also help to increase the mechanical properties of hybrid RVNRL-peroxide films due to the enhancement of the intraparticle crosslink density (chemical crosslinking) caused from monogeneity of the hybrid radiation and peroxide [vulcanizations [2][6][13].

Infra-red spectroscopy has been used to investigate the substances have been used in the reaction. Figure 2(a) to 2(d) show that there is no sign of residue HDDA and/or peroxide found in RVNRL samples because they are believe to be consumed by radiation polymerization and hydrolysis during the radiation vulcanization process. According to Figure 2(b) and 2(c), it can be clearly observed the presence of functional groups of C-O at wavelength ranges 1181-1225, while functional groups of C=O from HDDA and C=C (aromatic) from CHPO clearly observed at wavelength 1719 cm^{-1} and 1701 cm^{-1} respectively. However, there is no sign of functioning group of C=C (aromatic), C-O and C=O in the hybrid RVNRL-peroxide spectrums in Figure 2d. So it can be concluded that both HDDA and CHPO has been consumed during the vulcanization process.

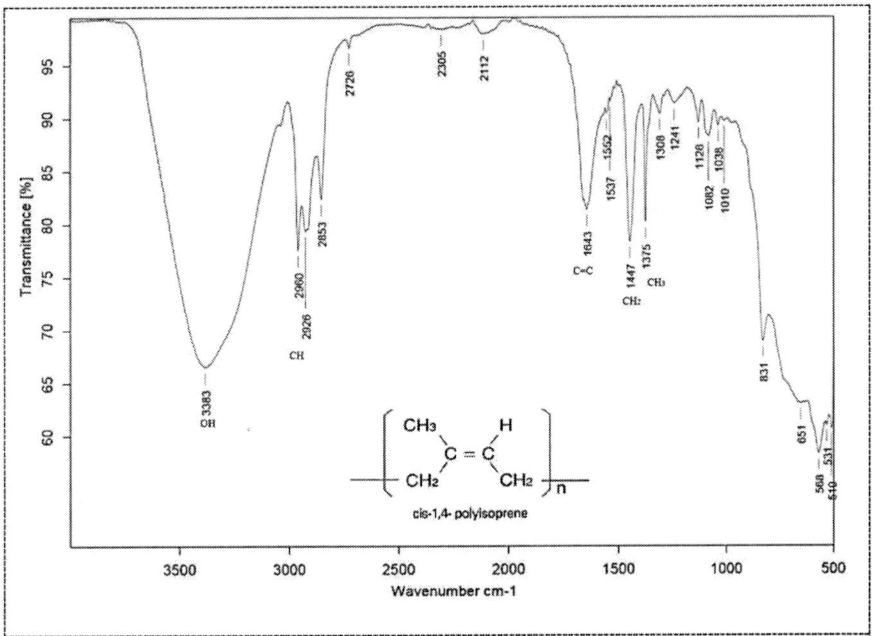

Figure 2 (a). Infra-red spectrum for polyisoprene.

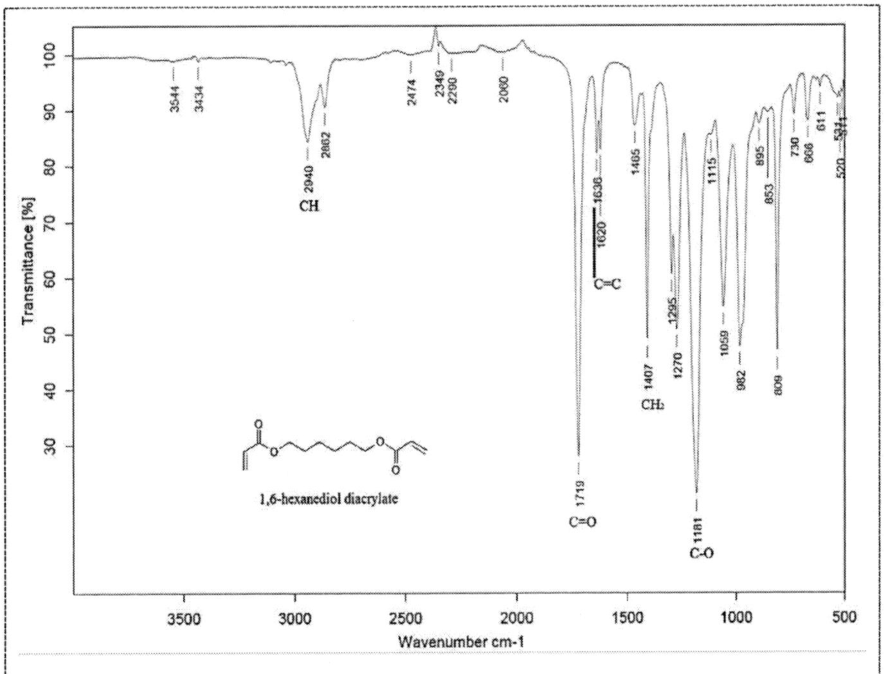

Figure 2(b). Infra-red spectrum for hexanedioldiacrylate

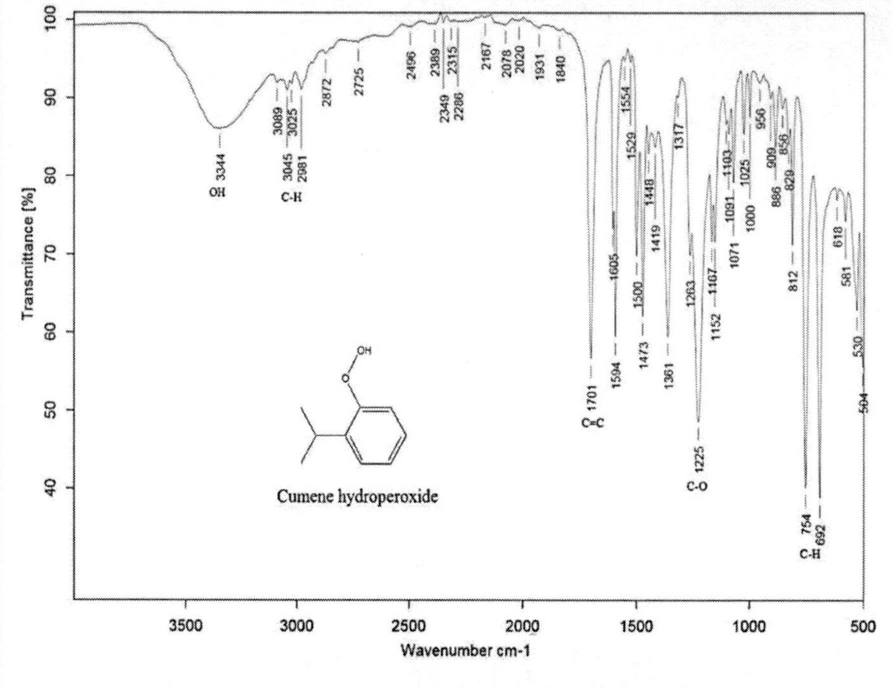

Figure 2(c). Infra-red spectrum for cumene hydroperoxide.

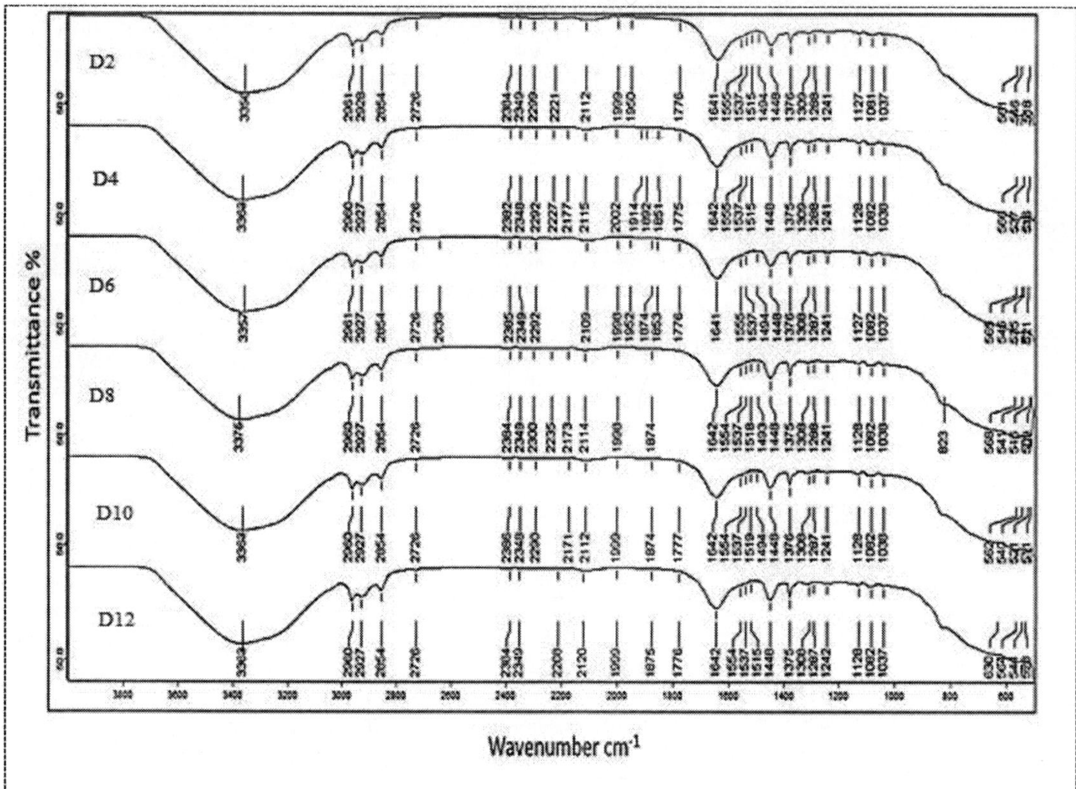

Figure 2 (d). Infra-red spectrums for hybrid RVNRL-peroxide at various irradiation doses

3.2. Effect of irradiation doses on gel content of hybrid RVNRL-peroxide

Soxhlet extractor is used to perform the extraction with toluene for the purpose of determining the gel fraction/content of RVNRL, which is defined as the percentage of a sample that does not dissolve in toluene. The crosslinked or gelled polymers only swell up and not dissolve in any solvent. The gel content is determined indirectly from the amount of soluble fraction and directly by weighing the dried gel [9, 14]. Figure 3 shows the relationship between gel content with irradiation doses. As shown, the extent of gel formation increases with the increasing of irradiation dose, indicating increases in crosslink density of the polymer. From the graph, it was observed that at 16 kGy of irradiation dose, hybrid RVNRL-peroxide film produced 97.7 % crosslink percentage and 7 % increment compared to the control sample at 90.7 %. However, it is also showed that the percentage of gel content decreased as the irradiation dose increased. This happens when the interparticle entanglement (physical crosslinking) starting to decreased when the irradiation dose increased. The interparticle entanglements are depended on the free rubber chain ends at the surface of each latex particle. These chains interpenetrate during film formation and contribute to the strength of the film by means of entanglements. The length of free rubber chain ends decreases with the increasing of irradiation dose because it is equivalent to molecular weight between crosslink [2]. Thus, the degree of crosslink in hybrid RVNRL-peroxide latex films increases up to maximum level and then will decrease with the increasing dose.

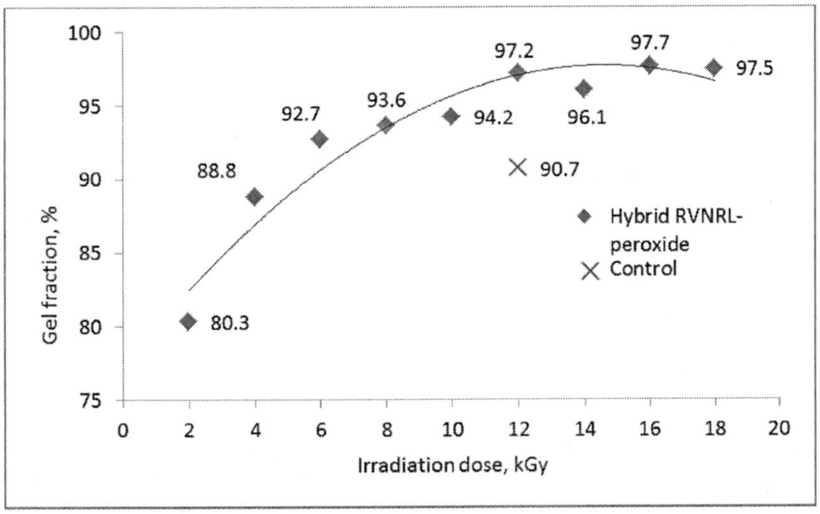

Figure 3. Effect of Irradiation Doses on Gel Fraction of Hybrid RVNRL-peroxide.

4. Conclusion

RVNRL with good tensile strength can be prepared by hybrid radiation and peroxide vulcanizations. Irradiation of latex compounding formulation based on 2.5 phr of hexanediol diacrylate (HDDA) as sensitizer, 0.1 phr of cumene hydroperoxide (CHPO) as co-sensitizer and 2.5 phr of Aquanox LP as an antioxidant at irradiation dose of 16 kGy can produced rubber film with tensile strength of 27.3 MPa with crosslink percentage of 97.7%. With that, it is concluded that the molecular structure of the peroxide has a great effect on the irradiation dose required for the vulcanization process, where the peroxides with aromatic structure will require higher irradiation dose compared to the aliphatic peroxides.

5. References

[1] Pairu I, Rusli D and Wan Manshol W Z 2016 *Radiation Effects and Defects in Solids* **171** p 1006-1015.

[2] Makuuci K 2003 *An Introduction to Radiation Vulcanization of Natural Rubber Latex* (T.R.I. Global Co. Ltd., Bangkok).

[3] Pairu I, Wan Manshol W Z, Chai C K, Sofian I, Mohd Noorwadi M L and Saadiah S 2017 *Jurnal Sains Nuklear Malaysia* **29**(1) p 28-36.

[4] ASTM D3577-01a 2001 *Standard Specification for Rubber Surgical Gloves* (ASTM International, West Conshohocken, PA).

[5] Sofian I, Wan Manshol W Z, Chai C K and Pairu I 2007 Improving On Tensile Strenght And Modulus Of Radiation Vulcanization Natural Rubber Latex (RVNRL) By Using t-butyl hydroperoxide (BHPO) As Co-sensitizer, *Research Note Malaysian Nuclear Agency* (Malaysia) p 04/07.

[6] Sofian I, Khairiah B, Chantara T R and Noor Hasni M A 2017 Radiation Vulcanization of Natural Rubber Latex with Organic Peroxide as The Co-Sensitizer *International Conference of Analytical Sciences 2017 (SKAM 30)* (Melaka, Malaysia, August 26-29, 2017).

[7] Wan Manshol W Z, Norjanah M, Jusoh H, Wan Khairul Amalina M N and Zulkifli J 1993 *Radiat. Phys. Chem.* **42** p 101-103.

[8] ASTM D412 2016 Standard Test Methods for Vulcanized Rubber and Thermoplastic Elastomers—Tension (ASTM International, West Conshohocken, PA).

[9] Jayasuriya M M, Makuuchi K and Yoshi F 2001 *European Polymer Journal* **37** p 93-98.

[10] Roslim R, Jefri J, Manroshan Singh J S, Siti Noor Suzila M U H and Amir Hashim M 2015 *Characterizing Physical Properties Of Peroxide Vulcanized Natural Rubber Latex Films* (Trans Tech Publications Ltd. Switzerland) **1134** p 236-242.

[11] Sofian I, Khairiah B, Chantara T R and Noor Hasni M A 2018 *Radiation Effects and Defects in Solids* **173**, 427-434.

[12] Lumen learning 2018 Organic compounds: https://courses.lumenlearning.com/boundless-chemistry/chapter/organic-compounds/.

[13] Siri Upathum C and Sonsuk M 1996 Development of an Efficient Process for Radiation Vulcanization of Natural Rubber Latex Using Hydroperoxide with Sensitizer *Proceedings of the Second International Symposium on RVNRL* (Kuala Lumpur, Malaysia, July 1996) p 15-17.

[14] ASTM D3616-95 2014 *Standard Test Method for Rubber—Determination of Gel, Swelling Index, and Dilute Solution Viscosity* (ASTM International, West Conshohocken, PA).

Acknowledgements

The authors would like to express their deepest appreciation to Revertex (M) Sdn. Bhd., Malaysia for s upplying HA latex to be used in this research and RAYMINTEX Plant, Malaysian Nuclear Agency for financial and technical supports of this work.

IOP Conf. Series: Materials Science and Engineering **555** (2019) 012027 doi:10.1088/1757-899X/555/1/012027

Angle resolve x-ray photoelectron spectroscopy analysis of La_2O_3 thin film of mixed carbon-glass and carbon-silicon substrate for micro-flexographic printing process

S. Hassan[1,a)], M.S.Yusof[1], Z. Embong[2]

[1]Faculty of Mechanical and Manufacturing Engineering, Universiti Tun Hussein Onn Malaysia, 86400 Batu Pahat, Johor, Malaysia.
[2]Faculty of Applied Sciences and Technology, Universiti Tun Hussein Onn Malaysia, Km. 1, Jln. Panchor, 84600 Panchor, Johor, Malaysia.

[a]suhaimihas@uthm.edu.my

Abstract. Adhesive property could be described as an interchangeably with some ink and substrate which was applied to one surface of two separate items that bonded together. Lanthanum oxide (La_2O_3) had been used as a rare earth metal candidate as printing ink. This metal deposit was printed on carbon-silicon substrate and carbon-glass substrate using Magnetron Sputtering technique. The choose of Lanthanum Oxide as a target is due to its wide application in producing micro-electronic devices such as thin film battery, electronic circuit and printed circuit board. The La_2O_3 was deposited on the surface of carbon-silicon and carbon-glass substrate were then analyzed using Angle Resolve X-Ray Photoelectron Spectroscopy (ARXPS). The position for each synthetic component in the narrow scan of Lanthanum (La) 3d and O 1s were referred to the electron binding energy (eV). This research was focused on the narrow scan regions which were O 1s and La 3d. Further discussion of the spectrum evaluation was discussed in detail. Here, it was proposed that from the adhesive and surface chemical properties of La was suitable as an alternative medium for micro-flexographic printing technique in printing multiple fine solid lines image at micro to nano scale feature. Micro-flexographic printing was developed in patterning technique from micron to nano scale range to be used for graphic, electronic and bio-medical device on variable substrates. Hence, this paper will describe the capability of this particular metal as rare earth metal in a practice of micro-flexographic printing process.

1. Introduction

Printing was a process reproducing an images using master form or printing plate. By combining both printing techniques which was micro-contact and flexographic also known as micro-flexographic printing had its own advantages and disadvantages, a new era of printing technology could be explored in producing micro to nano scale printing image[1]. The printed image was then analysis using Angle Resolve X-Ray Photoelectron Spectroscopy (ARXPS) to find the adhesive and chemical property between ink and substrate [2].

The attraction strenght between two adhesive or printed material was determined by the surface energy of the material. Adhesion was the molecular force of attraction betweeen unliked materials for example lanthanum oxide metal and mixed carbon-glass or carbon-silicon substrate. The higher surface energy, the molecular attraction will be greater. Particles size of La_2O_3 bulk were in the region of 1 μm which had been examined using Scanning Electron Microscopy (SEM) [3].

The chemical bonding between La metal deposited and substrate surface was due to a chemisorption and physisorption process. The adhesion property of La_2O_3 on Si substrate was successful when O 1s binding energy can be seen from La_2O_3 and SiO_2 components. There were four components that contributed to O 1s photoelectron which were La_2O_3, SiO_2, O bulk and water vapour [4].

The addition of pure La could give high denitriding rate compared with the experiment without La addition. La element can reduce the N contents in nitriding layers through the surface oxidation and attraction of La atoms [5]. The work and result had been analyzed and investigated by using X-ray diffraction (XRD), scanning electron microscope (SEM), X-ray photoelectron spectroscopy (XPS) and micro hardness tester.

Other than lanthanum material, graphene was the other example of rare earth metals that could be used as printing ink for micro-flexography printing. Graphene ink was a step to move forward in order to achieve high speed printing in electronic with simple, rapid, low cost method, less waste and roll-to-roll capability. This type of ink was practically used in electronic printing industries that aimed on printing multiple micro to nano solid lines [6].

Previous research had shown that printing components played an important roles in producing micro to nano scale printing image with a good adhesive property. The example of printing components were ink, printing plate, substrate, anilox and others. The ink properties could be ink chemistry, viscosity, rheological behavior, solvent evaporation rate and drying [7].

Previous research in fine solid line printing, the researcher [8] managed to print out 50μm line width and 50μm line gap by using carbon graphic inks as printing inks. The author used web press industrial machine as a printing method. This technique used photopolymer as a mould to transfer the ink from plate roller to substrate. While, in micro-contact printing (μCP), Perl showed that μCP could produce fine solid line below 1μm [9] which was smaller than flexography printing technique. The combination of both printing techniques hope will produce better image in micro to nano scale features.

Lanthanum was a suitable material to use in printing process because it will oxidize when exposed to air. Lanthanum material had been used as $La_{1-x}Sr_xFeO_3$ (LSFO) film in screen printing process in powders element synthesize by sol-gel. The lanthanum film had good sintering properties and fair compatibility with alumina ceramic substrates. The value of electrical conductivity was also high [10].

Lanthanum oxide characteristics was the most important issue that must be understood either it was suitable to be used as a printing ink for combination of flexography printing and μCP which known as micro-flexographic [11]. This study was investigated on the possibility of using lanthanum oxide ink in micro-flexographic to achieve the printing fine solid lines and good surface adhesion on mixed carbon-silicon and carbon-glass substrate.

2. Research Methodology

The micro-flexographic printing process was began from the preparation of printing plate. A pattern of fine solid line image was produced on printing plate [12] which was attached at plate cylinder. Micro-flexographic printing process was almost same as flexographic printing process [13]. Ink was transferred to the printing plate by using an engraved cylinder known as an anilox roll as shown in figure 1. Here, the thin film analysis had been performed into two categories which were deposition process of lanthanum using magnetron sputtering techniques and surface characterization using ARXPS. The take-off angle had been performed within 0° to 45° in subsequence order. At 0° take off-angle, analysis was reflected to the surface analysis for the bulk of the substrate while at grazing take-off angle, 45°, the others surface composition was determined for the uppermost oxide layer of mixed carbon-silicon and carbon-glass substrate.

X-ray Photoelectron Spectroscopy (XPS) was an advanced surface analytical techniques that used to characterize the elemental composition of the lanthanum oxide printing ink. The method was favourable for several nanometer thickness oxide or deposition [14]. XPS spectrum were obtained on an Omicron Nanotechnology spectrometer equipped X-ray source powered at 300W (15kVe20 mA) and operated under ultra-high vacuum (UHV) condition at 10^{-11} Torr. The kinetic energies of the photoelectrons were measured using a hemispherical electron analyser working in the fixed analyser transmission (FAT) pass energy mode, of 180 eV and 20 eV for wide and narrow scanning respectively.

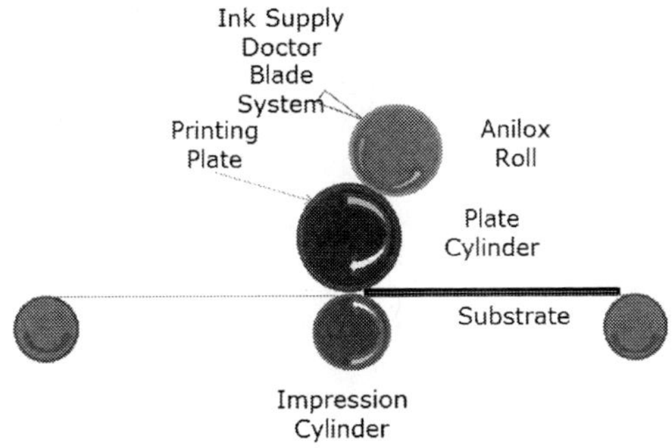

Figure 1. Schematic description of the micro-flexographic printing process

Spectrum analysis of variable angles were deconvoluted using Casa XPS software. The synthetic spectrum for the photoelectron peaks and its background signal was optimized using a specific model. Here, the appropriate background model used was Shirley method, while the synthetic photoelectron signal assigned using mixed of Gaussian-Lorentzian method with ratio 70% and 30% respectively. In this work, the data deconvolution was performed for element of interest on the substrate by referring to the narrow scan. There region of interest which were Lanthanum, La 3d, and oxygen, O1s.

Generally, lanthanum oxide had chemical properties that meet the requirements for applications in various electronic fields. The base material for the samples used in this study was commercially available lanthanum oxide target. La_2O_3 was known to spontaneously react with humidity in air to form lanthanum hydroxide $La(OH)_3$ [15]. Surface chemistry of lanthanum oxide could be produced, but the high reactivity of this compound will cause a rapid roughening of the surface due to the formation of hydroxides or carbonates.

RF magnetron sputtering was a suitable technique to print rare earth metal or metal oxide liked lanthanum oxide on variable substrates [16]. Recently, the fabrication of lanthanum oxide (La_2O_3) on substrate had been investigated intensively in the global scale [15]. Since there were few researchers correlated the properties of depositing La_2O_3 on carbon-silicon and carbon-glass substrate, the understanding to produce high-quality La_2O_3 printed using magnetron sputtering was poor.

3. Result and Analysis

The position for each synthetic component in the narrow scan of La 3d and O 1s were referred to the electron binding energy (eV). The spectrum analysis was deconvoluted by Casa XPS software. There were 4 series of take-off angle involved in these experiments as mentioned in the previous methodology section.

Firstly, the experiment analysis for the bulk of the carbon-silicon substrate was carried out at the angle of 0°. The deconvolution of O 1s photoelectron spectrum for La_2O_3 indicated that the oxygen species were comprises different peaks at 528.9, 530.3, 531.7 and 533.2 eV associated with SiO_2, bulk, La_2O_3 and OH^- as illustrated in figure 2(a). Then, the experiment analysis was continued for the bulk of the carbon-glass substrate which also carried out at the angle of 0°. The deconvolution of O 1s photoelectron spectrum for La_2O_3 indicated that the oxygen species were comprises different peaks at 528.9, 530.2, 531.5 and 532.7eV associated with SiO_2, bulk, La_2O_3 and OH^- as illustrated in figure 2(b).

At take-off angle 15°, the deconvolution of O 1s photoelectron spectrum for La_2O_3 on carbon-silicon substrate indicated that the oxygen species were comprises different peaks at 528.5, 530.2, 532.5 and 534.0 eV associated with SiO_2, bulk, La_2O_3 and OH^- as illustrated in figure 3(a). While, the deconvolution of O 1s photoelectron spectrum on carbon-glass substrate indicated that the oxygen species were comprises different peaks at 528.0, 530.0, 531.2 and 532.6eV as illustrated in figure 3(b).

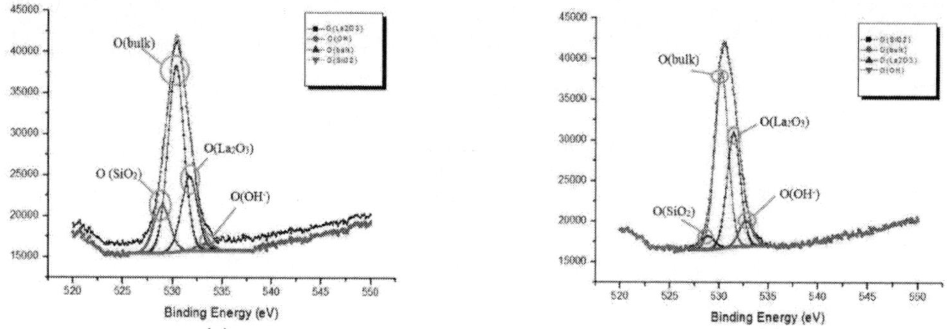

(a) (b)

Figure 2. XPS spectrum deconvolution for O 1s (a) on carbon-silicon substrate and (b) on carbon-glass substrate in 0° take-off angle

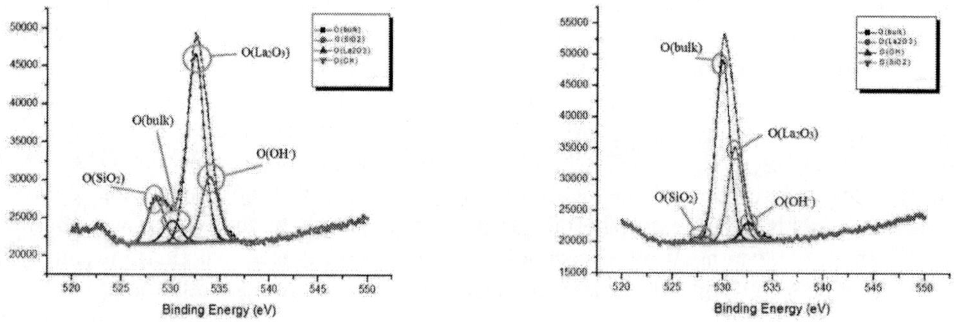

(a) (b)

Figure 3. XPS spectrum deconvolution for O 1s (a) on carbon-silicon substrate and (b) on carbon-glass substrate in 15° take-off angle

Then, experiment analysis of mixed carbon-silicon substrate was carried out at the angle 30°. The deconvolution of O 1s photoelectron spectrum for La_2O_3 indicated that the oxygen species were comprises different peaks at 527.7, 529.1, 533.6 and 535.1eV as illustrated in figure 4(a). The deconvolution of O 1s photoelectron spectrum for La_2O_3 on carbon-glass substrate indicated that the oxygen species were comprises different peaks at 528.0, 530.2, 531.5 and 533.0eV as illustrated in figure 4(b) which associated with SiO_2, bulk, La_2O_3 and OH^- respectively.

The last analysis for the bulk of the mixed carbon-silicon substrate was carried out at the angle of 45°. The deconvolution of O 1s photoelectron spectrum for La_2O_3 indicated that the oxygen species were comprises different peaks at 527.8, 529.2, 530.3 and 531.5eV as illustrated in figure 5(a). The deconvolution of O 1s photoelectron spectrum for La_2O_3 on carbon-glass substrate indicated that the oxygen species were comprises different peaks at 529.1, 530.8, 532.2 and 533.9eV as illustrated in figure 5(b) which associated with SiO_2, bulk, La_2O_3 and OH^- respectively.

From both carbon-silicon and carbon-glass substrate overall result, the high insensitive of photoelectron signal from both oxides at 45° take-off angle had become a clear evidence to support this finding. It means that most of the lanthanum oxide deposited was residual at the upper most layer of the substrate which at the bulk of the substrate. The information of oxygen species, O^{2-} component from O 1s narrow scan indicated there were 4 species of oxygen associated with the O 1s signal which were contributed by SiO_2, bulk oxide, La_2O_3 and hydroxide components. Here, two surface chemical process involved in oxide formation bonding which was chemisorption and physisorption process. It was cleared that the metal oxide bonding La_2O_3 was formed from the chemisorption process whiles the hydroxide component from physisorp process. This was evidenced that the deposited of Lanthanum metal with carbon-silicon substrate and carbon-glass substrate were bonded together with the substrate surface.

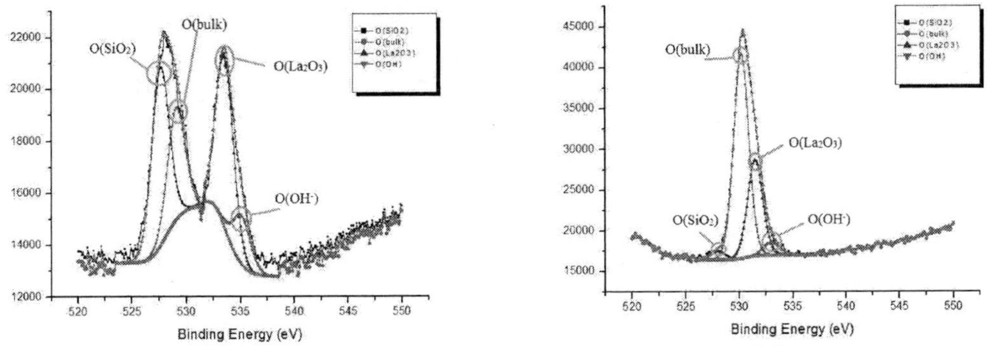

(a) (b)

Figure 4. XPS spectrum deconvolution for O 1s (a) on carbon-silicon substrate and (b) on carbon-glass substrate in 30° take-off angle

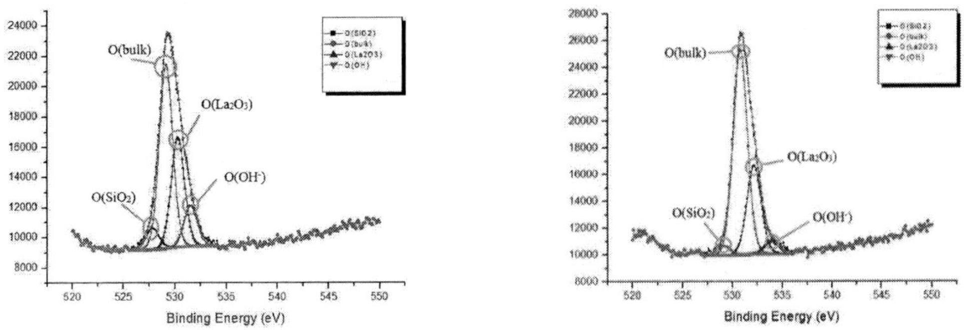

(a) (b)

Figure 5. XPS spectrum deconvolution for O 1s (a) on carbon-silicon substrate and (b) on carbon-glass substrate in 45° take-off angle

Therefore, it could be concluded that the surface adhesion of the Lanthanum metal deposited was better on carbon-glass substrate compared to carbon-silicon substrate. At take-off angle 45° of O 1s photoelectron spectrum, binding energy position at La_2O_3 on carbon-glass substrate (532.2) was higher than carbon-silicon substrate (530.3) as shown in table 1. The formation of this bonding was respected to a good adhesion between metal deposited and substrate due to the anion and cation interaction under a thermodynamic influenced. Hence most of the signal at this 45° take-off angle was mainly contributed by photoelectron from La_2O_3.

Table 1. A comparison at O 1s component between carbon-silicon and carbon-glass substrate.

Substrate	Binding Energies Position (eV) At Take-Off Angle 45°			
	O (SiO_2)	O (bulk)	O (La_2O_3)	O (OH^-)
Carbon-silicon	527.8	529.2	530.3	531.5
Carbon-glass	529.1	530.8	532.2	533.9

4. Conclusion

Surface adhesion study of La_2O_3 deposits on carbon-glass substrate compared to carbon-silicon substrate was successfully done. There were two types of oxide species on the substrates which were La_2O_3 and SiO_2. The oxide components that contribute to O 1s photoelectron were SiO_2, bulk, La_2O_3 and OH^- (water vapour). The chemical bonding formation between the La metal deposited and the surface was due to a chemisorption and physisorption process. A successful observation of La_2O_3 adhesion property on carbon-glass substrate was better than carbon-silicon substrates which clearly seen from the O 1s binding energy from La_2O_3 component. For conclusion, the micro-flexographic was good candidate for printing electronic with rare earth metal inks property, substrates and process parameters were main role to success the implementation.

5. References

[1] Hassan S, Yusof M S, Ding S, Maksud M I, Nodin M N, Mamat K A, Sazali M S and Rahim M Z 2017 Investigation of Carbon Nanotube Ink with PDMS Printing Plate on Fine Solid Lines Printed by Micro-flexographic Printing Method, *IOP Conf. Series: Materials Sci. and Eng.* **203.**

[2] Hassan S, Yusof M S, Embong Z, Ding S, Maksud M I 2018 Surface study of graphene ink for fine solid lines printed on BOPP Substrate in micro-flexographic printing using XPS analysis technique. *IOP Conf. Series: Materials Sci. and Eng.* **298(1)** p 012012.

[3] Balusamy B, *et al.* 2012 Characterization and bacterial toxicity of lanthanum oxide bulk and nanoparticles *J. of Rare Earths* **30(12)** p 1298-1302.

[4] Hassan S, *et al.* 2016 Angle resolved x-ray photoelectron spectroscopy (ARXPS) analysis of lanthanum oxide for micro-flexography printing *AIP Conf. Proc.* **1704** p 040002.

[5] Zhang C S, Yan M F, Sun Z 2013 Experimental and theoretical study on interaction between lanthanum and nitrogen during plasma rare earth nitriding *App. Surface Sci.* p 381-388.

[6] Hassan S, *et al.* 2015A Feasibility study of roll to roll printing on graphene *App. Mech. and Materials* **799-800** p 402-406.

[7] Hrehorova E, Pekarovicova A, Fleming P D 2006 Gravure printability of conducting polymer inks *NIP & Digital Fabrication Conf.* (Soc. for Imaging Sci. and Tech.).

[8] Yusof M S 2011 Printing fine solid lines in flexographic printing process (School Of Engineering, Swansea University) p 201.

[9] Perl A, Reinhoudt D N, Huskens J 2009 Microcontact printing: limitations and achievements *Adv. Materials* **21(22)** p 2257-2268.

[10] Cui X, Li S, Zhu X 2014 Microstructure and electrical properties of $La1{-}xSrxFeO3(x=0–0.6)$ film by a screen-printing method *Materials Lett. 130(0)* p 267-270.

[11] Hassan S *et al.* 2017 A Study of PDMS printing plate for fine solid lines image in micro-flexographic printing process *J. of Eng. and App. Sci.* **12** p 5644.

[12] Hassan S, *et al.* 2015 A study of nano structure by roll to roll imprint lithography *Int. Symp. on Tech. Management and Emerging Tech.* (ISTMET) (IEEE).

[13] Deganello D *et al.* 2010 Patterning of micro-scale conductive networks using reel-to-reel flexographic printing *Thin Solid Films* **518(21)** p. 6113-6116.

[14] Venkata R K *et al.* 2011 Heat capacity measurements and XPS studies on uranium–lanthanum mixed oxides, *J.of Alloys and Compounds* **509(7)** p 3229-3237.

[15] Sunding M F *et al.* 2011 XPS characterisation of in situ treated lanthanum oxide and hydroxide using tailored charge referencing and peak fitting procedures J. of Elec. Spect. and Related Phenomena **184(7)** p 399-409.

[16] Nayan N 2010 Optimization of RF magnetron sputtering plasma using Zn target *Int. Conf. on Enabling Sci. and Nanotech. (ESciNano)* (IEEE).

Acknowledgments
This project was supported by TIER 1 Grant (H197) and Contract Grant (H278) Research Management Center (RMC) and Faculty of Mechanical and Manufacturing Engineering (FKMP), University Tun Hussein Onn Malaysia (UTHM), Ministry of Education Malaysia.

Ludlum configuration for the measurement of mass attenuation coefficients at high energy photons by using Compton scattering method

M F Mohd Yusof[1a], N Mohd Isa[2], A B Abd Kadir[2], H Salleh[2]

[1]School of Health Sciences, Universiti Sains Malaysia, 16150, Kelantan, Malaysia
[2]Medical Physics and Secondary Standard Dosimetry Laboratory, Malaysian Nuclear Agency, 43000 Selangor, Malaysia

[a]mfahmi@usm.my

Abstract. The measurements of mass attenuation coefficients by using Compton scattering technique was carried out by using the Ludlum configuration. A ^{137}Cs sealed sourced was used and attenuated at angles between 30 and 75° to provide scattered gamma energies between 337.72 and 564.09 keV. The mass attenuation coefficients of solid water and perspex were measured and compared to the calculated value of water by using XCOM. The results showed that the measured mass attenuation coefficients of solid water and Perspex® phantoms were in agreement to the values of water within 6.84 and 7.20% average percentage of discrepancies. The results indicated the suitability of the Ludlum configuration for the measurement of mass attenuation coefficients by using the Compton scattering method.

1. Introduction

The value mass attenuation coefficients become the most important parameter to determine the absorption and scattering characteristics collectively known as attenuation properties of a material towards ionizing radiations [1]. Different types of materials would have different values of mass attenuation coefficients as it is proportional to the mass density, effective atomic numbers and electron density of the materials. Therefore two materials are postulated to have similar attenuation properties towards ionizing radiations when the values of mass attenuation coefficients are similar.

The values of mass attenuation coefficients of materials can be theoretically calculated by using the photon cross section database known as XCOM [2]. This program enables the user to calculate the mass attenuation coefficients of known element and compounds at very wide range of photon energies and had been used by many researchers. The mass attenuation coefficients of a material at low energy photons on the other hand can be experimented based on the transmission of photons through the material based on the Beer-Lambert law [3-5]. A previous study suggested that the measurement of mass attenuation coefficients of materials at high energy photons can be measured by using the Compton scattering method by using a high energy gamma emitter such as ^{137}Cs [6]. This method provides better accuracy of measurement at experimented gamma energy ranges as the scattering of the incident scattering gamma to an attenuator is commonly inelastic at specific gamma energies.

The Ludlum configuration had been used for the measurement of lead equivalent of materials and several quality control measurements in the Malaysian Nuclear Agency (Nuklear Malaysia). A previous study showed that the Ludlum configuration can be used for the measurement of mass attenuation coefficients at high energy photons using the transmission method [7]. Therefore, this study focused on the use of the Ludlum configuration for the measurement of mass attenuation coefficients at high energy photons by using the Compton scattering method.

2. Methodology

2.1. Experimental Set Up and System Calibration

The Ludlum configuration used in this study consists of the gas type Ludlum detector connected to a computer analysis and a ^{137}Cs sealed source that provided gamma peak energy of 662 keV. The ^{137}Cs sealed source was encapsulated in a lead container with collimation size of 0.1 cm to simulate the line source projection. An Al plate with approximate thickness of 0.1 cm was used as an attenuator to obtain the scattered photons. The Al plate was placed between the detector and the source at 20 cm distances from both the detector and the source as shown in Figure 1. The Ludlum Detector was placed at angles of 30°, 45°, 60° and 75° to measure the scattered gamma *energies. The scattered gamma energies was calculated based on the previous work by Limkitjaroenporn et* al. [6] by using the equation of,

$$E_{\gamma'} = \frac{E_\gamma}{1 + (1-\cos\theta)E_\gamma/mc^2} \quad (1)$$

where E_γ and $E_{\gamma'}$ is the incident and scattered gamma energy respectively, θ is the angle of scattered gamma and m is the electron rest mass [8][9]. This equation easily derived by assuming a relativistic collision between gamma ray and an electron initially at rest [6].

The energy calibration was performed on the Ludlum configuration to determine the linearity of signal in term of count per minute (CPM) of the detector at different photon energies. The calibration was performed by using ^{133}Ba, ^{137}Cs and ^{60}Co sealed sources that provided gamma peak energies of 0.36, 0.662 and 1.3 MeV respectively.

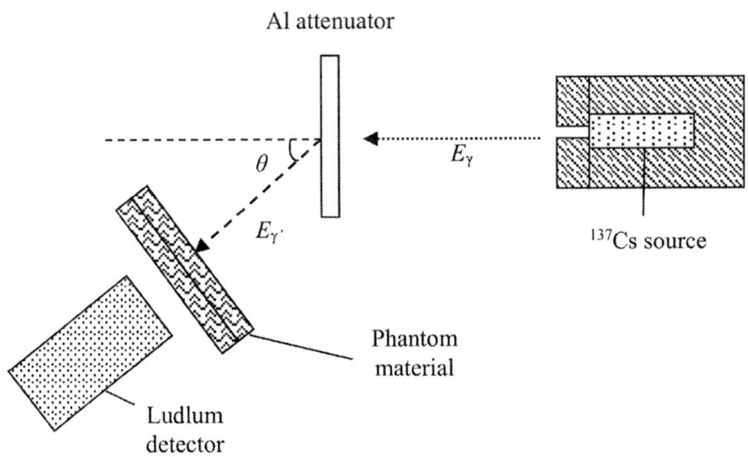

Figure 1. The experimental set up of the Ludlum configuration for the measurement of mass attenuation coefficients by using Compton scattering method.

Table 1. The scattered gamma energies at different angles of measurements.

Scattering angle, θ	Incident gamma energy, E_γ (keV)	Scattered gamma energy, $E_{\gamma'}$ (keV)
0	662	662.00
30		564.09
45		479.90
60		401.76
75		337.72

2.2 Experimented Materials

Two types of water equivalent phantom materials, the solid water and Perspex® phantoms were used in this study. The solid water phantoms is an epoxy resin-based materials commonly used for dosimetric and quality control measurements in radiotherapy. The Perspex® phantoms on the other hand is a methyl methacrylate-based materials commonly used as phantoms for diagnostic imaging. Both materials are made water equivalent with density close to water (1.0 g/cm^3). The elemental compositions, density and effective atomic number of solid water and Perspex® phantoms is presented in Table 2.

Table 2. The chemical compositions, density and effective atomic number of the phantom materials used in the study.

Phantom	Material	Chemical composition	Density (g/cm^3)	Effective atomic number, Z_{eff}
Solid water	Epoxy resin	$C_{21}H_{25}ClO_5$[1]	1.00[1]	7.35[1]
Perspex®	Methyl Methacrylate	$(C_5O_2H_8)_n$[1]	1.18[1]	6.48[1]

[1]Khan [8]

2.3 Measurement of Linear and Mass Attenuation Coefficients

The linear attenuation coefficient of the phantom samples was measured based on the transmission of photon through the samples based on the Beer-Lambert equation of,

$$I = I_o e^{-\mu x} \tag{2}$$

with I_o and I is the initial and transmitted photon respectively, μ is the linear attenuation coefficient of the sample medium and x is the thickness of the sample medium. The linear attenuation coefficient can be calculated by rearranging Equation 2 into the equation

$$\mu = \frac{\ln I_o/I}{x} \tag{3}$$

The mass attenuation coefficient, μ/ρ of the phantom materials can be calculated by dividing the value of linear attenuation coefficient with the density of the phantom material. The theoretical value of mass attenuation coefficients can be calculated by using the XCOM software. The molecular formula of the phantom materials and the range of photon energies were inserted into the calculation software and the tabulated data on the mass attenuation coefficients of the phantom materials were obtained. The theoretical values of mass attenuation coefficients of the phantom materials were compared to their measured values from the Ludlum configuration measurement.

3. Results and discussions

3.1. Calibration Curve of the System

The energy calibration curve of the Ludlum configuration is illustrated in Figure 2. The energy calibration graph showed an excellent linearity of signal detected by the Ludlum configuration shown by the linear regression value $R^2 = 0.999$. The result indicated the suitability of the Ludlum set up for measurement of mass attenuation coefficients at high energy photons.

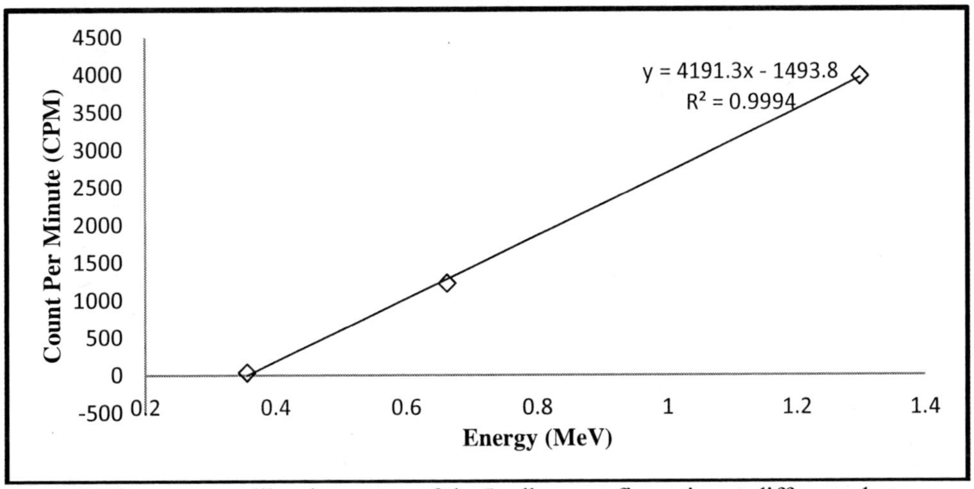

Figure 2. The energy calibration curve of the Ludlum configuration at different photon energies

3.2. Linear and Mass Attenuation Coefficients of Materials using Compton Scattering Methods

The measured linear and mass attenuation coefficients of the solid water and Perspex® phantoms in comparison to their respective XCOM values at all experimented scattered gamma energies are presented in Table 3 and 4 respectively. The comparison of mass attenuation coefficients of solid water and Perspex® phantoms to their respective XCOM values are illustrated in Figure 3 and 4 respectively. The results showed that the linear and mass attenuation coefficients of both solid water and Perspex® phantoms decreased at increased photon [6][7]. The results also showed that the measured mass attenuation coefficients of the phantom materials were in agreement to their respective theoretical values measured using XCOM calculations within 2.72 and 4.35% percentage of differences all experimented scattered photon energies in solid water and Perspex® respectively. The measurement of mass attenuation coefficient at transmitted gamma energies (angle = 0°) also showed an agreement between the measured and theoretical values by XCOM within 4,47 and 6.18% percentage of differences by solid water and Perspex® respectively. The paired sample *t*-test was calculated between measured and theoretical values of mass attenuation coefficients and presented in Table 5. The results showed that there was no significant different between the theoretical values to the measured values of mass attenuation coefficients of the phantom samples. This indicated the consistency between the experimental values of the mass attenuation coefficients to the theoretical values.

Table 3. The mass attenuation coefficients of solid water phantoms measured using Ludlum configuration and XCOM.

Energy	Solid water $_{XCOM}$	Solid water$_{meas}$		Percentage Difference
	μ/ρ (cm^2/g)	μ (cm^{-1})	μ/ρ (cm^2/g)	(%)
662	0.067	0.060	0.070	4.47
564.09	0.086	0.085	0.085	1.16
479.90	0.094	0.096	0.096	2.13
401.76	0.103	0.105	0.105	1.94
337.72	0.107	0.110	0.110	2.72

Table 4. The mass attenuation coefficients of Perspex® phantoms measured using Ludlum configuration and XCOM.

Energy	Perspex®$_{XCOM}$ (cm^2/g)	Perspex®$_{meas}$		Percentage Difference (%)
		μ (cm^{-1})	μ/ρ (cm^2/g)	
662	0.097	0.107	0.091	6.18
564.09	0.103	0.117	0.099	3.88
479.90	0.109	0.124	0.105	3.67
401.76	0.115	0.130	0.110	4.35
337.72	0.118	0.135	0.114	3.39

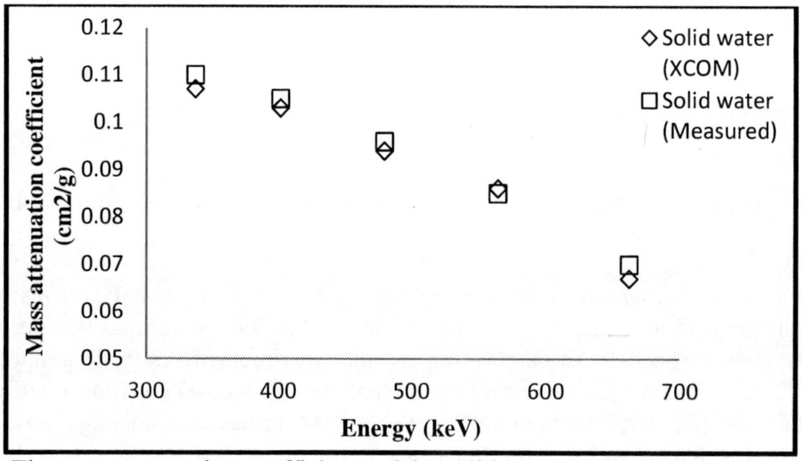

Figure 3. The mass attenuation coefficients of the solid water phantoms by using the XCOM calculation and measurements.

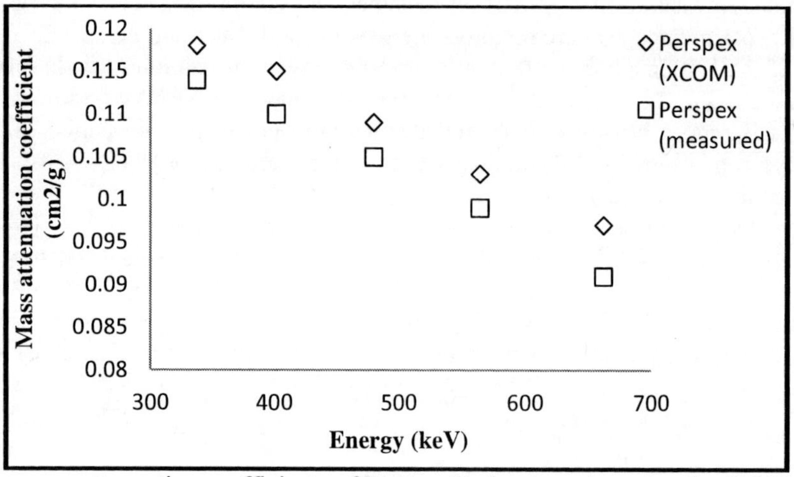

Figure 4. The mass attenuation coefficients of Perspex® phantoms by using the XCOM calculation and measurements.

Table 5. The mass attenuation coefficients of solid water and Perspex® phantoms in comparison to the theoretical values by using XCOM.

Paired samples	Mean	t	df	Sig. (2-tailed)
Solid water$_{meas}$-Solid Water$_{XCOM}$	0.096	1.457	3	0.241
Perspex$_{meas}$-Perspex$_{XCOM}$	0.111	17	3	0.040

4. Conclusion

The gamma energies reduced at increased scattered angles. The measured mass attenuation coefficients of the phantom materials decreased when measured at higher scattered gamma energies. The measured mass attenuation coefficients by using the Ludlum configuration by using the Compton scattering method showed good agreement to the theoretical values measured by using the XCOM software at all experimented gamma energies. The results indicated the suitability of the Ludlum configuration for the measurement of mass attenuation coefficients of materials by suing the Compton scattering method.

5. References

[1] Cevik U and Baltas H 2007 Measurement of the mass attenuation coefficients and electron densities for BiPbSrCaCuO superconductor at different energies *Nucl. Instr. Meth. Phys. Res. B* **256** p 619-625.

[2] Berger M J and Hubbell J H. 1987 XCOM: Photon cross-sections on a personal computer, *NBSIR*, 1987, 87-3597 *The Phys. of Radiation Therapy* ed M F Khan (Philadelphia: Lippincott Williams & Wilkins, 4th ed 2010).

[3] Hill R F, Brown S, and Baldock C 2007 Evaluation of the water equivalence of solid phantoms using gamma ray transmission measurements. *Radiat. Maes.* **43** p 1258-1264.

[4] Shakhreet B Z, Bauk S, Tajuddin A A, and Shukri A 2009 Mass attenuation coefficients of natural *Rhizophora* spp. wood for X-rays of 15.77-25.27 keV range, *Radiat. Prot. Dosim.*, **135** p 47-53.

[5] Marashdeh M W, Bauk S, Tajuddin A A, and Hashim R 2012 Measurementof mass attenuation coefficients of *Rhizophora* spp. binderless particleboards in the 16.59–25.26 keV photon energy range and their density profile using x-ray computed tomography. *Appl. Radiat Isot.* **70** p 656–662.

[6] Limkitjaroenporn P, Kaewkhao J, Chewpraditkul W, and Limsuwan P 2012 Mass attenuation coefficient and effective atomic number of Ag/Cu/Zn alloy at dfferent energy by Compton scattering technique *Procedia Eng.* **32** p 847-854.

[7] Mohd Yusof M F, Tajuddin A A, Hashim R, Bauk S, Isa N M, and Md Isa M J 2017 Mass attenuation coefficients of tannin-added *Rhizophora* spp. particleboards at 16.59-25.26 keV photons, and 137Cs and 60Co gamma energies. *Radiol. Phys. Technol.* **10(3)** p 331-339.

[8] Knoll G F 2000 *Radiation Detection and Measurement* (New York: John Wiley and Sons Inc. 3rd ed)

[9] Trousfanidis N 1983 *Measurement and Detection of Radiation* (New York: Hemisphere Publishing).

Acknowledgement

The author would like to thank the financial support by the Bridging Grant scheme no. 304.PPSK.6316145 by Universiti Sains Malaysia.

IOP Conf. Series: Materials Science and Engineering **555** (2019) 012029 doi:10.1088/1757-899X/555/1/012029

Measurement of mass attenuation coefficients of the fabricated *Rhizophora* spp. particleboards at high energy photons by using Compton scattering method

M F Mohd Yusof[1,a)], P N K Abd Hamid[2], N Mohd Isa[2], A B Abd Kadir[3], H Salleh[3]

[1]School of Health Sciences, Universiti Sains Malaysia, 16150, Kelantan, Malaysia
[2]School of Physics, Universiti Sains Malaysia, 11800 Penang, Malaysia
[3]Medical Physics and Secondary Standard Dosimetry Laboratory, Malaysian Nuclear Agency, 43000 Selangor, Malaysia

[a]mfahmi@usm.my

Abstract. The mass attenuation coefficients of the fabricated *Rhizophora* spp. particleboards were measured by using the Compton scattering technique and Ludlum configurations. The mass attenuation coefficents of the particleboards were measured at scattered gamma energies of ^{137}Cs between 337.72 and 564.09 keV and compared to the its theoretical values and the values of water by using XCOM calculation. The results showed that the measured mass attenuation coefficients of the fabricated *Rhizophora* spp. particleboards were in good agreement to its theoretical values and water (XCOM) within 3.8 and 5.9% discrepancies. The results showed the near attenuation properties of the fabricated *Rhizophora* spp. particleboards to water at scattered gamma energies and the suitability of the Ludlum configuration for the measurement of mass attenuation coefficients of materials.

1. Introduction

The potential of *Rhizophora* spp. particleboards as phantom had been studied by many researchers in the previous works [1-5]. The *Rhizophora* spp. particleboards showed good attenuation and dosimetric properties in comparison to water and other phantom materials such Perspex® and solid water phantoms at photons of the various energy ranges [6]. The mass attenuation coefficient is one of the most important parameter to determine attenuation properties of materials towards the ionizing radiations. The mass attenuation coefficient is determined based on the linear attenuation coefficients of a material when a line source passing through the material taking into considerations of the mass density of the material [7]. The measurement of mass attenuation coefficient shall be made in photons of various energy ranges to determine its proportionality at different photon [8].

The measurement of mass attenuation coefficient of materials with known elemental compositions can be determined theoretically by using computer program of photon cross-section database named XCOM [9][10]. This program enables the calculation of mass attenuation coefficient at various ranges of photon energies and has been used by many researchers. The mass attenuation coefficient of substances that consists of more than one element can also be calculated by inserting the molecular formula of the respective elements into the calculation program.

Several studies had been conducted for the measurements of mass attenuation coefficients of *Rhizophora* spp. particleboards based on the transmission of photons by using the X-ray fluorescent (XRF) and sealed sources [1-4][11]. A previous study suggested that the measurement of mass attenuation coefficient at high energy photons can be measured using the Compton scattering method by using ^{137}Cs gamma energy [12]. This method provided better accuracy of experimented gamma energy range as the scattering of incident gamma to the attenuator is commonly inelastic at specific angles and scattered gamma energies. This study measures the mass attenuation coefficients of the fabricated *Rhizophora* spp. particleboards at scattered gamma energies from the ^{137}Cs sealed source

2. Methodology

2.1. Rhizophora spp. Particleboard Phantoms

The fabricated *Rhizophora* spp. particleboards used in this study was based on the previous work by Mohd Yusof *et al.* [5] and Mohd Yusof *et al.* [11] as shown in Figure 1. The *Rhizophora* spp. particleboards were fabricated with the addition of tannin as adhesive material to increase their physical and mechanical properties [5]. Several evaluations on their physical properties including the percentage of elemental compositions, calculated effective density and effective atomic number were investigated and presented in Table 1. A previous work indicated the close values of mass attenuation coefficients of the particleboards to the values of water at ^{137}Cs and ^{60}Co gamma energies by using the transmission method [6]. Therefore, this study focused on the measurement of mass attenuation coefficients of the particleboards at scattered gamma energies of ^{137}Cs by using the Compton scattering method.

Figure 1. The fabricated *Rhizophora* spp. particleboards used in the study.

Table 1. The physical properties of the fabricated *Rhizophora* spp. particleboards

Physical properties	Descriptions
Elemental compositions	C (51.25%), O (43.11%), F (5.64%)[a]
Density (g/cm^3)	1.03[a]
Effective atomic number, Z_{eff}	7.22[a]
Electron density ($\times 10^{23}$/g)	3.33[a]

[a]Mohd Yusof *et al.* [5]

IOP Conf. Series: Materials Science and Engineering **555** (2019) 012029 doi:10.1088/1757-899X/555/1/012029

2.2. Experimental Set Up

The Ludlum configuration was used in this study to provide the scattered gamma energies as shown in Figure 2. It consists of the gas type Ludlum detector connected to a computer analysis. A ^{137}Cs sealed source with gamma peak energy of 662 keV was used to provide the incident gamma energies. The ^{137}Cs sealed source was encapsulated in a lead container with collimation size of 0.1 cm to simulate the line source projection. An aluminum (Al) plate with approximate thickness of 0.1 cm was used as an attenuator to produce the scattered photons. The Al plate was placed between the detector and the source at 20 cm distances from both the detector and the source as shown in Figure 3. The Ludlum detector was placed at angles of 30°, 45°, 60° and 75° to measure the scattered gamma energies. The scattered gamma energies was calculated based on the previous work by Limkitjaroenporn *et al.* [12] by using the equation of,

$$E_{\gamma\prime} = \frac{E_\gamma}{1 + (1-\cos\theta)E_\gamma/mc^2} \tag{1}$$

where E_γ and $E_{\gamma\prime}$ is the incident and scattered gamma energy respectively, θ is the angle of scattered gamma and m is the electron rest mass [12]. This equation easily derived by assuming a relativistic collision between gamma ray and an electron initially at rest [6]. The calculated scattered gamma energies of the ^{137}Cs is presented in Table 2.

Figure 2. The Ludlum configuration used for the measurement of mass attenuation coefficient in the study.

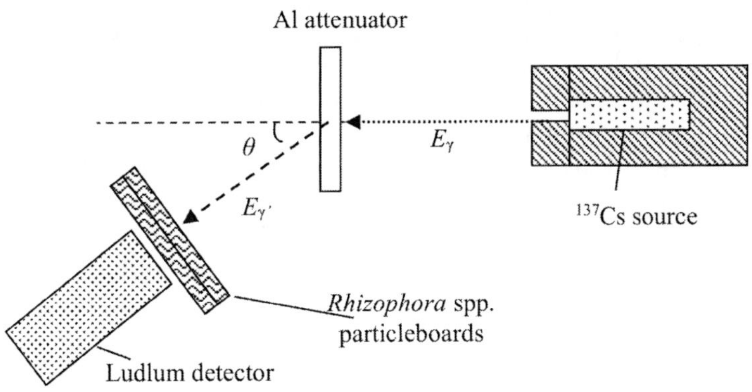

Figure 3. The experimental set up of the Ludlum configuration for the measurement of mass attenuation coefficients by using Compton scattering method.

Table 2. The scattered gamma energies at different angles of measurements.

Scattering angle, θ	Incident gamma energy, E_γ (keV)	Scattered gamma energy, $E_{\gamma'}$ (keV)
0	662	662.00
30		564.09
45		479.90
60		401.76
75		337.72

2.3 Measurement of Linear and Mass Attenuation Coefficients

The linear attenuation coefficient of the phantom samples was measured based on the transmission of photon through the samples based on the Beer-Lambert equation of,

$$I = I_o e^{-\mu x} \quad (2)$$

with I_O and I is the initial and transmitted photon respectively, μ is the linear attenuation coefficient of the sample medium and x is the thickness of the sample medium. The linear attenuation coefficient can be calculated by rearranging Equation 2 into the equation,

$$\mu = \frac{\ln I_o/I}{x} \quad (3)$$

The mass attenuation coefficient, μ/ρ of the phantom materials can be calculated by dividing the value of linear attenuation coefficient with the density of the phantom material. The measured mass attenuation coefficients of the particleboards were compared to the theoretical values of water by using the XCOM software calculation at similar photon energies [4]. The percentage differences of between the mass attenuation coefficients of the particleboards and water were calculated at all scattered gamma energies to determine the discrepancies between the materials.

3. Results and Discussions

The values of mass attenuation coefficients of the fabricated *Rhizophora* spp. particleboards at all experimented scattered gamma energies is shown in Table 3. The comparison of mass attenuation coefficients between the particleboards and the XCOM values of water is illustrated in Figure 4. The results showed that the mass attenuation coefficients of the fabricated *Rhizophora* spp. particleboards decreased exponentially when the scattered gamma energy increased [12]. The results also showed that the mass attenuation coefficients of the particleboards were in good agreement to the values of water (XCOM) at all experimented gamma energies. This indicated the similarities of attenuation properties of the fabricated *Rhizophora* spp. particleboards to water suggesting its potential to be used as water equivalent phantom material for high energy photons. The measured mass attenuation coefficients of the particleboards measured at transmitted and unattenuated ^{137}Cs gamma energies also showed good agreement to the scattered gamma energies [6]. This indicated that the scattered gamma energies can be used to determine the mass attenuation coefficients of a material.

The mass attenuation coefficients of the *Rhizophora* spp. particleboards in this study were measured based on the transmission of scattered gamma energies compared to the transmitted $K_{\alpha 1}$ photons used from the X-ray fluorescent (XRF) configuration on the previous works [1-5]. This study also measured the mass attenuation coefficients of lower density material compared to the high density material in the previous work [12]. The overall results indicated the suitability of the Compton scattering method for the measurement of mass attenuation coefficients of materials at high photons energies.

Table 3. The measured mass attenuation coefficients of the fabricated *Rhizophora* spp. particleboards at transmitted and scattered ^{137}Cs gamma energies.

Energy (keV)	Mass attenuation coefficients, μ/ρ (cm^2/g)		Percentage Difference (%)
	Water (XCOM)	*Rhizophora* spp. particleboards	
662	0.086[a]	0.080[a]	8.13
564.09	0.097	0.090[b]	7.21
479.90	0.105	0.095[b]	8.04
401.76	0.115	0.105[b]	8.69
337.72	0.118	0.115[b]	2.54

[a]Mohd Yusof et al., [11], [b]Current study

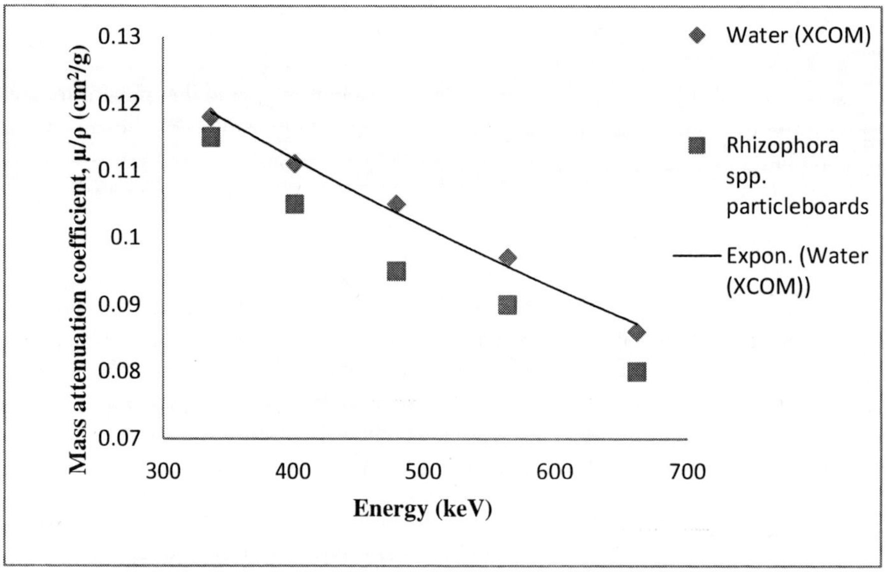

Figure 4. The mass attenuation coefficients of the fabricated *Rhizophora* spp. particleboards in comparison to water (XCOM) at scattered gamma energies.

4. Conclusion

The mass attenuation coefficients of the fabricated *Rhizophora* spp. particleboards decreased at higher scattered gamma energies. The values of mass attenuation coefficients of the fabricated *Rhizophora* spp. particleboards also showed good agreement to the theoretical values of water at all scattered gamma energies. The study also indicated the suitability of the Ludlum configuration for the measurement of mass attenuation coefficients by using the Compton scattering method.

5. References

[1] Marashdeh M W, Bauk S, Tajuddin A A, and Hashim R 2012 Measurement of mass attenuation coefficients of *Rhizophora* spp. binderless particleboards in the 16.59–25.26 keV photon energy range and their density profile using x-ray computed tomography *Appl Radiat Isot.*, **70** p 656–662.

[2] Abu Arra A., Hashim R, Bauk S, Kandaiya S and Tousi E T 2014 Fabrication and characterization of gum Arabic bonded *Rhizophora spp.* particleboards *Mater. Des.* **60** p 108-115.

[3] Tousi E T Z, BaukS, Hashim R, Jaafar M S, Abuarra A 2014 Measurement of mass attenuation coefficients of *Eremurus-Rhizophora spp.* particleboards for X-ray in the 16.63-25.30 keV energy range *Radiat. Phys. Chem.* **103** p 119-125.

[4] Ababneh B, Tajuddin A A, Hashim R, Shuaib I L 2016 Investigation of mass attenuation coefficient of almond gum bonded *Rhizophora* spp. particleboard as equivalent human tissue using XRF technique in the 16.6–25.3 keV photon energy, *Australas. Phys. Eng. Sci. Med.* **39** p 871-876.

[5] Mohd Yusof M F, Hashim R, Tajuddin A A, Bauk, S and Sulaiman O 2017 Characterization of tannin-added *Rhizophora* spp. as phantom materials for photon beams *Ind. Crop Prod.* **95** p 467-474.

[6] Mohd Yusof M F, Abd Hamid P N K, Tajuddin A A, Hashim R, Bauk S, Mohd Isa N, Md Isa M J 2017b Fabrication and characterisation of phantom material made of Tannin-added *Rhizophora spp.* particleboards for photon and electron beams, *J. of Phys. Conf. Series* **851** p 012036.

[7] Khan F M 2010*The Phys. of Rad. Therapy 4th ed.* (Philadelphia: Lippincott Williams and Wilkins).

[8] Cevik U and Baltas H 2007 Measurement of the mass attenuation coefficients and electron densities for BiPbSrCaCuO superconductor at different energies. *Nucl. Instr. Meth. Phys. Res. B* **256** p 619-625.

[9] Hubbell J H 1982 Photon Mass Attenuation and energy-absorption coefficients *Int. J. Appl. Radiat. Isot.* **33(11)** p 1269-1290.

[10] Berger M J and Hubbell J H 2010 1987 XCOM: Photon cross-sections on a personal computer *NBSIR The Physics of Radiation Therapy* ed M F. Khan (Philadelphia: Lippincott Williams and Wilkins) p 87-3597.

[11] Mohd Yusof M F, Abd Hamid P N K , Tajuddin A A , Hashim R, Bauk S, Mohd Isa N, Md Isa M J 2017c Mass attenuation coefficient of tannin-added *Rhizophora* spp. particleboards at 16.59–25.56 keV photons, and ^{137}Cs and ^{60}Co gamma energies *Radiol. Phys. Technol.,* 10(3), 331-339.

[12] Limkitjaroenporn P, Kaewkhao J, Chewpraditkul W and Limsuwan P 2012 Mass attenuation coefficient and effective atomic number of Ag/Cu/Zn alloy at different energy by Compton scattering technique *Procedia Eng.* **32** p 847-854.

Acknowledgement

The author would like to thank the financial support by the Bridging Grant scheme no. 304.PPSK.6316145 by Universiti Sains Malaysia.

IOP Conf. Series: Materials Science and Engineering **555** (2019) 012030 doi:10.1088/1757-899X/555/1/012030

X-ray fluorescent (XRF) configuration for the measurement of mass attenuation coefficients at low energy photons

Abdullah S N A[1,a)], Mohd Yusof M F[2], Kabir N A[1], Tajuddin A A[1] and Mohd Rodin M N M[1]

[1]School of Physics, Universiti Sains Malaysia, 11800 USM, Penang, Malaysia
[2] School of Health Sciences, Universiti Sains Malaysia, 16150 Kota Bharu, Kelantan, Malaysia

[a]sitinor_azizah87@yahoo.com

Abstract. An X-ray fluorescent (XRF) configuration was used to measure the mass attenuation coefficient of materials at low energy photons. An Americium-241 (^{241}Am) annular source was used in conjunction with niobium, palladium, molybdenum and tin plates that provided $K_{\alpha 1}$ photons with energies between 16.59 and 25.26 keV. The transmitted photons were measured by using a low energy Germanium (LEGe) detector and analyzed by using MAESTRO software. A calibration with ^{241}Am showed good linearity between photon energies and peak channels with R^2 value near to 1. The measurement of mass attenuation coefficient was carried out on Al plates and Perspex based on the transmitted photons on the samples. The measured mass attenuation coefficients were compared to the XCOM calculations. The results showed that the measured mass attenuation coefficients of aluminum and Perspex® were in good agreement to their XCOM values within 10% percentage of discrepancies at all experimented photon energies. The overall results indicated the suitability of the XRF configuration for the measurement of mass attenuation coefficients at low energy photons.

1. Introduction

The accurate measures of radiation interaction parameters with matter plays a significant value in many applied fields of sciences. The essential parameter, for example the attenuation coefficient of a material is extensively evaluated in medical, nuclear and radiation physics, radiation protection and dosimetry, as well as industrial, environmental and agricultural studies [1].

The probability of photon interaction by a physical process or another per unit distance travelled is called linear attenuation coefficient, μ [2]. However, mass attenuation coefficient (μ/ρ) is more important in characterization of a material. This is because it offers the reduction of photon intensity as a function of mass instead of path length of the travelled material [3]. Besides, attenuation of γ-rays and X-rays is closely related to density and atomic number of a material. Thus, it provides many other photon interaction parameters such as atomic cross-section, effective atomic number and electron density. [4]

Theoretically, XCOM computer program is commonly used for obtaining the attenuation coefficient values that can be compared experimentally [5]. However, the XCOM has a few limitations such as it does not consider the molecular and solid-state effects which can change the element's cross section. Therefore, several measurements have been reported on the experimental determination of attenuation coefficients. Some of these are using proton-induced X-ray emission (PIXE) [6], high energy gamma photon [7][8] and X-ray fluorescent (XRF) techniques [9][10].

X-ray fluorescence (XRF) spectrometry is widely used for quantitative analysis because it is inexpensive and requires less technical effort to run the system [11]. It is a non-destructive analytical technique used to determine the elemental composition of a sample by measuring the fluorescent (or secondary) X-ray emitted from the sample when it is excited by a primary radiation source. This characteristic X-ray is unique and usually defined as the fingerprint for the specific element.

This study was conducted to measure the mass attenuation coefficients of two materials: aluminum (Al) and Perspex® using XRF configuration at effective energy range of 16.59 keV to 25.26 keV. The values were then compared to the mass attenuation coefficients calculated from XCOM simulation software.

2. Materials and methods

2.1. Materials Used

Pure aluminum and Perspex® plates were used as the absorber in this experiment as shown in Figure 1. The dimension of aluminum plates are 7.6 cm x 7 cm and four thicknesses of 0.06 cm, 0.125 cm, 0.15 cm and 0.2 cm with the density of 2.70 g/cm^3. The Perspex®, with the density of 1.18 g/cm^3 has 5 cm x 5 cm dimension with thicknesses of 0.5 cm, 1 cm, 1.5 cm and 2 cm.

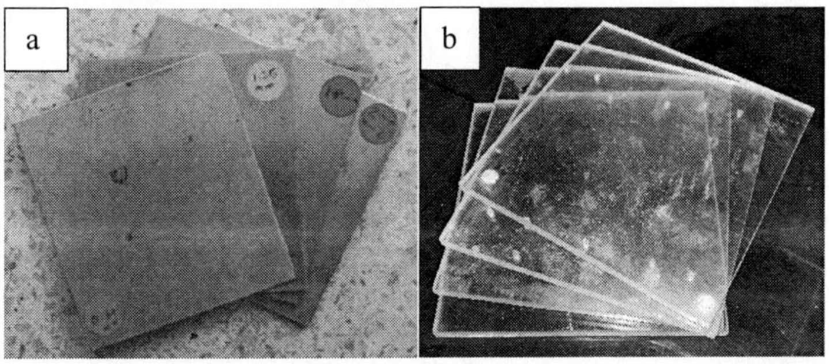

Figure 1. a) Aluminum plates and b) Perspex® used in the study.

2.2. Determination of mass attenuation coefficient using ^{241}Am source

The XRF instrument at Biophysics Laboratory, School of Physics, Universiti Sains Malaysia (USM) was used for the measurements of mass attenuation coefficient of the absorbers. A schematic drawing of the experimental arrangement is presented in Figure 2. A 59.54 keV of 100 mCi ^{241}Am was used to irradiate high purity metal plates to produce the X-ray fluorescence photons. The photons transmitted through the absorber were detected in a narrow beam geometry by the CANBERRA low energy Germanium (LEGe) detector. The detector output pulses were amplified by an ORTEC 572 amplifier and the spectrum was collected by a multichannel analyser (MCA-3 series) for a period of 60 seconds.

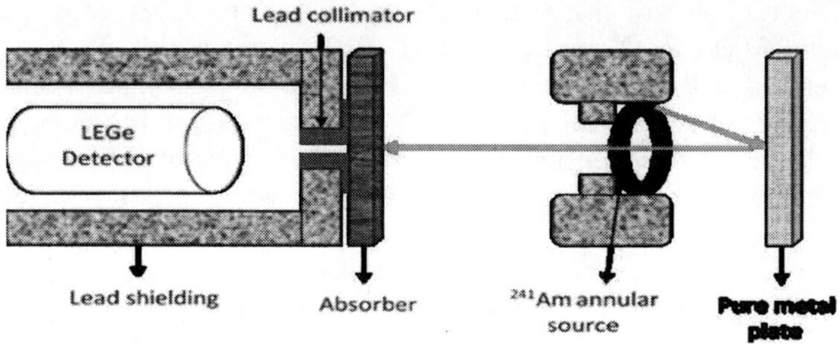

Figure 2. The schematic diagram of the experimental set-up for attenuation measurement.

In this experiment, 4 types of high purity metal plates were used: niobium (Nb), molybdenum (Mo), palladium (Pd) and tin (Sn) having $K_{\alpha 1}$ fluorescent X-rays of 16.59, 17.74, 21.21 and 25.26 keV respectively. Table 1 presents the details of the metal plates [12].

Table 1. The details of metal plates used in the X-ray fluorescent (XRF) configuration.

Plate	Atomic Number (Z)	Thickness (mm)	Purity (%)	$K_{\alpha 1}$ energy (keV)
Niobium (Nb)	41	0.14	99.8	16.61
Molybdenum (Mo)	42	0.11	99.9	17.47
Palladium (Pd)	46	0.10	99.9	21.17
Tin (Sn)	50	0.28	99.999	25.27

Next, the net area under the $K\alpha_1$ peak was measured, the total intensity of incident photons (N_o) and those that are transmitted without interaction (N) were calculated in unit of counts. From that, linear attenuation coefficient (μ) expressed in cm^{-1} can be calculated by

$$\mu = \frac{1}{x} \ln \frac{N_0}{N} \quad (1)$$

with x is the thickness of the absorber [13]. However, since a range of thicknesses were used, the linear attenuation coefficient was determined by the gradient of ln $[N_o/N]$ against thickness curve.

Yet, the probability of interaction is proportional to the number of atoms per volume. This dependency can be overcome by normalizing the linear attenuation coefficient to the density of the material, defined as mass attenuation coefficient, μ/ρ (cm^2/g). After the mass attenuation coefficient of the two types of absorbers were calculated experimentally, the results were then compared to the theoretical values obtained from XCOM simulation software.

2.3. Evaluation of Statistical Analysis Using Paired Sample t-test

The paired sample *t*-test was carried out to compare the means between two sets of data and show how significant the difference was from each other. The two tailed paired-sample *t*-test was chosen for this study to prove the similarity of the mass attenuation coefficients of the absorbers obtained by XRF method and XCOM computer program. The null hypothesis was the mean difference of the mass attenuation coefficient between both methods is equal to zero, H_0: $\mu_d = 0$ [14]. In brief, the *p*-value obtained must be higher than 0.05 (5%) to show that the mass attenuation coefficients from both methods were considered statistically equivalent.

3. Results and discussion

3.1. Mass attenuation coefficient for Aluminium and Perspex® using XRF method and XCOM

The mass attenuation coefficients of Aluminium samples were determined and presented in Figure 3. The graphs showed that the values of mass attenuation coefficients decreased when the photon energies increased [1,2]. The comparison between the experimental values using XRF system and calculated values from XCOM software. Accordingly, it was found that the mass attenuation coefficient values measured using the XRF configuration were very close to the theoretical values calculated in XCOM computer program [5]. The reasonable agreement between the two measurements gave confidence in the technique adopted during the experiment.

Similar results were also observed in Perspex® as shown in Figure 4. The XCOM values of Perspex® were obtained by inserting the percentage of elemental compositions of Perspex® into the XCOM calculation software. The values of measured mass attenuation coefficients of Perspex® however showed larger differences to its respective values in XCOM at higher photon energies compared to that in the lower photon energies.

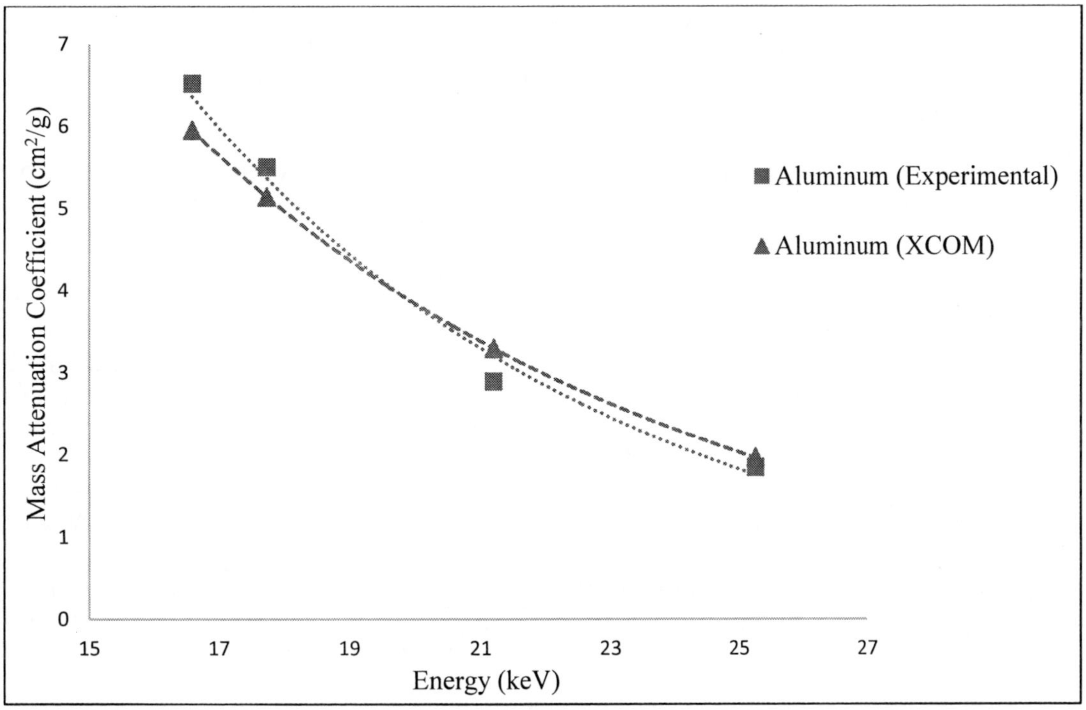

Figure 3. Mass attenuation coefficients of aluminium determined by XRF method and XCOM against photon energy.

IOP Conf. Series: Materials Science and Engineering **555** (2019) 012030 doi:10.1088/1757-899X/555/1/012030

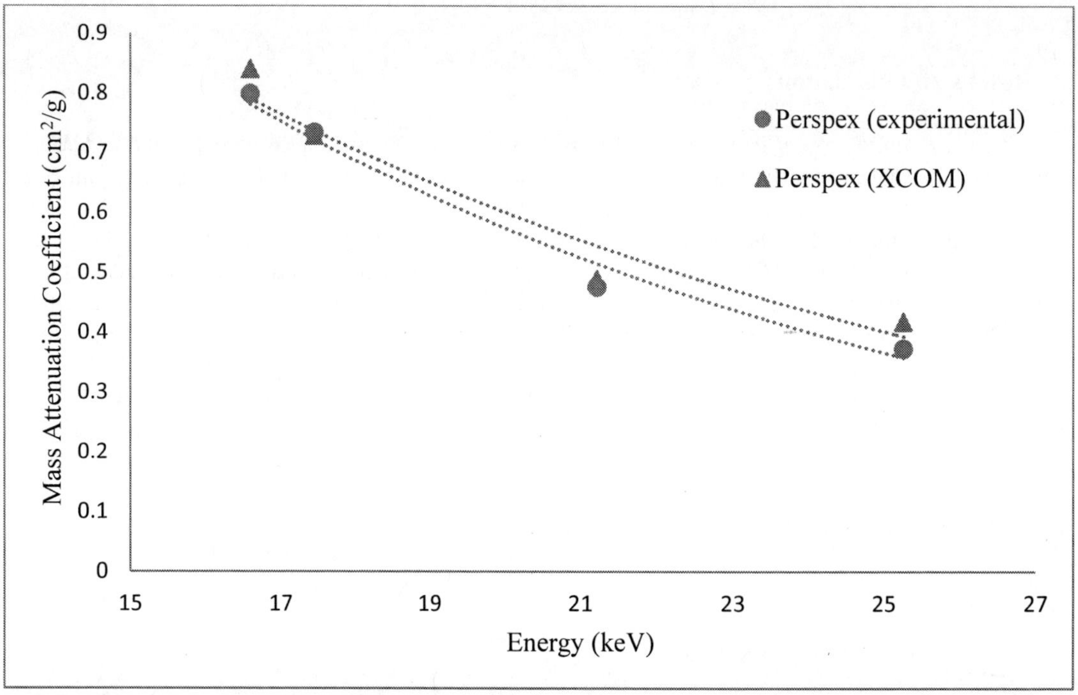

Figure 4. Mass attenuation coefficients of Perspex® determined by XRF method and XCOM against photon energy.

3.2. *Paired-sample t-test analysis for mass attenuation coefficient of Aluminium and Perspex*

Table 2 presented the results of the two tailed paired-sample *t*-test of mass attenuation coefficient of aluminium and Perspex®. The *p*-value for both absorbers calculated were higher than 0.05, indicating that the pairs of mass attenuation coefficient between the two methods had no significant difference. Thus, the null hypothesis was accepted. This showed the similarity between mass attenuation coefficient obtained by experiment and calculated by XCOM. Consequently, the XRF technique used in this study produced an accurate and reliable results for mass attenuation coefficient measurement.

Table 2. The paired-sample *t*-test of mass attenuation coefficients of the absorbers between experimental and XCOM values.

Pair	**Paired Differences**		**df**	**t**	***p*-value (two-tailed)**
	Means	**Standard Deviation**			
Aluminum XRF - XCOM	0.102	0.385	4	0.462	0.676
Perspex XRF - XCOM	0.023	0.005	4	1.953	0.146

4. Conclusions

The comparison of the mass attenuation coefficient for both absorbers: aluminium and Perspex® showed that our experimental results were, in general, close to the theoretical values calculated using XCOM, represented by the solid curve. Furthermore, the paired-sample *t*-test indicated that the mass attenuation coefficients obtained by both methods were considered statistically equivalent. Hence, the XRF configuration in our biophysics lab is suitable and reliable to be used for mass attenuation coefficient measurement at low energy range.

5. References

[1] Marashdeh M W, Al-Hamarneh I F, Munem E M A, Tajuddin A A, Ariffin A and Al-Omari S 2015 Determining the mass attenuation coefficient, effective atomic number, and electron density of raw wood and binderless particleboards of *Rhizophora* spp. by using Monte Carlo simulation. *Results in Phys.* **5** p 228-234.

[2] Tekin H O and Manici T 2017 Simulations of mass attenuation coefficients for shielding materials using the MCNP-X code *Nucl. Sci. T*ech. **28(7)** p 95.

[3] Tousi E T, Hashim R, Bauk S and Jaafar M S 2017 Evaluation of the mass attenuation coefficient and Effective Atomic Number of the *Eremurus* spp. root in mammography energy range *IOSR-JOP* **09(01)** p 100-104.

[4] Levet A and Özdemir Y 2017 Determination of effective atomic numbers, effective electrons numbers, total atomic cross-sections and buildup factor of some compounds for different radiation sources *Radiat. Phys. Chem.* **130** p 171-176.

[5] Berger M J and Hubbell J H 1987 XCOM: Photon cross-sections on a personal computer *NBSIR* **87** p 3597.

[6] Abdullah K K, Ramachandran K K, Karunakaran N, Babu B R S, Josephm A, Thomas R and Varier K M 2010 X-ray attenuation around K-edge of Zr, Nb, Mo and Pd: A comparative study using proton-induced X-ray emission and ^{241}Am gamma rays. *Pramana J. of Phys*, **75(3)** p 459-469.

[7] Yusof M F M, Abd Hamid P N K, Tajuddin A A, Hashim R, Bauk S, Mohd Isa N. and Md Isa M. J 2017 Characterization and attenuation study on tannin-added *Rhizophora* spp. particleboard at high energy photon and electron *IOP Conf. Series: J. of Phys.* **851** p 012036.

[8] Yusof M F M, Abdullah A, Tajuddin A A, Hashim R, Bauk S 2016 Attenuation properties and percentage depth dose of tannin-based *Rhizophora* spp. particleboard phantoms using computed tomography (CT) and treatment planning system (TPS) at high energy x-ray beams *AIP Conf. Series* **1704** p 040003.

[9] Marashdeh M W, Bauk S, Tajuddin A A, and Hashim R 2012 Measurement of mass attenuation coefficients of *Rhizophora* spp. binderless particleboards in the 16.59–25.26 keV photon energy range and their density profile using x-ray computed tomography *Appl. Radiat. Isot.* **70** p 656–662.

[10] Alshipli M and Kabir N A 2017 Measurement of attenuation coefficients and CT numbers of epoxy resin and epoxy-based Rhizophora spp particleboards in computed tomography energy range. *Radiat. Phys. Chem.* **149** p 41-48.

[11] Kacal M R, Han I and Akman F 2015 Determination of K shell absorption jump factors and jump ratios of 3d transition metals by measuring K shell fluorescence parameters *Appl. Radiat. Isot.* **95** P 193-199.

[12] Yusof M F M, Abd Hamid P N K, Tajuddin A A, Hashim R and Bauk S 2015 Mass attenuation coefficient of binderless, pre-treated and tannin-based *Rhizophora* spp. particleboards using 16.59 – 25.26 keV photon energy range *AIP Conf. Series* **1659** p 040007.

[13] Bushberg J T *et al.* 2012 *The essential physics of medical imaging* (Philadelphia: Lippincott Williams & Wilkins).

[14] Abd Hamid P N K, Mohd Yusof M F, Tajuddin A A and Hashim R 2017 Measurement of mass attenuation coefficients for corn starch bonded *Rhizophora* spp. particleboards at 16-59-25.26 keV photons using X-ray fluorescence configuration *Jurnal Sains Nuklear Malaysia*, **29(2)** p 47-54.

Acknowledgements

The authors would like to thank the financial support of Incentive Grant no. 1001/PFIZIK/823048 and Bridging Grant no. 304/PPSK/6316145 by Universiti Sains Malaysia.

IOP Conf. Series: Materials Science and Engineering **555** (2019) 012031 doi:10.1088/1757-899X/555/1/012031

Evaluation of image quality and recovery coefficient of corn starch-bonded *Rhizophora* spp. particleboards as phantom for SPECT/CT imaging

P N K Abd Hamid[1,a)], M F Mohd Yusof[2], A A Tajuddin[1], R Hashim[3], R Zainon[4]

[1]School of Physics, Universiti Sains Malaysia, 11800, Penang, Malaysia
[2]School of Health Sciences, Universiti Sains Malaysia, 16150, Kelantan, Malaysia
[3]School of Industrial Technology, 11800, Penang, Malaysia
[4]Advanced Medical and Dental Institute, Universiti Sains Malaysia, 13200, Penang, Malaysia

[a]pnkah14_phy063@student.usm.my

Abstract. The aim of this study was to evaluate the performance of corn starch-bonded *Rhizophora* spp. particleboards as phantom for SPECT/CT including the image contrast, source dimension ratio and recovery coefficient values. Phantom set made of the *Rhizophora* spp. particleboards was constructed according to the Jaszczak phantom commonly used in SPECT/CT with external dimension of 22 and 18 cm diameter and length respectively. Cylindrical vials with different diameter sizes filled with ^{99m}Tc unsealed source were inserted in drilled holes of the constructed particleboards phantom. The particleboards phantom was scanned by using SPECT/CT scanning mode and the SPECT images were reconstructed using clinical protocol. The results showed that the contrast value, source dimension ratio and recovery coefficient of *Rhizophora* spp. particleboards in good agreement with Jaszczak phantom. The overall results showed an excellent agreement of performance between corn starch-bonded *Rhizophora* spp. particleboards to the Jaszczak (water) as standard phantom material. The overall results indicated that the corn starch-bonded *Rhizophora* spp. can be highly recommended to be used as tissue-equivalent phantom material for Quality Assurance and dosimetry works in SPECT/CT imaging.

1. Introduction

Nuclear medicine images are formed based on the localization of radionuclide in the target organ. The reconstruction of tomographic images by detecting the gamma energies around the patient body using the gamma camera is known as single photon emission computed tomography (SPECT). The hybrid between SPECT and computed tomography (CT) known as SPECT/CT system increased the sensitivity and specificity of nuclear medicine studies.

Water has been used as standard tissue equivalent material had been widely used as phantom for dosimetric studies and quality control (QC) involving ionizing radiations [1]. Several solid-type materials made of water equivalent such as acrylic and Perspex® had been introduced as phantom material to replace water in many diagnostic imaging procedures but still failed to give accurate dosimetry in comparison to water and soft tissues.

Previous studies indicated the suitability of particleboards made of *Rhizophora* spp. as phantoms in many applications in medical physics [2-4]. However, there is a limited data on the potential of the

Rhizophora spp. particleboards for *in-vivo* detection of ionizing radiations such as in the nuclear imaging [5]. Therefore, this study focused on the evaluation of SPECT/CT images in the *Rhizophora* spp. particleboards in comparison to the standard phantom material.

2. Methodology

2.1. Preparation of Phantom for SPECT Imaging

A total of eighteen pieces of *Rhizophora* spp. particleboards were fabricated based on the study by Abd Hamid *et al.* [6][7]. A total of two sets of 6 cylindrical-shaped vials made of acrylic with 6 cm length and different diameters of 1.3, 1.6, 1.9, 2.5, 3.0 and 3.5 cm were constructed for the Jaszczak and the particleboards phantoms as shown in Figure 1 [7].

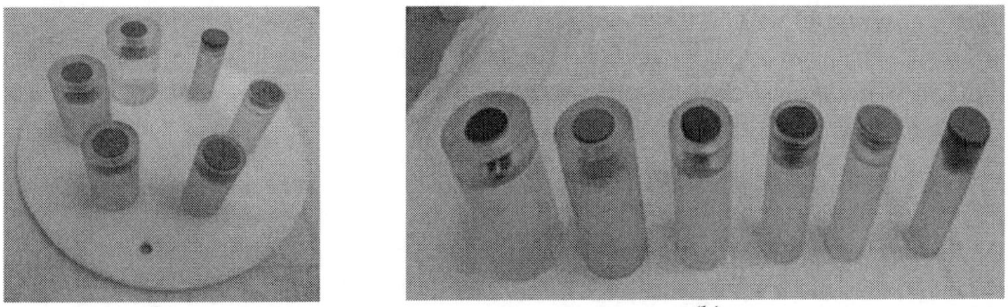

(a) (b)

Figure 1. Two sets of constructed radionuclide vials for (a) Jaszczak phantom and (b) for corn starch-bonded *Rhizophora* spp. particleboards phantom.

A ^{99m}Tc unsealed source with approximate activity concentration of 59 200 Bq/ml was injected into vials using a syringe to prevent the presence of air bubbles. The vials containing ^{99m}Tc radionuclide were positioned in its inserts within the particleboards phantom as shown in Figure 2(a). A number of eighteen pieces of particleboards were stacked to form the cylindrical SPECT/CT phantom with the diameter of 22 cm and thickness of 18 cm. Four plastic rods were used to fix and align the particleboards together and the extended rods at the top surface of particleboard were fixed with plastic screws to minimize the air gaps between the particleboard pieces as shown in Figure 2(b). Another set of radiopharmaceutical vials were positioned in the Jaszczak phantom filled with water to simulate the attenuation of human soft tissues as shown in Figure 2(c).

(a) (b) (c)

Figure 2. (a) The cylindrical vials were positioned in the inserts of the constructed corn starch-bonded *Rhizophora* spp. particleboards, (b) the 18 pieces of *Rhizophora* spp. particleboards were stacked together and (c) the Jaszczak phantom inserted with its set of radiopharmaceutical vials.

IOP Conf. Series: Materials Science and Engineering **555** (2019) 012031 doi:10.1088/1757-899X/555/1/012031

2.2 Image Quality Study of SPECT Images

The regions of interest (ROI) were drawn at the centre of the 'hot spot' area within the circular regions of vial on the SPECT images defined by using the CT images. Another ROI with similar size was also drawn on the background region that outside the region of vial images in the middle region of the phantom. The similar area size of ROI for each vial was used on all image slices. The image contrast of each vial was calculated by using equation [8]:

$$\text{Contrast} = \frac{C_{cyl} - C_{bgd}}{C_{bgd}} \tag{1}$$

with c_{cyl} is the average count per pixels in the cylindrical region and c_{bgd} is the average count per pixels in the background region. The apparent source dimension was evaluated by measuring the full width at half maximum from the plotted intensity profiles of radionuclide distributions.

The apparent source dimension was calculated by multiplying the measured full width half maximum (FWHM) with scale factor (mm/pixel) given by the gamma camera manufacturer as equation:

$$\text{Apparent Source Dimension}, d = FHWM \times SF \tag{2}$$

with *SF* is the scale factor given by the gamma camera manufacturer with the value of 4.42 mm/pixel. The ratio between the measured apparent source dimensions in the SPECT/CT image to the actual dimension given by the inner diameter of the vials containing the radionuclide was calculated by using equation:

$$\text{Ratio of source dimension} = \frac{d}{D} \tag{3}$$

with *d* and *D* is the measured apparent and actual dimension of the radionuclide respectively.

2.3 Analysis of Recovery Coefficient Values

The SPECT images of *Rhizophora* spp. particleboards and Jaszczak phantoms were reconstructed using Butterworth filter with scatter and attenuation correction applied. An ROI was drawn over the area of radionuclide distribution of each vial in the SPECT image and the average counts per pixel were measured. The activity concentration of apparent image were calculated by using following equation 3.25 [9]:

$$\text{Apparent Activity Concentration}, \mathrm{A_a} = \frac{R_{ROI}}{S_{vol}} \times \frac{VS}{T_{acq}} \tag{4}$$

with R_{ROI} is the count per pixel (counts/pixel) in drawn ROI, *VS* is the voxel size with the values of $4.42 \times 4.42 \times 4.42$ mm^3 given by the manufacturer and T_{aqq} is the total scanning time (s). The actual activity of ^{99m}Tc source was calculated according radioactive decay law by using equation:

$$\text{True Activity Concentration, } \mathrm{A_t} = A_0 e^{-\lambda t} \tag{5}$$

with A_t and A_0 is the current and initial activity of the radionuclide source (Bq) respectively, λ is decay constant of the radionuclide with the value of 3.21×10^{-5} s^{-1} and t is the time elapsed from the initial activity (s). Therefore, the recovery coefficient (RC) values of SPECT images of the radionuclide measured in the constructed corn starch-bonded *Rhizophora* spp. particleboards phantom was calculated by using following equation [8]:

$$\text{Recovery Coefficient, } RC = \frac{A_a}{A_t} \tag{6}$$

3. Results and Discussions

3.1 Evaluation of SPECT Image Contrast

The contrast evaluation of SPECT images of the phantoms scanned at 10 s, 20 s and 30 s per projection are illustrated in Figures 3. The images contrast of corn starch-bonded *Rhizophora* spp. particleboards shows a consistency to the contrast of images by Jaszczak phantom at 10 s, 20 s and 30 s scan time per projection.

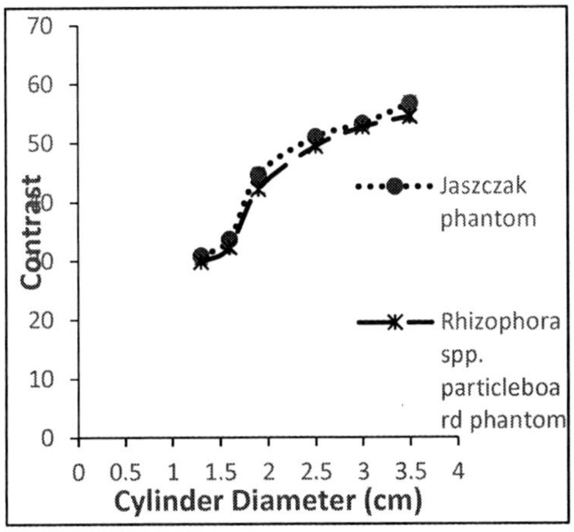

(a)

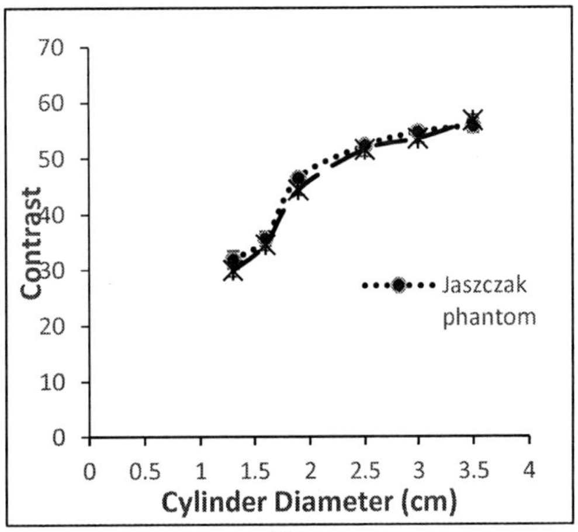

(b)

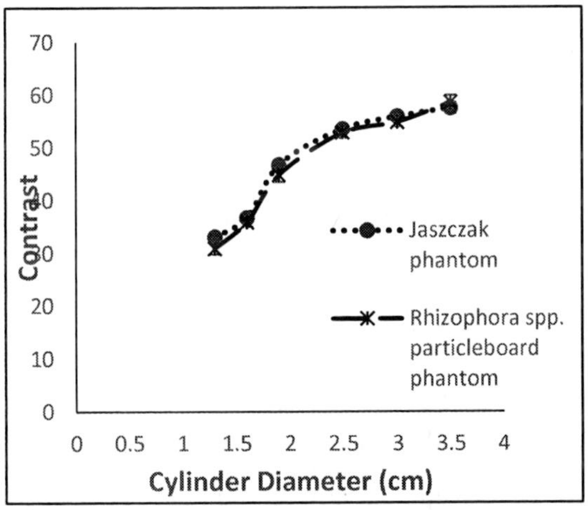

(c)

Figure 3. The contrast of the SPECT images for corn starch-bonded *Rhizophora* spp. particleboards in comparison to Jaszczak phantom at (a) 10 s/projection, (b) 20 s/projection and, (c) 30 s/projection of time setting.

The *p*-values obtained from paired sample *t*-test for pair of Jaszczak – *Rhizophora* spp. particleboards phantom as shown in Table 1. There are no significant difference between image contrast of Jaszczak phantom and corn starch-bonded *Rhizophora* spp. particleboards scanned at 20 s and 30 s time per projection shown by the *p*-values 0.079 and 0.113 respectively.

Table 1. The result of paired sample *t*-test for SPECT image contrast between Jaszczak phantom and corn starch-bonded *Rhizophora* spp. particleboards.

Time per projection (s)	Paired Differences		df	t	*p*-value (2-tailed)
	Means	St. Deviation			
10	1.537	0.697	5	5.400	0.003
20	0.978	1.090	5	2.198	0.079
30	0.755	0.965	5	1.916	0.113

3.2. Evaluation of Source Dimention Ratio

Table 2 shows the ratio of apparent diameter to actual diameter of cylinder size for corn starch-bonded *Rhizophora* spp. particleboards and Jaszczak phantom. The nearest apparent size to actual size was obtained by 2.5 cm of cylinder diameter shown by the lowest value of ratio 1.35 and 1.41 for Jaszczak phantom and *Rhizophora* spp. particleboards respectively followed by cylinder diameter with 3.0 cm. The ratio of dimension for corn starch-bonded *Rhizophora* spp. particleboards was in agreement to values of Jaszczak phantom within 4.44% of percentage differences.

Table 2. The ratio apparent to actual source dimension for each cylinder diameter for corn starch-bonded *Rhizophora* spp. particleboards and Jaszczak phantoms.

Cylinder Diameter (cm)	Jaszczak Phantom	*Rhizophora* spp. Particleboards	Difference (± %)
1.3	2.05	2.11	2.93
1.6	1.68	1.74	3.57
1.9	1.49	1.54	3.36
2.5	1.35	1.41	4.44
3.0	1.38	1.42	2.90
3.5	1.51	1.51	0

3.3. Evaluation of Recovery Coefficient

The recovery coefficient (RC) is the ratio of the apparent activity concentration to the true activity concentration of a region of interest. The similar pattern of RC curves can be observed for Jaszczak phantom and *Rhizophora* spp. particleboards as illustrated in Figure 4. The result showed the rapidly improvement of RC to the value of 1.0 at increasing diameter of cylinder between 1.3 and 2.5 cm. The RC of the SPECT images at cylindrical diameter sizes of 2.5, 3.0 and 3.5 cm on the other hand were greater than 1.0 for both phantoms. The ideal value of RC would be close to 1.0 [10]. In some situation, RC values can be greater than 1.0 due to spill-over effect or spill-in effect. Spill-over effect occurred due to contamination of activity from the neighbouring tissue or spot to these hot areas [8].

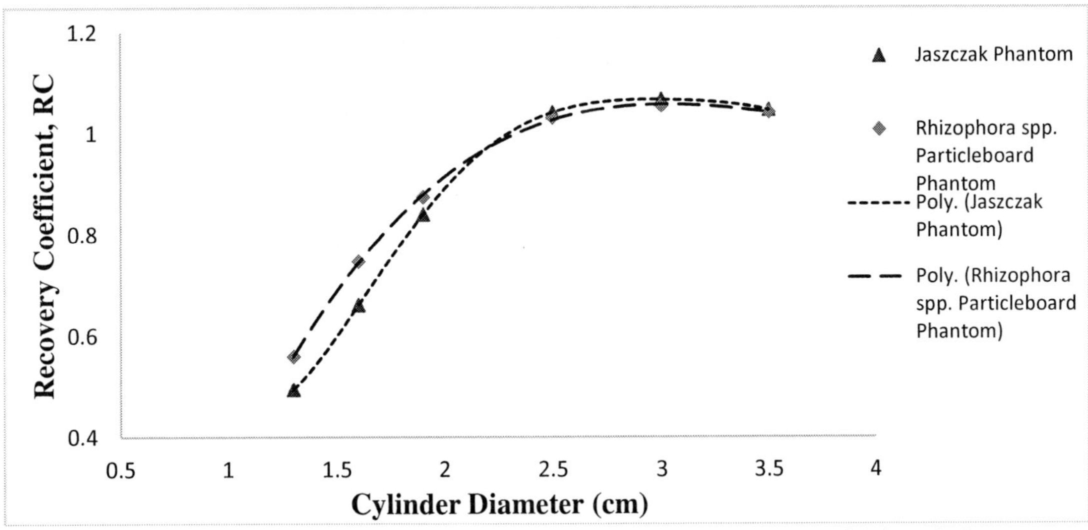

Figure 4. The recovery coefficient curves of Jaszczak and *Rhizophora* spp. particleboard phantoms at different diameter of radionuclide vial sizes.

4. Conclusion

The contrast values of SPECT image by the corn starch-bonded *Rhizophora* spp. particleboards were consistent with the Jaszczak phantom. The calculation of ratio between apparent and actual source dimension for corn starch-bonded *Rhizophora* spp. particleboards were consistent to the Jaszczak phantom. The evaluation of RC show an excellent agreement of RC curves between corn starch-bonded *Rhizophora* spp. particleboards to Jaszczak phantom with RC values close to ideal values of 1.0. The RC curve of the corn starch-bonded *Rhizophora* spp. particleboards was also consistent to the Jaszczak phantom. The overall results indicated that the corn starch-bonded *Rhizophora* spp. can be highly recommended to be used as new phantom material for Quality Assurance and dosimetry works in SPECT/CT. imaging.

5. References

[1] Khan F M and Gibbons J P 2014. *Khan's The Phys. of Radiat. Therapy* (5th ed) (Philadelphia: Lippincott Williams and Wilkins).

[2] Marashdeh M W, Bauk S, Tajuddin A A, and Hashim R 2012 Measuremen tof mass attenuation coefficients of *Rhizophora* spp. binderless particleboards in the 16.59–25.26 keV photon energy range and their density profile using x-ray computed tomography *Appl. Radiat. Isot,* **70** p 656–662.

[3] Tousi E T, Hashim R, Bauk S, Jaafar M S, Abuarra A and Ababneh B 2014 Some properties of particleboards produced from *Rhizophora* spp. as a tissue-equivalent phantom material bonded with *Eremurus* spp. *Radiat. Meas*. **54** p 14-21.

[4] Mohd Yusof M F, Hashim R, Tajuddin A A and Bauk S 2017 Characterization of tannin-added *Rhizophora* spp. particleboards as phantom materials for photon beams. *Ind. Crop Prod.* **9**, p 467-474.

[5] Abuarra A, Hashim R, Bauk S, Kandaiya S and Tousi E T 2014b Fabrication and characterization of gum Arabic bonded *Rhizophora* spp. particleboards *Mater. Design* **60** p 108-115.

[6] Abd Hamid P N K, Mohd Yusof M F, Tajuddin A A and Hashim R 2017 Measurement of mass attenuation coefficients for corn starch bonded *Rhizophora* spp. particleboards at 16-59-25.26

keV photons using X-ray fluorescence configuration *Jurnal Sains Nuklear Malaysia* 29(2), 47-54.

[7] Abd Hamid P N K, Mohd Yusof M F, Tajuddin A A, Hashim R 2017 and Zainon R 2018 Design and evaluation of corn starch-bonded *Rhizophora* spp. particleboard phantoms for SPECT/CT imaging *IOP Conf. Series: Mater. Sci. and Eng.* **298** p 012041.

[8] Cherry S R, Sorenson J A and Phelps M E 2012 *Phys. in Nuclear Medicine 4th ed* (Philadelphia: Saunders Elsevier).

[9] IAEA Human Health Reports No. 9 2014. Quantitative nuclear medicine imaging: concepts, requirements and methods. (Vienna: International Atomic Energy Agency).

[10] Gopal B S 2010 *Basics of PET Imaging 2nd ed* (New York: Springer).

Acknowledgement

The author would like to thank the financial support by the Bridging Grant scheme no. 304.PPSK.6316145 by Universiti Sains Malaysia.

IOP Conf. Series: Materials Science and Engineering **555** (2019) 012032 doi:10.1088/1757-899X/555/1/012032

Future of nuclear medicine- an overview

Siti Amira Othman[1,a)], Nurul Fathihah Abu Bakar[1], Nor Farah Amirah Nor Azman[1] and Nurin Saqinah Jasrin[1]

[1]Faculty of Applied Sciences and Technology, Universiti Tun Hussein Onn Malaysia, Pagoh Educational Hub, KM 1, Jalan Panchor, 84000, Pagoh, Muar, Johor.

[a]sitiamira@uthm.edu.my

Abstract. The development of nuclear medicine has led to an entire subset of extremely effective strategies in diagnosis and disease therapy. The emerging advancements in nuclear medicine are vast and diverse. Nuclear medicine tests and treatments involve the use of small amounts of radioactive material in order to diagnose and determine the severity of a variety of conditions and diseases. Basically, nuclear medicine involves the use of the latest imaging facilities such as magnetic resonance imaging, computed tomography scans and positron emission tomography which will find in many modern medical institutions today. One significant advantage of using nuclear medicine is the ability to make complex medical procedures simpler and safer for patients. With this technology, physicians and medical professionals are able to examine in great detail the most sensitive areas. All examinations can happen without subjecting patients to possibly dangerous and invasive procedures and surgeries. This medical technology has enabled the development of treatment options for patients suffering from serious illnesses, such as cancer either through radiation or chemotherapy. Health care professionals now face many challenges in their profession such as new rules and the latest practices that enable them to learn new skills and new ways of practicing their daily tasks in nuclear medicine areas. As a result, they are confronted with new skills and new ways of practicing in nuclear medicine area every day.

1. Future of Nuclear Medicine

Nuclear Medicine is one of dominant part in medicine world that use unsealed radioactive substances for therapy and diagnosis. The substances consist small amount of radionuclides which known as radiopharmaceuticals. Based on history, radiopharmaceuticals have been existed in the middle of 1920-s. During that time German scientists had been working hard on one experiment on rats using radionuclides. Phosphorus-32 was the first radionuclides that have been used in the late 1930 to treat leukaemia disease. After 1946, Iodine-131 began to take a dominant lead in the field which it's has been used to treat thyroid cancer. The differences between therapy and diagnosis is diagnosis (radionuclides imaging) is one of the technique in medical instrument or device that has an ability to emits radiation from the injected radionuclides which the output result is detected by a special indicator or detector such as scintillation cameras (gamma cameras or positron emission tomography (PET) cameras). Usually the output result will be in detail or high resolution image to give real 3D image of the body. The result of image is due to the distribution of medically injected, inhaled, or ingested radionuclides within the body examples include bone scans, thyroid scans, renal scans, and many others as well. For therapeutic purposes, radiopharmaceuticals are administered to treat disease with curative (able to cure or cause to get better) or palliative intent (providing relief from the symptoms and stress of a serious illness). It relies on killing abnormal cells within body by high but localized radiation exposure. Examples include I-131 therapies for thyroid disorders (hyperthyroidism and thyroid cancer), Strontium and Samarium bone palliation for osseous metastases and etc.

Radiopharmaceutical has developed into more crucial field of medicine. The application of global pharmaceutical in market is estimated to increase from $4.9 billion dollar in 2010 to 7.9 billion dollar in 2011. There is a significant increase in the global demand in radiopharmaceutical field, with the increase of disease related to cardiac, neurological and cancer disease. Radiopharmaceutical consists of drug component and a radioactive component or substances such as radioisotope and molecules labelled with radioisotopes, which are currently use in both diagnosis and therapy. Radionuclides refer to substances that have same number of protons but different number of neutrons. Unstable radionuclide causes their nuclei to undergo rearrangement while changing to a stable state and energy is give off. Most radionuclides contain a component that emits gamma radiation. Radiopharmaceuticals are the main component of nuclear medicine practice, where it was administered in diagnosing, managing and treating various types of diseases.

The obvious distinction between radiopharmaceuticals and traditional drugs is the lack of effect of pharmacological activity on human bodies as the part of pharmaceutical. Various radiopharmaceuticals have been developed that have different targeting mechanisms or routes. Some are given in the simplest salt form as non-invasive diagnostic imaging agents to provide functional and structural information about organs and diseased tissues. This became their main advantage, which is by allowing non-invasive external monitoring or targeted therapeutic irradiation, the effect on the biological processes in the body was minimized. Radiopharmaceutical was recorded as having an excellent safety record and their incidence of side effects is extremely low. They may be given to the patients in several ways such as orally or placed into the eyes or the bladder. Radioisotopes may be used internally or externally in intensive purposes of radiopharmaceutical, which have been used widely as tracers of physiologic processes. If the radioisotope were implanted in sealed capsules in a tissue, was applied externally; the dose could be terminated by the removal of the sources. If the radioisotope were implanted in an unsealed capsule, was applied internally; the dose cannot be stopped by the removal of the sources, and the dose tend to possess higher energy. Radiopharmaceutical are widely used in diagnostic and therapeutic area of human diseases, nearly 95% of pharmaceutical are used for diagnostic purposes, while the other 5% are usually used as therapeutic purposes. Since radiopharmaceutical was usually used as tracer, it has no pharmacological effects.

There are several examples that prostrate the use of radiopharmaceutical in therapeutic procedures such as Non-Hodgkin's Lymphoma Therapy. In therapeutic treatments for Non-Hodgkin's Lymphoma Therapy, a radioisotope called radioimmunotherapy (RIT) is given and attached it to an antibody in order to deliver radioactivity to specific cells. Rhenium 186 (3.8 d) is used for pain relief in bone cancer. Beta emitter with weak gamma for imaging Iridium 192 (74 d) is supplied in wire form for use as an internal radiotherapy source for cancer treatment (used then removed). Palladium 103 (17 d) is used to make brachytherapy permanent implant seeds for early stage prostate cancer. Yttrium 90 (64 h) is used for cancer brachytherapy and as silicate colloid for the relieving.

2. Application of nuclear medicine

2.1. Nitrogen-13

In producing nitrogen-13-ammonia in usable quantities with least contamination, a method has been developed in order to realise the statement. The system involves a device to produce a proton beam which travels along the selected path and strikes the target material in a target chamber. The process will produce nitrogen-13 atoms and alpha particles. In the early production of N-13-ammonia it was produced from alpha particle irradiation and heating of boron nitride in sodium hydroxide [1]. Then, it was produced by deuteron irradiation of methane gas or metal carbides, but somehow gave low yields. This shows that, this method was not too efficient. Thus another method was developed which is the proton irradiation of natural water target that involves oxygen bombarded with proton that producing nitrogen-13 and alpha particle [2][3]. This is due to the hydrogen abstraction from the target water matrix.

Another method to produce N-13 was found around year 1991 by Wieland *et al.* They produced the N-13 directly in the water by adding free radical scavengers like ethanol, acetic acid and hydrogen to avoid the formation of the oxo anions. The method is involving the use of pressurized, dilute aqueous solutions of acetic acid and ethanol. Previous study has stated that combination of hydrogen and ethanol was more effective rather stand alone at high beam doses [6][7]. However, it has limiting factor which is the nuclear reaction has very small cross-section of the reaction which requires accelerators or cyclotrons with energies that higher than 11 MeV. By using cyclotron, it produced radionuclide by irradiation reaction with 11 MeV protons. It bombards oxygen-16 with protons and emits alpha particles and nitrogen-13 as final product [5].

The nitrogen-13-ammonia is produced from the reduction process of nitrogen-13-nitrates and nitrites. It is converted into N-13-ammonia in aqueous medium which involving exothermic reaction. The application of nitrogen-13 as radiotracer is very essential in myocardial perfusion imaging. It is a nuclear medicine procedure that illustrates the function of heart muscle (myocardium) [1]. It is first developed by Dr. Herrmann Blumgart to measure cardiac strength by injecting patients with a radioactive compound known as Radium C. The substance was injected into venous system and pass through the right heart into the lungs then into left heart and out into arterial system where it then detected by Wilson chamber. The chamber monitor the circulation time of the radioactivity produced. It is concluded from the procedures that the longer the circulation time, the weaker the heart. The role of radioactive substance, like nitrogen-13 is important in order to determine the cardiac physiology (function).

2.2. Technetium- 99m

Technetium-99m is extensively used in medical application. The higher energy of gamma ray about 140keV is appropriate to be used in diagnosis as radioactive tracer. The techenetium-99m is not only used for skeleton image but also able to apply for brain, thyroid, lungs, liver, spleen, gall bladder and bone marrow. Technetium-99m is a radioisotope which emitting gamma rays. This allows Technetium-99m to be used for single photon emission computed tomography (SPECT). SPECT is a nuclear imaging technique which used gamma to be detected by the gamma camera. When the radioisotope injected in the patient body and the concentrated will attach to the organ of interest and give a signal for gamma camera to detect it. The gamma camera needs to be rotated every three to six degree of patient body until reach 360° for recording the detectable data. This projection takes about 15-20 minutes to finish.

Technetium-99m also has been used widely in bone and brain scans. Bone scans can be used directly to heal the skeletal injury and some cases are allowing the Techenetium-99m to react with tumour of the bones. It is also able to detect strokes by brain scanning [7]. The property of Technetium-99m radioisotope is able to identify the lymph nodes drain cancer such as breast cancer. The radioisotope will be used together with antimony sulphide colloid as radioactive maker in our body to detect the abnormal activity of lymph nodes. This radioactive maker is injected by shielding syringe to lymph node drain area and the area spotted with abnormal activity as the area of cancer growth.

Next, the immunoscintigraphy is a process to detect hard-to-find cancers by injected the Technetium-99m into monoclonal antibody. Monoclonal antibody is one of the immune system proteins that are capable to binding with cancer cells. Medical equipment will detect the gamma ray that is emitted by Technetium-99m and diagnose that the higher concentration indicates the location of tumours. It usually relates with intestine organ treatment. Other than that, Technetium-99m is incorporate with tin compound to detect gastrointestinal bleeding sites. The detection occurs by binding them with red blood cells and maps the circulatory system disorders.

2.3 Rubidium-82

Biomarkers have the potential to aid in the risk assessment, diagnosis, early detection, prognosis and prevention of the breast cancer which is important and, numerous reports have been made on biomarkers, but only a few have been practiced and used in clinical. For the breast cancer risk especially for woman, shows a genetic risk factors. The food content of Rubidium was inversely associated with cancer incidence in early 1974 was found among Hopi Indians [8]. Tumour volume was significantly reduced in vivo study after the administration of Rubidium Carbonate in mice. Levels of Rubidium in tumour tissue were significantly higher than in the normal tissue or particularly breast tumour tissues. Due of that, Rubidium is used in tumour through intake of rubidium in form of food with control dose.

In human, Rubidium is natural substances and micronutrient like Selenium, for evaluating its anticancer efficacy, it would be meaningful to compare rubidium levels in biological samples between cancer patients and normal control patients and it is also important to know the application of the Rubidium as a biomarker in the risk assessment. Through urine, Rubidium is excreted and urine concentration is a good indicator for the exposure of the Rubidium. The acid toxins leaking out of the tumour mass were neutralized and render them to become nontoxic. This is because the presence of Rubidium salts in the body fluids neutralized the toxin. In the colorectal patients, the concentrations of the Rubidium in whole blood are much lower than the patients in healthy controls. The Rubidium was sequestered by the tumour so the levels of rubidium in patient's serum and urine is low.

Since the late 1950, Rubidium used in the context of myocardial perfusion have been studied [9]. In relation to myocardial perfusion, the Rubidium is extracted from blood and is taken up myocardium which requires high of energy. Similar to conventional single photon emission computed tomography (SPECT) myocardial perfusion imaging (MPI), Rubidium-82 can generate a clear perfusion image because it is an extractable tracer and it can identify viable myocardium when the cell membrane was disruption during the rapid washout of Rb-82 from the tissue. The radiation exposures for patients is limited around (5.5mSv for 60mCi) because it has short half-life, the total study time can be reduced as the image acquisition requires is short after tracer administration. Rb-82 has lower extract fraction which may affect the hyperemic flow measurements and this will make the image resolution reduced.

However, data support with flow quantification of Rb-82 will be feasible, accurate and reproducible because of Rubidium activity can be infused over a longer interval to reduce the peak dead time-losses and the PET imaging will be displayed in three dimensional modes. When using Rubidium in PET, quick reconstruction of the rest uptake frame can often detected while the patient is still on the bed, and can be resolved by instructing the patient to eat or drink water before repeating the image acquisition [9]. PET also provides reliable images, particularly in large patients even though the patient with large breast and arm present in the field of view. By using Rubidium in PET, we can see our heart's image and show how much blood enters the heart muscle either during rest or exercise.

3. Conclusion

Nuclear Medicine is a promising area in the foreseeable future. The incorporation of molecular imaging and radiopharmaceutical will represent a significant paradigm shift for the specialty of this area.

4. References

[1] Boogers M J., Fukushima K, Bengel F M and Bax J J 2011 The role of nuclear imaging in the failing heart: myocardial sympathetic innervation and future applications *Heart Failure Reviews* p 411- 423.

[2] Joliot F, Curie I 1934 Artificial production of a new kind of radio-element *Nature* **133** p 201-202.

[3] Vaalburg W, Kamphuis J A, Beerling-Vander Molen H D *et al.* 1975 An improved method for the cyclotron production of 13n-labelled Ammonia. Int. J. Appl. Radiat. Isot. p 316-318.

[4] Wieland B, Bida G, Padgett H *et al.* 1991 In target production of [13n] Ammonia via proton irradiation of dilute aqueous ethanol and acetic acid mixtures *Int. J. Rad. Appl. Instrum. A* p 1095-1098.

[5] Berridge M S, Landmeier B J 1993 In-target production of [13n] Ammonia: target design, products, and operating parameters *Appl. Radiat. Isot.* p 1433-1441.

[6] Monahan W G, Tilbury R S, Laughlin J S 1972 *Uptake of 13n-labeled Ammonia* J. Nucl. Med. p 274-277 (Pub: Med).

[7] Smith E M 1964 Properties, uses, radiochemical purity and calibration of Tc-99m *J. of Nucl. Med.* p 871- 882.
[8] Brewer A K, Clarke B J, Greenberg M and Rothkopf N 1979 The effects of Rubidium on mammary tumour growth in C57blk/ 6J mice *Cytobios* p 99- 101.
[9] Yoshinaga K, Klein R and Tamaki N 2010 Generator produced Rubidium-82 positron emission tomography myocardial perfusion imaging- from basic aspects to clinical applications *J. of Cardiology* p 163- 173.

Acknowledgements
The authors would like to thank the Applied Physics Programme, Faculty of Applied Sciences and Technology Universiti Tun Hussein Onn Malaysia for facilities provided and gratefully acknowledged the financial support a research grant (H074) that make the research possible.